Food and Beverage Fermentation Technology

Food and Beverage Fermentation Technology

Contributors

Rashid Abafita Abawari et al.

AURIS
Reference

www.aurisreference.com

Food and Beverage Fermentation Technology

Contributors: Rashid Abafita Abawari et al.

Published by Auris Reference Limited

www.aurisreference.com

United Kingdom

Food and Beverage Fermentation Technology

ISBN: 978-1-78154-809-7

British Library Cataloguing in Publication Data
A CIP record for this book is available from the British Library

Printed in the United Kingdom

Exclusively distributed by CBS Publishers & Distributors Pvt. Ltd.

Sales & Distribution Rights only for India, Pakistan, Bangladesh, Sri Lanka, Nepal and Bhutan.This book is not to be sold outside these territories.

Contents

vi

List of Abbreviations

LAB	Lactic Acid Bacteria
AMB	Aerobic Mesophilic Bacteria
ASF	Aerobic Spore-formers
EMB	Eosin Methylene Blue
ACA	American coolship ale
BAS	Barcoded amplicon sequencing
TRFLP	Terminal restriction fragment length polymorphism
LAB	Lactic acid bacteria
OTU	Operational taxonomic unit
RDP	Ribosomal Database Project
WLD	Wallerstein Differential Agar
PCoA	Principal coordinates analysis
LDLR	Density lipoprotein receptor
LDA	Linear discriminant analysis
FID	Flame ionization detection
AF	Alcoholic fermentation
MSL	Maximum residue levels
MSM	Model synthetic medium
SEM	Scanning electron microscope
TEM	Transmission electron microscopy
MRS	Mann Rogosa Sharpe
MRS	Mann Rogosa Sharpe
BEV	Bovine enterovirus
VFA	Volatile fatty acids
YNB	Yeast Nitrogen Base
FVs	Fermentation vessels
PM	Phenotypic Microarray
DAD	Diode array detection
HPLC	High-performance liquid chromatography
OWLS	Optical waveguide lightmode spectroscopy
ANN	Artificial Neural Network
NN	Neural Network
MIMO	Multiinput and multi-output
MRE	Mean relative error
CNN	Computational neural networks
GCP	Green Chilli Powder
FFNN	Feed forward neural networks
DLF	Dielectric loss factor
PCA	Principal Component analysis
GRNNs	Generalized regression neural networks

List of Contributors

Rashid Abafita Abawari
Soil and water Process, Bio-fertilizer Development Project Jimma agricultural research Institute, Jimma, Ethiopia.

Nicholas A. Bokulich
Department of Viticulture and Enology, Robert Mondavi Institute of Wine and Food Science, University of California Davis, Davis, California, United States of America,
Department of Food Science and Technology, Robert Mondavi Institute of Wine and Food Science, University of California Davis, Davis, California, United States of America

Charles W. Bamforth
Department of Food Science and Technology, Robert Mondavi Institute of Wine and Food Science, University of California Davis, Davis, California, United States of America

David A. Mills
Department of Viticulture and Enology, Robert Mondavi Institute of Wine and Food Science, University of California Davis, Davis, California, United States of America,
Department of Food Science and Technology, Robert Mondavi Institute of Wine and Food Science, University of California Davis, Davis, California, United States of America

Yunhye Kim
Department of Food and Nutrition, College of Human Ecology, Yonsei University, Seoul, South Korea

Sun Yoon
Department of Food and Nutrition, College of Human Ecology, Yonsei University, Seoul, South Korea

Sun Bok Lee
Department of Food and Nutrition, College of Human Ecology, Yonsei University, Seoul, South Korea

Hye Won Han,
Department of Food and Nutrition, College of Human Ecology, Yonsei University, Seoul, South Korea

Hayoun Oh
Department of Food and Nutrition, College of Human Ecology, Yonsei University, Seoul, South Korea

Wu Joo Lee
Department of Food and Nutrition, College of Human Ecology, Yonsei University, Seoul, South Korea

Seung-Min Lee
Department of Food and Nutrition, College of Human Ecology, Yonsei University, Seoul, South Korea

Carole Camarasa
INRA, UMR1083, Montpellier, France,
SupAgro, UMR1083, Montpellier, France,
Universite´ Montpellier 1, UMR1083, Montpellier, France

Isabelle Sanchez
INRA, UMR1083, Montpellier, France,
SupAgro, UMR1083, Montpellier, France,
Universite´ Montpellier 1, UMR1083, Montpellier, France

Pascale Brial
INRA, UMR1083, Montpellier, France,
SupAgro, UMR1083, Montpellier, France,
Universite´ Montpellier 1, UMR1083, Montpellier, France

Fre´de´ ric Bigey
INRA, UMR1083, Montpellier, France,
SupAgro, UMR1083, Montpellier, France,
Universite´ Montpellier 1, UMR1083, Montpellier, France

Sylvie Dequin
INRA, UMR1083, Montpellier, France,
SupAgro, UMR1083, Montpellier, France,
Universite´ Montpellier 1, UMR1083, Montpellier, France

Marion Favier
University of Bordeaux, ISVV, Unit of Enology EA 4577, Villenave d'Ornon, France,
SARCO, research subsidiary of the Laffort group, BP 40, Bordeaux, France,

Eric Bilhe` re
University of Bordeaux, ISVV, Unit of Enology EA 4577, Villenave d'Ornon, France,

Aline Lonvaud-Funel
[1] University of Bordeaux, ISVV, Unit of Enology EA 4577, Villenave d'Ornon, France,

Virginie Moine
Laffort, BP 17, Bordeaux, France

Patrick M. Lucas
University of Bordeaux, ISVV, Unit of Enology EA 4577, Villenave d'Ornon, France,

Xiang-yu Sun
College of Food Science and Nutritional Engineering, China Agricultural University, Beijing, 100083, P.R. China,

Yu Zhao
Faculty of Science, University of Copenhagen, København S, Denmark

Ling-ling Liu
College of Food Science and Nutritional Engineering, China Agricultural University, Beijing, 100083, P.R. China,

Bo Jia
College of Food Science and Nutritional Engineering, China Agricultural University, Beijing, 100083, P.R. China,

Fang Zhao
College of Food Science and Nutritional Engineering, China Agricultural University, Beijing, 100083, P.R. China,

Wei-dong Huang
College of Food Science and Nutritional Engineering, China Agricultural University, Beijing, 100083, P.R. China,

Jicheng Zhan
College of Food Science and Nutritional Engineering, China Agricultural University, Beijing, 100083, P.R. China,

Yantyati Widyastuti
Research Center for Biotechnology, Indonesian Institute of Sciences, Cibinong, Indonesia;

Rohmatussolihat
[1]Research Center for Biotechnology, Indonesian Institute of Sciences, Cibinong, Indonesia;

Andi Febrisiantosa
Technical Implementation Unit for Development of Chemical Engineering Processes, Indonesian Institute of Sciences, Yogyakarta, Indonesia.

Hendrik A. Scheinemann
Institute of Bacteriology and Mycology, University of Leipzig, Faculty of Veterinary medicine, An den Tierkliniken 29, 04103 Leipzig, Germany,
Gesellschaft zur Förderung von Medizin-, Bio- und Umwelttechnologien e. V. Erich-Neuß-Weg 5, 06120 Halle (Saale), Germany

Katja Dittmar
Institute of Parasitology, University of Leipzig, Faculty of Veterinary medicine, An den Tierkliniken 35, 04103 Leipzig, Germany,

Frank S. Stöckel
Institute of Parasitology, University of Leipzig, Faculty of Veterinary medicine, An den Tierkliniken 35, 04103 Leipzig, Germany,

Hermann Mülle
Institute of Virology, University of Leipzig, Faculty of Veterinary medicine, An den Tierkliniken 29, 04103 Leipzig, Germany,

Monika E. Krüger
Institute of Bacteriology and Mycology, University of Leipzig, Faculty of Veterinary medicine, An den Tierkliniken 29, 04103 Leipzig, Germany,

Cyprian E. Oshoma
Bioenergy and Brewing Science Building, School of Biosciences, University of Nottingham, Sutton Bonington Campus, Loughborough, Leics, United Kingdom,

Darren Greetham
Bioenergy and Brewing Science Building, School of Biosciences, University of Nottingham, Sutton Bonington Campus, Loughborough, Leics, United Kingdom,

Edward J. Louis
Centre for Genetic Architecture of Complex Traits, University of Leicester, Leicester, United Kingdom,

Katherine A. Smart
3 SAB Miller PLC, Surrey, United Kingdom,

Trevor G. Phister
PepsiCo Int. Beaumont Park, Leycroft Road, Leicester, United Kingdom,

Chris Powell
Bioenergy and Brewing Science Building, School of Biosciences, University of Nottingham, Sutton Bonington Campus, Loughborough, Leics, United Kingdom,

Chenyu Du
Bioenergy and Brewing Science Building, School of Biosciences, University of Nottingham, Sutton Bonington Campus, Loughborough, Leics, United Kingdom,
School of Applied Sciences, University of Huddersfield, Queensgate, Huddersfield, United Kingdom

Marina C. N. Assohoun
Department of Biotechnology and Food Microbiology, University Nangui Abrogoua, Abidjan, Côte d'Ivoire;
Department of Nutrition and Food Security, University Nangui Abrogoua, Abidjan, Côte d'Ivoire.

Théodore N. Djeni
Department of Biotechnology and Food Microbiology, University Nangui Abrogoua, Abidjan, Côte d'Ivoire;

Marina Koussémon-Camara
Department of Biotechnology and Food Microbiology, University Nangui Abrogoua, Abidjan, Côte d'Ivoire;

Kouakou Brou
Department of Nutrition and Food Security, University Nangui Abrogoua, Abidjan, Côte d'Ivoire.

Cs. Csutorás
Institute of Food Science, Egerfood Regional Knowledge Center, Eszterhazy Karoly University, Eger, Hungary

K. Rácz
Eger Crown Winehouse Ltd., Kerecsend, Hungary

G. Z. Nagy
Institute of Food Science, Egerfood Regional Knowledge Center, Eszterhazy Karoly University, Eger, Hungary

O. Hudák
Institute of Food Science, Egerfood Regional Knowledge Center, Eszterhazy Karoly University, Eger, Hungary

L. Rácz
Institute of Food Science, Egerfood Regional Knowledge Center, Eszterhazy Karoly University, Eger, Hungary

Maria Rosa Alberto
Facultad de Bioquímica, Química y Farmacia, Universidad Nacional de Tucumán (UNT), Tucumán, Argentina;
Centro Científico Tecnológico Tucumán—Consejo Nacional de Investigaciones Científicas y Tecnológicas (CCT-CONICET), Tucumán, Argentina.

Maria Francisca Perera
[1]Facultad de Bioquímica, Química y Farmacia, Universidad Nacional de Tucumán (UNT), Tucumán, Argentina;

Mario Eduardo Arena
Facultad de Bioquímica, Química y Farmacia, Universidad Nacional de Tucumán (UNT), Tucumán, Argentina;
Centro Científico Tecnológico Tucumán—Consejo Nacional de Investigaciones Científicas y Tecnológicas (CCT-CONICET), Tucumán, Argentina.

Madhukar Bhotmange
Laxminarayan Institute of Technology, Rashtrasant Tukadoji Maharaj Nagpur University, Nagpur 440033. India

Pratima Shastri
Laxminarayan Institute of Technology, Rashtrasant Tukadoji Maharaj Nagpur University, Nagpur 440033. India

Preface

Over the past decade, new applications of genetic engineering in the fermentation of food products have received a great deal of coverage in scientific literature. The book Food and Beverage Fermentation Technology covers numerous fermentation procedures by reviewing the type of raw material used, its pretreatment, the temperature/climate, and the conditions of the fermentation process. The aim of first chapter is to document the microbiology of the product and antibiotic susceptibility patterns of LAB. Second chapter demonstrates that ACA exhibits a conserved core microbial succession in absence of inoculation, supporting the role of a resident brewhouse microbiota. Third chapter aims to investigate whether in vitro fermentation of soy with L. plantarum could promote its beneficial effects on lipids at the molecular and physiological levels. The aim of fourth chapter is to investigate how the adaptation to a broad range of ecological niches may have selectively shaped the yeast metabolic network to generate specific phenotypes. Fifth chapter aims to determine whether O. oeni contains plasmids of technological interest. Sixth chapter aims to understand copper tolerance of wine yeast and establish the mechanism by which yeast decreases copper in the must during fermentation. Seventh chapter focuses on the role of lactic acid bacteria in milk fermentation. Eighth chapter describes a simple, easy-to-use, low-tech hygienization method which conserves nutrients and does not require large investments in infrastructure. Ninth chapter emphasizes on screening of non- saccharomyces cerevisiae strains for tolerance to formic acid in bioethanol fermentation. In eleventh chapter, we focuses on industrial scale investigations modelling real fermentation processes for the best. Twelfth chapter focuses on lactic acid fermentation of peppers. Application of artificial neural networks to food and fermentation technology is discussed in last chapter.

Chapter 1

MICROBIOLOGY OF KERIBO FERMENTATION: AN ETHIOPIAN TRADITIONAL FERMENTED BEVERAGE

Rashid Abafita Abawari

Soil and water Process, Bio-fertilizer Development Project Jimma agricultural research Institute, Jimma, Ethiopia.

ABSTRACT

Keribo is an indigenous traditional fermented beverage and is being served on holidays, wedding ceremony and also used as sources of income of many households in Jimma zone. The aim of this study was to document the microbiology of the product and antibiotic susceptibility patterns of LAB. Samples of Keribo were collected from Jimma town and four of its districts. Keribo was fermented in the laboratory following the traditional techniques for microbial succession monitored at 6 h intervals. Finally, dominant LAB was evaluated for their antibiotic susceptibility patterns against eight antibiotics. Samples of Keribo from open markets and households in Jimma zone showed average Lactic Acid Bacteria (LAB), Aerobic Mesophilic Bacteria (AMB), Aerobic Spore-formers (ASF) and yeasts with mean counts of (log CFU mL^{-1}) 2.70 ± 2.07, 2.34 ± 2.37, 4.96 ± 2.80 and 2.01 ± 0.60, respectively. The mean counts of Enterobacteriaceae, staphylococci and moulds were below detectable levels. The early stage was dominated by AMB and ASF. However, the mean counts of LAB increased exponentially for the first 30 h and remain constant thereafter. Leuconostoc mesenteroides, identified as the most dominant LAB, were found to be susceptible to penicillin G, gentamicin, ampicilin, chloramphenicol, amikacin, bacitracin and norfloxacin but resistant to vancomycin.

INTRODUCTION

Traditional fermented foods and beverages are those traditionally fermented products based on the skills of the household occupants by indigenous knowledge systems and is produced from a variety of locally available cereal ingredients using traditional techniques by the people of that area themselves (Abegaz et al., 2002). They became part of the cultural and traditional norm among the indigenous communities in rural areas. Ashenafi (2002) indicated that fermented foods and beverages constitute a major portion of peoples' diet in all parts of the world. In Ethiopia, like in many developing countries, fermented food products constitute a major portion of peoples' diet. Ethiopia is a country rich in cultural diversity and hence, varieties of foods and beverages are processed and consumed among the various ethnic groups. The fermented products are, however, produced on fairly small scale and usually for local consumption (Ashenafi, 2006).

The indigenous natural fermentation takes place by the microorganisms involved in the process but particular microbial community will succeeded the end product in any food fermentation. Isolation of such microbes should not only be confined to dominant organisms but also it should include other microbes found in lower numbers which might have an important function in the process. Microbiological, identification of the role of each organism, nutritional and technical investigation should be carried out on each of the fermentation processes. The various microorganisms involved in each fermentation process should be isolated, characterized, studied and preserved (Aidoo et al., 1992). Microbiologically, fermentation of traditional fermented products relies on the microorganisms (LAB) present in the substrates, fermenting vats and equipments. Lactic acid bacteria are the most frequently encountered groups in almost all fermented products. They are known to produce varieties of chemical compounds. Inhibition of the growth of pathogenic microorganisms by lactic acid bacteria in some fermented products is accounted to organic acids, low pH, hydrogen peroxide, diacetyl, competition and nutrient depletion, altered redox potential, CO_2, ethanol, crowding and production of antibiotics like substance such as bacteriocins (Adams and Nicolaides, 1997).

Keribo is an indigenous traditional fermented beverage produced and consumed in different parts of the country, including Jimma zone. It is produced mainly from barley and sugar. Fermented Keribo constitutes a major part of the beverages being served on holidays, wedding ceremony and also as sources of income of many households in Jimma zone. The popularity of this traditional fermented beverage is more reflected among the religious groups and those do not like alcoholic drinks. Being considered as a non- or low-alcoholic beverage, Keribo is popular among both adults and children. It has poor keeping quality with shelf-life of not more than a day or two and it has a pronounced characteristic of the deteriorating beverage at the end of 48 h of fermentation.

The traditional indigenous technology of cheese, bread, beer and wine is well documented. The microbiology and fermentation processes of some of the traditional Ethiopian fermented foods, condiments and beverages are also studied and documented. A review on the microbiology of Ethiopian foods and beverages (Ashenafi, 2006) has revealed the availability of scientifically documented information on the microbiology of a number of traditional Ethiopian fermented foods and beverages. Among the traditional Ethiopian fermented beverages, the fermentation processes and microbial dynamics during fermentation of 'Tella'(Samuel and Berhanu, 1991), 'Borde' (Bacha et al., 1998) and 'Shamita' (Bacha et al., 1999) are described. Moreover, the safety consideration of Ethiopian foods and beverages has shown the possibility of isolating some food-borne pathogens from some fermented products (Ashenafi, 2002). However, there is no scientifically documented information both on the microbiology and safety of 'Keribo' preparation. The aim of this study was, therefore, to document and analyze its microbiological flora with emphasis on the fermenting lactic acid bacteria, antibiotic susceptibility patterns of the isolates and safety of 'Keribo'. The present study reports on microbial succession occurred during traditional fermentation of Keribo.

MATERIALS AND METHODS

Enumeration of microbial groups: The counting and characterization of microbial groups was carried out following standard microbiological methods (Mugula et al., 2001). Samples of Keribo (25 mL each) were drawn aseptically at 6 h intervals during laboratory fermentation of Keribo and 25 mL each collected from local venders were mixed with 225 mL sterile peptone water (1.5%), homogenized and were spread-plated in duplicate on pre-dried agar plates of Plate Count(PCA) (CDH, India) for Aerobic Mesophilic Bacteria (AMB), Eosin Methylene Blue (EMB) (CDH, India) for Enterobacteriaceae (EB) , Mannitol Salt (MSA) (CDH, India) for Staphylococci, De man, Rogasa, Sharpe (MRS) (OXOID) for Lactic Acid Bacteria (LAB) and Chloramphenicol-bromophenol Blue (CBB) for yeasts/ moulds, CBB consisted of (g L^{-1} distilled water) yeast extracts 5.0 g, glucose 20 g, choramphenicol 0.1 g, Bromophenol-blue 0.01 g, agar 15 g, pH 6.0-6.4. The total Aerobic Mesophilic Bacteria (AMB) was enumerated on PCA plates after incubation at 30-32°C for 48 h. After incubation at 30-32°C for 20-24 h after which purplish red colonies surrounded by reddish zone of precipitated bile were counted as coliforms and large mucoid colonies that are pink to purple because of their lactose fermentation were enumerated as members of Enterobacteriaceae (EB). All snow white colonies of LAB were counted on MRS agar plates after anaerobic incubation using anaerobic jar (anaerobic Gas pack System, Oxoid) at 30-32°C for 48 h. Aerobic spore formers were counted on Plate Count (PC) agar after appropriate dilution was heat-treated at 80°C for 10 minutes in water bath and spread- plated. The numbers of AMB, EB, Staphylococci, LAB, yeasts/moulds or ASF from their respective duplicate countable plates are reported as log CFU mL^{-1} calculated from the arithmetic mean of total samples. After colony counting, 10 to 15 colonies were randomly picked from countable plates of MRS agar for further identification. Colonies of LAB were transferred into about 5 mL MRS broth (HIMEDIA, India) and purified by repeated streaking on MRS agar. The pure cultures of LAB were streaked on slants of MRS agar and were stored at 4°C for further characterization.

Characterization of the dominant LAB: Lactic acid bacteria isolated from Keribo samples were identified on the basis of key characteristics and tests (Facklam and Elliott, 1995; Ricciardi et al., 2005; Yousif et al., 2005; Bahiru et al., 2006). Morphological characterization of the pure culture was conducted microscopically using oil emulsion objectives.

The preparation was observed under Digital Olympus spectro-microscopy connected to Screen display. Cell grouping, motility, presence or absence of endospores and cell shape were the basic features to be evaluated during morphological observation. Gram-reaction was tested based on the KOH test of Gregersen (1978). Production of the enzyme oxidase was tested according to Kovacs (1956) and formation of catalase was determined by flooding young colonies with 3% solution of H_2O_2 and Oxidative or fermentative utilization of glucose by each isolate was assessed by the O/F test (Hugh and Leifson, 1953). Gas production from glucose was assigned in MRS broth containing inverted Durham tubes. The broth was inoculated with two colonies from fresh grown MRS agar plate. The broth was then sealed with melted petroleum jelly and the tube is incubated at 30-32°C for 48 h. Gas production was indicated when the inverted Durham tubes pushed upward. Salt tolerance was done using MRS broth containing 4, 6.5, 8, 15 and 20% (w/v) NaCl. Heat tolerance at 15, 40 and 45°C in MRS broth (Oyewole and Odunfa, 1990), deamination of arginine (Pilone et al., 1991), acid production from carbohydrates (Nair and Surendran, 2005) and production of yellow pigment (Farrow et al., 1989) in MRS broth was conducted. Coagulase test was done by placing isolates from pure culture on a clean microscopic slide and mixed with blood plasma. Agglutination or clumping of cocci within 5 to 10 seconds was taken as positive result. Nitrate reduction in nitrate broth and indole production in tryptone broth was evaluated (Facklam et al., 1989).

Source of dominant LAB involved in Keribo fermentation: To evaluate the dominant LAB involved in Keribo fermentation, sample of table sugar (25 g), which is commonly used for Keribo preparation, was mixed with

225 mL sterile peptone water (1.5%) and homogenized using vortex mixer. The homogenate was serially diluted (10^{-1} to 10^{-2}) and 0.1 mL aliquot of appropriate dilution was spread-plated in duplicate on pre-dried plates of De-Mann Rogossa and Sharpe (MRS) agar for counts of LAB. The inoculated plates were incubated anaerobically using anaerobic jar (anaerobic Gas pack System, Oxoid) at 30-32°C for 48 h. Repeatedly purified colonies of LAB were subjected to morphological and biochemical analysis.

Physico-chemical analysis: The pH of samples was determined by dipping an electrode of a digital pH meter (HANNA-211 meter, Portugal) into 10 mL aliquot sample drawn during laboratory fermentation. The pH meter was calibrated against standard buffer solutions at pH 4.0 and 7.0 (Merck). The total amount of lactic acid present in each of the sample drawn was determined by titration against a 0.1 N NaOH (Byaruhanga, 1998). The percent of lactic acid present in the sample was calculated using the formula:

$$\text{Lactic acid (\%)} = \frac{\text{Amount of NaOH x Normality of NaOH x 9}}{\text{Volume of sample (mL)}}$$

Determination of antibiotic susceptibility patterns of Isolates: Susceptibility of the LAB to 8 types of antibiotics was performed by the disc diffusion method as described by Bauer et al. (1966) and Liasi et al. (2009) using commercially available antibiotic disc (Oxoid). The commercial antibiotics used were penicillin G (Pen, 10 unit), ampicillin (Amp, 10 μg), amikacin (Amk, 30 μg), norfloxacin (Nx, 10 μg), chloramphenicol (Chl, 30 μg), Vancomycin, (Van, 10 μg), gentamycin (Gen, 30 μg) and bacitracin (B, 10 μg). After incubation of the plates, inhibition zone diameters were measured inclusive of the diameter of the discs. The isolates were classified as sensitive S ($\geq$21 mm); intermediate, I (16-20 mm) or resistant, R ($\leq$15 mm), respectively according to Vlkova et al. (2006). For purpose of data analysis, the intermediates were considered as sensitive (Ferraro, 2000; Rojo-bezares et al., 2006).

Statistical analysis: To see if there was significant difference in microbial counts among study areas was analyzed by analysis of variance (ANOVA) and means were separated by Duncan's test at $\alpha = 0.05$.

RESULTS

Microbial load of Keribo samples and identification of LAB: The minimum counts of Aerobic Spore Forming bacteria (ASF) was over 5 log CFU mL^{-1} with the mean count of 4.96 log CFU mL^{-1} and maximum counts of 7.97 log CFU mL^{-1} in Keribosamples. Similarly, the mean count of LAB in collected samples was 2.70 log CFU mL^{-1} with maximum count of 6.89 log CFU mL^{-1}

(Table 1). The counts of staphylococci, Enterobacteriaceae and Molds were below detectable level ($<\log 2$ CFU mL^{-1}).

LAB, AMB and yeasts were among the commonly isolated microbial groups in Keribosamples next to ASF. The Keribo samples from the five study areas did not have significant differences in counts of LAB and yeasts ($p>0.05$) (Table 2). However, samples from areas 1, 2 and 5 had significantly higher counts of AMB than samples from areas 3 and 4 ($p<0.05$). Similarly, samples from areas 1, 2, 3 and 5 had significantly higher counts of ASF than samples from area 4 ($p<0.05$) (Table 2).

About 52% (n = 26) of Keribo samples had counts of LAB ranging from Log 2 to Log 6 CFU mL^{-1}. However, about 48% (n = 24) of the sample had counts of LAB below Log 2 CFU mL^{-1}.

Table 1: Microbial count (log CFU mL-1) of different microbial groups detected in Keribo samples

Microbial group	Mean±SD	%CV	Minimum	Maximum
LAB	2.70±2.07	76.66	0.0	6.89
AMB	2.34±2.37	101.28	0.08	8.31
ASF	4.96±2.80	56.45	0.0	7.97
Yeasts	2.01±0.60	29.85	0.81	3.10
EB	<2	-	-	-
Staph	<2	-	-	-
Molds	<2	-	-	-

LAB: Lactic acid bacteria, AMB: Aerobic mesophilic bacteria, ASF: Aerobic spore former, Staph, staphylococci, EB: Enterobacteriaceae

Table 2: Microbial counts (CFU mL-1) of Keribo samples from five areas (n = 50)

Microbial Groups	Log CFU mL^{-1} (Mean±S.D)				
	Area 1	Area 2	Area 3	Area 4	Area 5
LAB	2.47±2.02ᵃ	2.69±1.80ᵃ	3.07±2.12ᵃ	2.56±2.65ᵃ	2.69±2.06ᵃ
AMB	2.27±1.82ᵃ	3.04±0.56ᵃ	1.26±1.99ᵇ	1.46±0.94ᵇ	3.69±3.34ᵃ
ASF	5.69±2.17ᵃ	5.37±1.89ᵃ	6.73±2.24ᵃ	2.44±2.58ᵇ	4.54±3.35ᵃ
Yeasts	2.48±0.44ᵃ	2.08±0.49ᵃ	2.10±0.26ᵃ	1.60±0.67ᵃ	1.87±0.77ᵃ

*Averages in rows followed by the same letters are not significantly different ($p>0.05$)

Even though about 28% (n = 14) of the samples had counts of mesophilic spore formers below detectable level, majorities of the samples (72%, n = 36) had these counts ranging from Log 2 to Log 7 CFU mL-1. Contrary to this, majorities of keribo samples (64%, n = 32) had counts of Aerobic mesophilic bacteria below Log 2 CFU mL-1 although 36% (n = 18) of the samples had counts ranging from Log 2 to Log 8 CFU mL-1. All samples had mould counts below detectable level. About 56% (n = 28) of the samples had yeast counts = Log 2 CFU mL-1. Only 2% (n = 1) of samples had counts of Enterobacteriaceae at levels >log 2 CFU mL-1 (Fig. 1).

Physico-chemical change during laboratory Keribo fermentation: Changes in pH and TA during Laboratory keribo fermentation are as shown in Fig. 2. The initial pH of unfermented Keribo at 0 h was around 5.75. The pH dropped from 5.75 to around 4.5 within the first 6 h of fermentation. The pH further dropped gradually to as low as 4.0 after 30 h of fermentation with maximum drop down to 3.7 in 48 h fermentation (Fig. 2). Generally, titratable acidity increased from 0.07 to 0.13% during the first 6 h of fermentation. Thereafter, the amount of lactic acid increased gradually up to about 20% during 36 h of fermentation. The amount of lactic acid produced reached a value of 25% at 48 h of fermentation (Fig. 2).

Microbial change during Laboratory Keribo fermentation: At early stage of fermentation, the mean counts of LAB, yeasts and moulds were below detectable level. Both yeasts and moulds remained below detectable level throughout fermentation with no significant rise in the counts of yeasts and a total elimination of molds at the end of 48 h fermentation. The counts of LAB, however, increased exponentially for the first 24 h of fermentation followed by gradual increment thereafter (Fig. 3). Likewise, counts of ASF increased exponentially for the first 30 h of fermentation with gradual decline up to 48 h of fermentation.

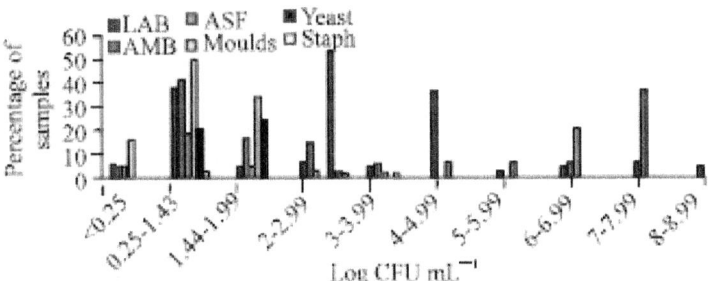

Figure 1: Distribution of counts of major microbial groups in Keribo samples. AB, lactic acid bacteria; AMB, Aerobic mesophilic bacteria; ASF, Aerobic spore former; EB, Enterobacteriaceae; Staph, staphylococci

On the other hand, the counts of AMB decreased rapidly in the course of fermentation for the first 24 h and began to decrease gradually for the rest of the fermentation period. In the later part of fermentation both AMB and ASF had related count.

Grouping of the dominant LAB: Lactic acid bacteria isolated from Keribo samples were classified into the genus Leuconostoc and were further classified into the species level on the basis of their physiological and other biochemical characteristics (Table 3). The identified isolates of Leuconostoc species were non-spore-forming, single cocci, gram-positive, catalase positive and non-motile. When they are cultivated in MRS broth, cells precipitated rapidly and growth did not occur on the surface. The determination of fermentation products showed that the isolates could convert glucose to lactic acid and CO_2 through a typical hetero-fermentative pathway. It produced ammonia from arginine and hydrolyzed bile esculine, but indole production was negative in tryptone broth. These bacteria also produced a characteristic yellow pigment in MRS broth. Moreover, tests for temperature tolerance showed the organism could tolerate temperature of up to 40°C. Prolonged incubation in MRS broth at 4, 6.5, 8, 15 and 20% NaCl resulted in turbidity growth that shows increased in the number of bacteria. Based on these characteristics the isolates were identified as L. mesenteroids.

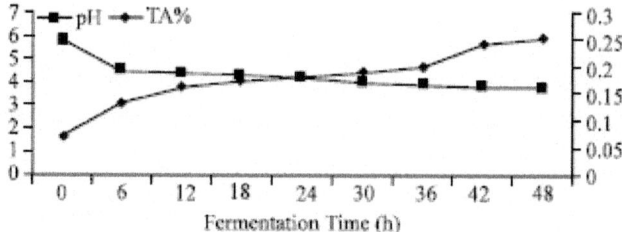

Figure 2: pH and TA Changes in traditionally fermenting Keribo

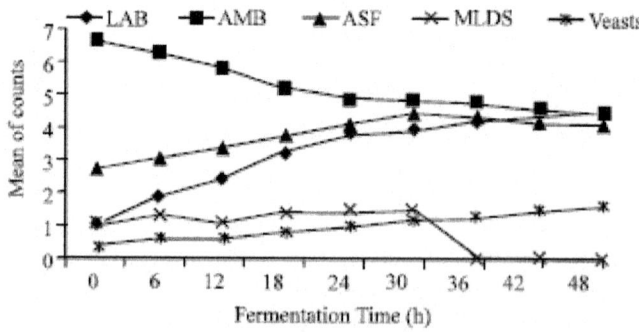

Figure 3: Dynamic of LAB, AMB, ASF and Mould/ Yeast during traditional Laboratory fermentation of Keribo, Where, LAB: Lactic acid bacteria, AMB: Aerobic mesophilic bacteria, ASF: Aerobic spore formers, MLDS: Molds

Table 3: Morphological, physiological and biochemical characteristics of LAB

Characteristics	Isolates S_1	S_2	S_3	S_4	S_5	S_6	S_7
Cell morphology	Cocci	Cocci	Cocci	Cocci	Cocci	cocci	Cocci
Cellular arrangement	Single	Single	Single	Single	Single	single	Singe
Motility test	-	-	-	-	-	-	-
Catalase activity	+	+	+	+	+	+	+
Coagulase test	-	-	-	-	-	-	-
Gas from glucose	+	+	+	+	+	+	+
Yellow color from MRS broth	+	+	+	+	+	+	+
Growth at temperature (°C)							
15	+	+	+	+	+	+	+
40	+	+	+	+	+	+	+
45	-	-	-	-	-	-	-
Tolerance to NaCl (%)							
4	+	+	+	+	+	+	+
6.5	+	+	+	+	+	+	+
8	+	+	+	+	+	+	+
15	+	+	+	+	+	+	+
20	+	+	+	+	+	+	+
Indole production in Tryptone broth	-	-	-	-	-	-	-
Arginine hydrolysis	+	+	+	+	+	+	+
Bile esculine hydrolysis	+	+	+	+	+	+	+
Acid from							
Lactose	+	+	+	+	+	+	+
Mannitol	+	+	+	+	+	+	+
Nitrate broth	-	-	-	-	-	-	-

Source of dominant lab involved in keribo fermentation: The data gathered in this study showed, during traditional keribo preparation, barley was deeply roasted and boiled in boiling water at 65 to 70°C for 10 to 20 min. Thereafter, it was filtered and table sugar and yeast was added to the filtrate. The microbiological analysis indicated that the dominating LAB isolated from both collected Keribo sample and Keribo prepared by experienced woman showed catalase positive and all were hetero-fermentative. The result showed that only a single genus and/or a single species of LAB involves in Keribo fermentation. Result obtained from microbiological analysis of table sugar revealed that the isolate was gram-positive, catalase-positive, fermentative, non-oxidative, non-spore- forming, hetero-fermentative and single cocci that considered as Leuconostoc species, mainly Leuc. mesenteroides.

Determination of antibiotic susceptibility patterns of Isolates: Results of the sensitivity studies of the LAB isolates tested against 8 different types of antimicrobial agents are shown in Table 4. All isolates were susceptible to penicillin G, Norfloxacin, gentamycin, ampicillin chloramphenicol, amikacin and norfloxacin. However, the isolates were resistant to Vancomycin.

Besides sharing similarity in morphological and physiological characteristics, the isolates shared similarity in antibiotics resistance/sensitivity. The isolates could be species of the same strains.

Table 4: Antibiotic sensitivity of Leu. Mesenteroids

	Groups of LAB isolated from different localities				
Antimicrobial agents	Leum₁	Leum₂	Leum₃	Leum₄	Leum₅
Penicilin G	+	+	+	+	+
Vancomycin	-	-	-	-	-
Amikacin	+	+	+	+	+
Bacitracin	+	+	+	+	+
Ampicilin	+	+	+	+	+
Norfloxacin	+	+	+	+	+
Chloramphenicol	+	+	+	+	+
Gentamicin	+	+	+	+	+

-: Resistant, +: Susceptible, Leum: Leuconostoc mesenteroides

DISCUSSION

Keribo is a traditional, non-alcoholic, dark brown colored fermented beverage commonly consumed in rural and urban areas of Jimma zone, southwestern of Ethiopia, with some similarity to Boza of Bulgaria, Albania, Turkey and Romania (Blandino et al., 2003). It is produced by an over-night fermentation of cereal (barley) predominantly by activities of LAB like the fermentation of shamita (Bacha et al., 1999).

High count of LAB could account for acidification of the product with extension of fermentation periods. LAB have been involved in the natural fermentation of many traditional Ethiopian fermented foods and beverages (Bahiru et al., 2006).

Deep-roasting o f the cereal and boiling at about 65-70°C for 15 to 20 min during Keribo preparation must have eliminated most of the contaminant associated with the raw materials. As most of the isolates failed to tolerate temperature above 40°C, the single species of LAB that dominated in the final product must have joined the system from sugar used for fermentation. Efiuvwevwere and Akoma (1997)reported similar treatment of ingredients at 70°C for 30 min during preparation of pasteurized Nigerian beverage, Kunun-zaki, in which most of the microorganisms were destroyed except the Bacillus species and the thermo-tolerant lactic acid bacteria.

Since the cooking process (deep roasting and boiling at 65 to 70°C) and low pH inactivates the contaminants, contamination of Keribo with Staphylococcus and Enterobacteriaceae could be due to post production contamination. The occurrence of Staphylococcus (0.83%) and Enterobacteriaceae (0.75%) are evidence of poor hygienic conditions of some of the Keribo samples. These organisms may be contaminants from unsafe water used either to dilute the ready-to-consume Keribo or wash utensils. The utensils used for preparation of Keribo and serving are made of low quality plastic and necked-bottles that are difficult to be cleaned.

Although, there are no microbiological standards set for the traditional fermented foods/beverages of Ethiopia, the mean counts of staphylococci, Enterobacteriaceae, yeasts and molds observed among the samples of Keribo were on the lowest margin of the standards set for fruit juices served in the Gulf region, indicating the maximum count permitted for total colony count of coliforms, yeast and molds are $1x104$, 100 and $1x103$ CFU mL^{-1}, respectively (Gulf Standards, 2000). However, the means counts of aerobic spore-formers and aerobic mesophilic bacteria of the samples were 4.96 log CFU mL^{-1} (with the maximum count of 7.97 log CFU mL^{-1}) and 2.34 log CFU mL^{-1} (with maximum of 8.31 log CFU mL^{-1}), respectively. On the basis of the Gulf Standards, it is clear that the colony counts of LAB, AMB and ASF in our Keribo samples exceeded the standard by considerable margin. From long history of its safety, the high counts of LAB may not pose hazard to the health of consumers. The low mean counts of staphylococci also avoid the risk of enterotoxin production as toxin production among these groups is possible after the counts exceed or equals 106 CFU mL^{-1} (James, 2000). High counts of aerobic mesophilic bacteria may trigger health problems provided that there are potential pathogenic strains among the strains including E. coliand Salmonella species.

The microbiology of Keribo samples drawn an intervals during controlled laboratory fermentation were observed to have mean counts of Coliforms, Enterobacteriaceae, Enterococci and Staphylococci below detection level. The two steps heat treatment during Keribo preparation (deep roasting of barley and boiling of roasted barley in water to dissolve it) has contributed to eliminate these bacterial groups. Moreover, the drop in pH level in the course of fermentation due to rise in the level of percent lactic acid could account to the betterment and microbiological safety of the fermented product. Laboratory prepared Keribo had comparable microbial counts with samples obtained from local Keribo brewers in Jimma Zone. Although with steady increase and below detectable level at the end of fermentation, the mean counts of yeasts increased throughout fermentation (over a period of 48 h) of the laboratory prepared Keribo. Likewise, there was an increase in the number of LAB and aerobic spore formers. The growth of yeasts appeared not to be inhibited by the acidity developed by the activities of lactic acid bacteria and proliferation with ease (Etchells et al., 1943).

The positive reaction to catalase test of our LAB isolates is in contradiction to the common characteristics of LAB (Aguirre and Collins, 1993). The result of the present study is in agreement with the rare observations reported in the distant past (Johnson and Mccleskey, 1957; Whittenbury, 1964; Yousten et al., 1975; Lucey and Condon, 1986) and even some recent reports (Bayane et al., 2006; Azizpour et al., 2009).

In the present study utilization of mannitol in MRS broth which was not reported among other Leuconostoc spp, accumulation of H2O2 that did not affect better growth of the leuconostocs and showed catalase activity (Lucey and Condon, 1986), high salt concentration (20%NaCl), high temperature (40°C) tolerance (Oyewole and Odunfa, 1990) and vancomycin-resistant were the specific criteria to decide on isolated LAB to be Leu. mesenteroids.

Although, the first report from traditional Ethiopian fermented beverages, the dominance of Leuconostoc species was reported earlier during the cassava fermentation for gari production (Okafor, 1977). Besides dominating microflora of the final product, several studies have shown that Leu. mesenteroides could also initiate fermentation processes such as the fermentation of idli (Mukherjee et al., 1965), sauerkraut (Pederson and Albury, 1969; Steinkraus, 1992), teff (Gashe,1985), Cassava (Oyewole and Odunfa, 1990) and Burukutu (Kolawole et al., 2007).

Aerobic Mesophilic Bacteria (AMB) initiated Keribo fermentation at 0 h to 6 h as shown by their early leading rate of growth followed by the succession of LAB. The initial high pH 5.75 of the Keribo fermentation at 0 h would explain the reason for growth of Aerobic Mesophilic Bacteria (AMB) while the lower pH (pH = 4.47) at 6 h fermentation began to inhibit their growth. The high numbers of LAB attained after 6 h fermentation was responsible for a marked reduction of pH and increment in TA resulting in inhibition of most Aerobic Mesophilic Bacteria (AMB). Thus, fermentation for 24 h appeared to be a turning point for an accelerated reduction in number of aerobic mesophilic bacteria and stabilization of the maximum numbers of acid producing bacteria involved in Keribo fermentation. Thereafter, the LAB entered steady growth and showed relative decline during 36 h fermentation that corresponds with lactic acid production. This was because the cocci which would normally initiate fermentation were suppressed by rapid decrease in pH with accelerated increase in acidity followed by high growth rate of LAB responsible for end fermentation. This result was in agreement with the disappearance ofLeu. mesenteroides beyond the first 48 h of the cassava fermentation during fufu production due to its inability to tolerate the increasing acidity of the fermenting mash (Oyewole and Odunfa, 1990).

The microbiological analysis of common sugar used in Keribo preparation was showed LAB with the same morphological and physiological characteristic which was similar to LAB obtained from dynamics and sample collected from venders. Overall, the results obtained from analysis of Keribo samples, laboratory fermented Keribo (dynamics) and samples of sugar used for the making of Keribo were similar in morphological and physiological characteristics. Therefore, the addition of sugar and yeast to un-malted, deeply roasted and boiled barley to initiate fermentation of Keribo is the possible source of LAB responsible for Keribo fermentation.

Antimicrobial resistance has been increasing in many parts of the world; it becomes increasingly important to monitor the antimicrobial susceptibility of lactic acid bacteria isolated from food and drinks including Keribo. All Leu. mesenteroids isolates were resistant to vancomycin and susceptible to penicillin G, gentamicin, ampicilin as also reported by Bacha et al. (2010). The low resistance to the commonly used antibiotics of these strains could show low contribution of these strains in the dissemination of resistance genes to potential pathogens in the environment, including fermented foods. Thus, the observed Intrinsic resistance of LAB to vancomycin could be the result of natural resistance of the isolates (Salminen et al., 1998).

To sum-up, during production and sales, venders and local processors must always keep their personal hygiene to discourage contamination. Sellers should also ensure that they do not expose the fermented products during display because this may predispose them to contamination. Improving the processing condition and upgrading traditionally fermented food production could improve the food in-security problems of the community. In order to produce the desired amount of traditional fermented beverages, it calls for optimization of the production processes and/or techniques. Hence, future studies should include the selection of most suitable strains for starter culture development that may be used to scale up the production of Keribo from households- level to large scale production.

CONCLUSION

This study indicated that LAB inoculated from yeast and sugar added and the yeasts are involved in the fermentation process. The prevalence of isolation of potential pathogen in traditional Keribo was low as mean counts of coliforms, Staphylococci, Enterococci and Enterobacteriaceae were below detectable level. It requires further study on the effect of heat treatment on nutritional value of the final product.

REFERENCES

1.   Abegaz, K., F. Beyene, T. Langsrud and A.J. Narvhus, 2002. Indigenous processing methods and raw materials of borde, an Ethiopian traditional fermented beverage. J. Food Technol. Afr., 7: 59-64.

2. Adams, M.R. and L. Nicolaides, 1997. Review of sensitivity of different food borne Pathogens to fermentation. Food Control, 8: 227-239.

3. Aguirre, M. and M.D. Collins, 1993. Lactic acid bacteria and human clinical infection. J. Applied Bacteriol., 75: 95-107.

4. Aidoo, K.E., S. Angel, E.V.MA. Carpio, H.A. Dirar and S. Feresu et al., 1992. Research Priorities in Traditional Fermented Foods. In: Applications of Biotechnology to Traditional Fermented Foods, Gaden, J.R.E.L., M. Bokanga, S. Harlander, C.W. Hesseltine and K.H. Steinkraus (Eds.). National Academy Press, Washington, DC., pp: 3-17.

5. Ashenafi, M., 2002. The microbiology of Ethiopian foods and beverages. A review. SINET Ethiop. J. Sci., 25: 97-140.

6. Ashenafi, M., 2006. A review on the microbiology of indigenous fermented foods and beverages of Ethiopia. Ethiop. J. Biol. Sci., 5: 189-245.

7. Azizpour, K., A. Tkmechi and N. Agh, 2009. Characterization of lactic acid bacteria isolated from the intestines of common carp of West Azarbaijn, Iran. J. Anim. Vet. Adv., 8: 1162-1164.

8. Bacha, K., T. Mehari and M. Ashenafi, 1998. The microbial dynamics of borde fermentation, a traditional ethiopian fermented beverage. SINET: Ethiop. J. Sci., 21: 195-205.

9. Bacha, K., T. Mehari and M. Ashenafi, 1999. Microbiology of the fermentation of Shamita, a traditional Ethiopian fermented beverage. SINET: Ethiop. J. Sci., 22: 113-126.

10. Bacha, K., T. Mehari and M. Ashenafi, 2010. Antimicrobial susceptibility patterns of LAB isolated from Wakalim, a traditional Ethiopian fermented sausage. J. Food Saf., 30: 213-223.

11. Bahiru, B., T. Mehari and M. Ashenafi, 2006. Yeast and lactic acid flora of tej, an indigenous Ethiopian honey wine: Variation within and between production units. Food Microbiol., 23: 277-282.

12. Bauer, A.W., W.M. Kirby, J.C. Sherris and M. Turck, 1966. Antibiotic susceptibility testing by a standardized single disk method. Am. J. Clin. Pathol., 45: 493-496.

13. Bayane, A., D. Roblain, R.D. Dauphin, J. Destain, B. Diawara and P. Thonart, 2006. Assessment of the physiological and biochemical characterization of a Lactic acid bacterium isolated from chicken faeces in sahelian region. Afr. J. Biotechnol., 5: 629-634.

14. Blandino, A., M.E. Al-Aseeri, S.S. Pandiella, D. Cantero and C. Webb, 2003. Cereal-based fermented foods and beverages. Food Res. Int., 36:

527-543.

15. Byaruhanga,Y.B., 1998. Inhibition of Bacillus cereus by lactic acid bacteria in mageu, a sour maize beverage. M.Sc. Thesis, University of Pretoria, South Africa.

16. Efiuvwevwere, B.J.O. and O. Akoma, 1997. The effects of chemical preservatives and pasteurization on the microbial spoilage and shelf-life of Kunun-Zaki. J. Food Saf., 17: 203-213.

17. Etchells, J.L., I.D. Jones and M.A. Hoffman, 1943. Brine preservation of vegetables. Proceedings of the Institute Food Technology, June 2-4, 1943, USA., pp: 176-182.

18. Facklam, R. and. J.A. Elliott, 1995. Identification, classification and clinical relevance of catalase-negative, gram-positive cocci, excluding the streptococci and enterococci. Clin. Microbiol. Rev., 8: 479-495.

19. Facklam, R., D. Hollis and M.D. Collins, 1989. Identification of gram-positive coccal and coccobacillary vancomycin-resistant bacteria. J. Clin. Microbiol., 27: 724-730.

20. Farrow, J.A.E., R.R. Facklam and. M.D. Collins, 1989. Description of Leuconostoc citreum new species and Leuconostoc pseudomesenteroides new species. Int. J. Syst. Bacteriol., 39: 279-283.

21. Ferraro, M.J., 2000. Performance Standards for Antimicrobial Disk Susceptibility Tests. 7th Edn., NCCLS, Wayne, PA., USA., ISBN-13: 9781562383930, Pages: 26.

22. Gashe, B.A., 1985. Involvement of lactic acid bacteria in the fermentation of tef (Eragrotis tef), An ethiopian fermented food. J. Food Sci., 50: 800-801.

23. Gregersen, T., 1978. Rapid method for distinction of gram-negative from gram-positive bacteria. Eur. J. Applied Microbiol. Biotechnol., 5: 123-127.

24. Gulf Standards, 2000. Microbiological criteria for food stuffs-part 1. GCC. Riyadh, Saudi Arabia.

25. Hugh, R. and E. Leifson, 1953. The taxonomic significance of fermentative versus oxidative metabolism of carbohydrates by various gram negative bacteria. J. Bacteriol., 66: 24-26.

26. James, J.M., 2000. Modern Food Microbiology. 6th Edn., Aspen Publishers Inc., Maryland, USA., Pages: 268.

27. Johnson, M.K. and C.S. Mccleskey, 1957. Studies on the aerobic carbohydrate metabolism of Leuconostoc mesenteroides. J Bacteriol., 74: 22-25.

28. Kolawole, O.M., R.M.O. Kayode and B. Akinduyo, 2007. Proximate and microbial analyses of burukutu and pito produced in Ilorin, Nigeria. Afr. J. Biotechnol., 6: 587-590.

29. Kovacs, N., 1956. Identification of Pseudomonas pyocyana by the oxidase reaction. Nature, 178: 703-703.

30. Liasi, S.A., T.I. Azmi, M.D. Hassan, M. Shuhaimi, M. Rosfarizan and A.B. Ariff, 2009. Antimicrobial activity and antibiotic sensitivity of three isolates of lactic acid bacteria from fermented fish product, Budu. Malaysian J. Microbiol., 5: 33-37.

31. Lucey, C.A. and S. Condon, 1986. Active role of oxygen and NADH oxidase in growth and energy metabolism of Leuconostoc. Microbiology, 132: 1789-1796.

32. Mugula, J.K., S.A.M. Nnko and T. Sorhaug, 2001. Changes in quality attributes during storage of Togwa, a lactic acid fermented gruel. J. food Saf., 21: 181-194.

33. Mukherjee, S.K., M.N. Albury, C.S. Pederson, A.G. Van veen and K.N. Steinkraus, 1965. Role of Leuconostoc mesenteroids in leaving leavening batter of Idli, a fermented food of India. Applied Microbiol., 13: 227-231.

34. Nair, P.S. and P.K. Surendran, 2005. Biochemical characterization of lactic acid bacteria isolated from Fish and Prawn. J. Culture Collections, 4: 48-52.

35. Okafor, N., 1977. Microorganisms associated with cassava fermentation for garri production. J. Applied Bacteriol., 42: 279-284.

36. Oyewole, O.B. and S.A. Odunfa, 1990. Characterization and distribution of lactic acid bacteria in cassava retting during fufu production. J. Applied Bacteriol., 68: 145-152.

37. Pederson, C.S. and M.N. Albury, 1969. The Sauerkraut Fermentation. New York State Agr. Expt. Sta. Tech. Bull. Bulletin 824, Geneva.

38. Pilone, G.J., M.G. Claytone and R.J. Van Duivenboden, 1991. Characterization of wine lactic acid bacteria: Single broth culture for tests of heterofermentation, mannitol from fructose and ammonia from arginine. Am. J. Enol. Vatic., 42: 153-157.

39. Ricciardi, A., E. Parente, P. Piraino, M. Paraggio and P. Romano, 2005. Phenotypic characterization of lactic acid bacteria from sourdoughs for Altamura bread produced in Apulia (Southern Italy). Int. J. Microbiol., 98: 63-72.

40. Rojo-Bezares, B., Y. Saenz, P. Poeta, M. Zarazaga, F. Ruiz-larrea and C. Torres, 2006. Assessment of antibiotic susceptibility within lactic acid

bacteria strains isolated from wine. Int. J. Food Microbiol., 111: 234-240.

41. Salminen, S., A. von Wright, L. Morelli, P. Marteau and D. Brassart et al., 1998. Demonstration of safety of probiotics: A review. Int. J. Food Microbiol., 44: 93-106.

42. Samuel, S. and A.G. Berhanu, 1991. The microbiology of Tella fermentation. Sinet. Ethiop. J. Sci., 14: 81-92.

43. Steinkraus, K.H., 1992. Lacti Acid Fermentation. In: Applications of Biotechnology in Traditional Fermented Foods, Gaden, E.L. JR., M. Bokanga, S. Harlander, C.W. Hesseltine and K.H. Steinkraus (Eds.). National Academy Press, Washington, DC., pp: 43-51.

44. Vlkova, E., V. Rada, P. Popelarova, I. Trojanova and J. Killer, 2006. Antimicrobial susceptibility of bifidobacteria isolated from gastrointestinal tract of calves. Livestock Sci., 105: 253-259.

45. Whittenbury, R., 1964. Hydrogen peroxide formation and catalase activity in the Lactic acid bacteria. J. Gen. Microbiol., 35: 13-36.

46. Yousif, N.M.K., P. Dawyndt, H. Abriouel, A. Wijjaya and U. Schillinger et al., 2005. Molecular characterization, technological properties and safety aspects of Enterococci from Hussuwa. An African fermented sorghum product. J. Applied Microbiol., 98: 216-228.

47. Yousten, A.A., J.L. Johnson and M. Salin, 1975. Oxygen metabolism of catalase-negative and catalase-positive strains of Lactobacillus plantarum. J. Bacteriol., 123: 242-247.

Chapter 2

BREWHOUSE-RESIDENT MICROBIOTA ARE RESPONSIBLE FOR MULTI-STAGE FERMENTATION OF AMERICAN COOLSHIP ALE

Nicholas A. Bokulich[1,2], Charles W. Bamforth[2] , David A. Mills[1,2]

[1] Department of Viticulture and Enology, Robert Mondavi Institute of Wine and Food Science, University of California Davis, Davis, California, United States of America,

[2] Department of Food Science and Technology, Robert Mondavi Institute of Wine and Food Science, University of California Davis, Davis, California, United States of America

ABSTRACT

American coolship ale (ACA) is a type of spontaneously fermented beer that employs production methods similar to traditional Belgian lambic. In spite of its growing popularity in the American craft-brewing sector, the fermentation microbiology of ACA has not been previously described, and thus the interface between production methodology and microbial community structure is unexplored. Using terminal restriction fragment length polymorphism (TRFLP), barcoded amplicon sequencing (BAS), quantitative PCR (qPCR) and culture-dependent analysis, ACA fermentations were shown to follow a consistent fermentation progression, initially dominated by *Enterobacteriaceae* and a range of oxidative yeasts in the first month, then ceding to *Saccharomyces* spp. and *Lactobacillales* for the following year. After one year of fermentation, *Brettanomyces bruxellensis* was the dominant yeast population (occasionally accompanied by minor populations of *Candida* spp., *Pichia* spp., and other yeasts) and*Lactobacillales* remained dominant, though various aerobic bacteria became more prevalent. This work demonstrates that ACA exhibits a conserved core microbial succession in absence of inoculation, supporting the role of a resident brewhouse microbiota. These findings establish this core microbial profile of spontaneous beer fermentations as a target for production control points and quality standards for these beers.\

INTRODUCTION

American coolship ale (ACA) is a type of beer produced in the United States using production practices adopted from the lambic brewers of Belgium, in an attempt to create a similar style of sour ale. Traditional lambic is fermented entirely spontaneously by exposing the cooling, boiled wort to the atmosphere overnight in an open, shallow vessel known as a "coolship," during which time it becomes inoculated by autochthonous yeasts and bacteria that perform the fermentation, lasting 1–3 years in oak casks [1], [2]. During this time, several stages of succession occur in the microbial community profile that are likely responsible for the unique flavors exhibited by these beers. Previous culture-based studies revealed that*Enterobacteriaceae* and oxidative yeasts (*Kloeckera* spp.) dominate the early fermentation, lasting 1–2 months, after which *Saccharomyces* spp. take hold and are responsible for the main alcoholic fermentation and *Pediococcus* spp. proliferate, producing copious quantities of lactic acid [1], [3]. Finally, *Brettanomyces* spp. dominate the late stage of fermentation after 1 year, producing volatile phenols and other characteristic aroma compounds [3]. The combined metabolic activity of several microbial populations, the low pH, and the high ethanol environment result in the production of very high concentrations of aromatic esters, volatile phenols, and other compounds responsible for the unique sensory qualities of these beers [3],[4]. ACA, currently produced by a small number of breweries, is a direct adaption by American craft brewers to recreate a lambic-style beer, adopting the same recipes and production methods employed by the Belgian lambic breweries. However, to our knowledge, neither of these beers has ever been studied using culture-independent techniques, and how geographical location influences this fermentation microbial profile has yet to be determined.

In this study, ACAs from different batches were selected from one American brewery and followed using TRFLP to give a low-resolution view of bacterial community patterns as well as high-resolution taxonomic identification of yeasts and lactic acid bacteria (LAB) across samples and across time. Representative samples were then selected from clusters identified by TRFLP and sequenced using 16S rDNA barcoded amplicon sequencing (BAS) on the Illumina GAIIx to provide an in-depth look at bacterial community structure over time. Quantitative PCR (qPCR) and culture-based techniques were also used to characterize changes in the microbial community over time. Findings established that ACA fermentation involves a multiphase, core microbial profile, which is conserved batch-to-batch, supporting the presence of resident brewhouse microbiota responsible for conducting the fermentation. Additionally, this core profile displayed some notable similarities to the microbial profile of lambic, suggesting that the shared production methods exert a common selective niche environment for spontaneous beer fermentation.

MATERIALS AND METHODS

Sampling

Wort was prepared according to the standard protocol of the brewery studied, with a typical mash followed by at least 1 hr of kettle boiling, after which wort was transferred to an open coolship and left to cool overnight. The following morning, after the wort reached ~22°C, it was transferred to enclosed oak barrels and fermented at cellar temperature. Barrels were topped off as needed to minimize headspace exposure to air during fermentation. Prior to bottling, a fruit slurry was added to the beer and allowed to referment in bottles.

Samples were collected starting the morning after overnight exposure of the wort to the atmosphere in the coolship (wk 0) through up to 184 wk of fermentation in oak casks (Figure 1A;). Some bottle-refermented samples were also tested from batches one and two (wk 148). In order to collect samples representing the complete 3-yr fermentation of ACA, eight different batches were sampled representing an overlapping mosaic of time points : batches one and two covered wk 95–184; batches three and four from wk 60–149; batches five and six from wk 8–92; and batches seven and eight from wk 0–83. Replicate barrels from each batch were sampled, where possible, and the same barrels were followed throughout the sampling period (Figure 1A;). Samples of fresh wort from batches seven and eight were also collected prior to coolship exposure to ensure that populations detected by culture-independent methods were not due to residual DNA from grain-associated populations surviving kettle boil; no amplification could be achieved from any of these samples.

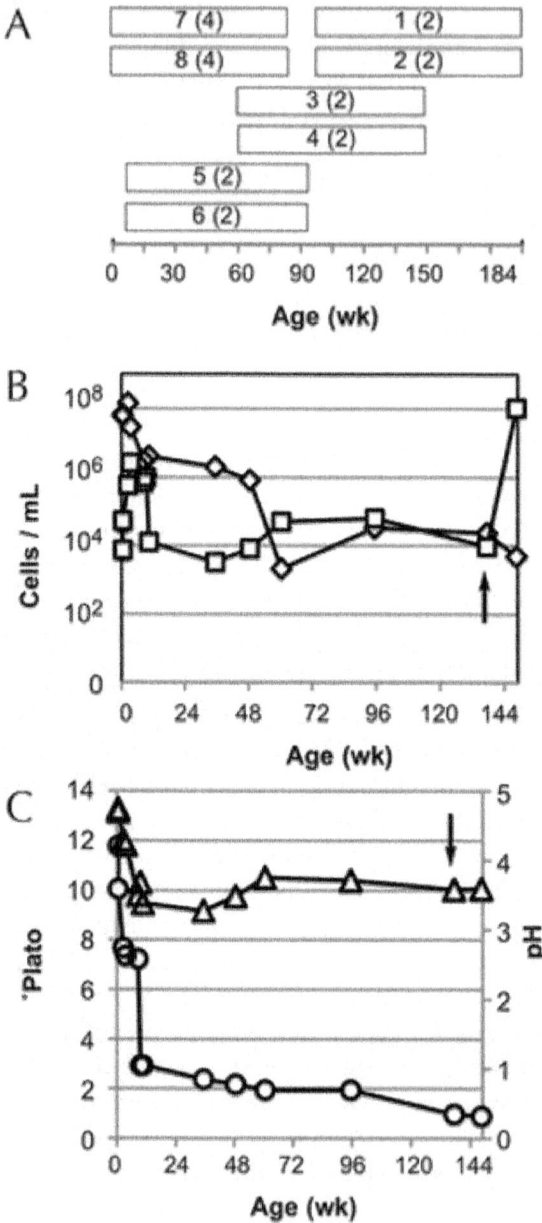

Figure 1: ACA fermentation profile and sampling regime. *Panel A*: Sampling regime employed representing 3 years of ACA fermentation. White bars represent the span of sampling times for each batch. Labels indicate batch number (number of barrel replicates in parenthesis). *Panels B/C*: Real-time PCR of total bacteria and total yeast populations (*Panel B*) and pH and °Plato (*Panel C*) across ACA fermentation. All values are

averages of multiple batches tested in duplicate (where possible). Error bars indicate ± 1 standard deviation. $\Diamond$, Total bacteria; $\square$, total yeasts; $\triangle$, pH; $\circ$, °Plato. Arrow indicates time at which fruit was added and beer was refermented in bottle. doi:10.1371/journal.pone.0035507.g001

Samples were collected aseptically from sample ports located in the head of the barrels, 5–6 in. above the lower rim. Approximately 200 mL were bled from the sampling port before sample collection. Fresh samples used for culture-dependent techniques were rush-shipped on ice, stored at 4°C until analysis, and processed within 24 hr. Samples for TRFLP analysis only were immediately (within 24 hr of sampling, immediately following shipping on ice) centrifuged at 4,000× g for 15 min at 4°C, decanted, and the remaining pellet stored at −20°C until analysis.

DNA Extraction

From the centrifuged samples, 100 μL of cell pellet were removed and washed 3 times by suspension in 1 mL ice-cold PBS, centrifugation at 8,000× g (5 min), and the supernatant discarded. The cell pellet was then suspended in 200 μL DNeasy lysis buffer (20 mM Tris-Cl [pH 8.0], 2 mM Sodium EDTA, 1.2% Triton X-100) supplemented with 40 mg/mL lysozyme and incubated at 37°C for 30 min. From this point, the extraction proceeded following the protocol of the Qiagen Fecal DNA Extraction Kit (Qiagen, Valencia, CA), with the addition of a bead beater cell lysis step of 2 min at maximum speed following addition of "buffer ASL" using a FastPrep-24 bead beater (MP Bio, Solon, OH). DNA extracts were stored at −20°C until further analysis. Duplicate extractions were made for all samples, except for those samples (very old samples still maturing in barrels) containing such low concentrations of cells that the entire sample needed to be used for one extraction to obtain a workable quantity of DNA.

TRFLP Analysis

PCR amplification was performed in 50-μL reactions containing 1 μL of DNA template, 25 μL 2× Promega GoTaq Green Master Mix (Promega, Madison, WI), 1 mM $MgCl_2$, and 2 pmol of each primer. Each PCR was performed in triplicate and the products combined prior to purification.

For amplification of the ITS1/ITS4 domain of yeast 26S rDNA genes (ITS-TRFLP) [5], the forward primer used was ITS1HEX (5′-[5HEX] TCCGTAGGTGAACCTGCGG-3′) and the reverse primer was ITS4 (5′-TCCTCCGCTTATTGATATGC-3′) [6]. The PCR conditions were an initial denaturation at 95°C for 2 min, followed by 30 cycles of denaturation at 95°C for 1 min, annealing at 50°C for 1 min, and extension at 72°C for 2 min, and with a final extension at 72°C for 7 min.

For amplification of universal bacterial 16S rDNA genes (16S-TRFLP), the forward primer used was Uni331F-FAM (5'-[5FAM] TCCTACGGGAG-GCAGCAGT-3') [7] and the reverse primer was 1492R (5'-GGTTACCTTGT-TACGACTT-3') [8]. The PCR conditions were an initial denaturation at 95°C for 2 min, followed by 30 cycles of denaturation at 95°C for 30 sec, annealing at 50°C for 30 sec, and extension at 72°C for 2 min, and with a final extension at 72°C for 5 min.

Lactic acid bacteria (LAB)-specific TRFLP (LAB-TRFLP) [9] was performed using the primers NLAB2F (5'-[5HEX]-GGCGGCGTGCCTAATACATGCAAGT-3') and WLAB1R (5'-TCGCTTTACGCCCAATAAATCCGGA-3') [9]. PCR conditions consisted of an initial denaturation at 95°C for 5 min, followed by 30 cycles of denaturation at 95°C for 45 sec, annealing at 66°C for 30 sec, and extension at 72°C for 45 sec, and with a final extension at 72°C for 5 min.

All samples were amplified in triplicate and combined prior to purification using QIAquick PCR Purification Kit (Qiagen), following the manufacturer's instructions. Restriction digests were performed according to the manufacturer's instructions for each individual enzyme. Digestions of ITS PCR products were performed using HaeIII, DdeI, and HinfI. Digestions of 16S PCR products were performed using AluI, MspI, HaeIII, and HhaI. Digestions of LAB-TRFLP products used MseI and Hpy118I. The digested DNA was submitted to the UC Davis College of Biological Sciences Sequencing Facility, for fragment separation via capillary electrophoresis. Traces were visualized using the program Peak Scanner v1.0 (Applied Biosystems, Carlsbad, CA) using a baseline detection value of 10 fluorescence units. Peak filtration and clustering were performed with R software using the scripts and analysis protocols designed by Abdo and colleagues [10]. Operational taxonomic unit (OTU) picking for both LAB-TRFLP and 16S-TRFLP was based on *in silico* digest databases generated by the virtual digest tool from MiCA [11] of good-quality 16S rDNA gene sequences compiled by the Ribosomal Database Project Release 10 [12], [13], allowing up to 3 nucleotide mismatches within 15 bp of the 5' terminus of the forward primer. Putative species assignments of ITS sequence fragments were made by comparing fragment peak data to a Wine Yeast ITS TRFLP database developed in-house [5]. Principal coordinates were computed from Bray-Curtis dissimilarity scores of raw TRFLP data (prior to taxonomic classification/grouping) using QIIME [14].

Illumina Sequencing Library Construction

For amplification of the V4 domain of bacterial 16S rRNA genes, we used primers F515 and R806 [15], both modified to contain an illumina adapter region for sequencing on the illumina GAIIx platform and, on the forward primer, an 8 bp Hamming error-correcting barcode to enable sample multiplexing [16]. A list of V4 primers and barcodes used is presented in. PCR reactions contained 5–100 ng DNA template, 1× GoTaq Green Master Mix (Promega), 1 mM MgCl$_2$, and 2 pmol of each primer. Reaction conditions consisted of an initial 94°C for 3 min followed by 35 cycles of 94°C for 45 sec, 50°C for 60 sec, and 72°C for 90 sec, and a final extension of 72°C for 10 min. All samples were amplified in triplicate and combined prior to purification. Amplicons were purified using the Qiaquick 96 kit (Qiagen), quantified using PicoGreen dsDNA reagent (Invitrogen, Grand Island, NY), mixed at equimolar concentrations, and gel-purified using the Qiaquick gel extraction kit (Qiagen) all according to respective manufacturers' instructions. Purified libraries were submitted to the UC Davis Genome Center DNA Technologies Core for cluster generation and 150 bp paired-end sequencing on the Illumina GAIIx platform.

Data Analysis

Raw Illumina fastq files were demultiplexed, quality-filtered, and analyzed using QIIME [14]. The 150-nt reads were truncated at any base receiving a quality score <1e-5, and any read containing one or more ambiguous base call was discarded, as were truncated reads of <75 nt. OTUs were assigned using the QIIME implementation of UCLUST [17], with a threshold of 97% pairwise identity, and representative sequences from each OTU selected for taxonomy assignment. OTUs were classified taxonomically using a QIIME-based wrapper of the Ribosomal Database Project (RDP) classifier [18] against the RDP 16S rDNA database core set [12], [13], using a 0.80 confidence threshold for taxonomic assignment. Any OTU representing less than 0.01% of the total sequences was removed to avoid inclusion of erroneous reads, leading to inflated estimates of diversity. Filtered sequences were aligned using PyNast against a template alignment of the RDP core set filtered at 97% similarity. Beta diversity estimates were calculated within QIIME using weighted UniFrac [19] distances between samples. From these estimates, principal coordinates were computed to compress dimensionality intro three-dimensional principal coordinate analysis (PCoA) plots. Observed-species alpha-rarefaction of filtered OTU tables was also performed in QIIME to confirm that sequence coverage was adequate to capture the species diversity observed in all samples (Figure S2).

Quantitative PCR (qPCR)

Quantitative PCR was performed in 20-µL reactions containing 10–100 ng of DNA template, 0.2 µM of each respective primer, and 10 µL of Takara

SYBR 2× Perfect Real Time Master Mix (Takara Bio Inc). For amplification of total bacteria, the primers Uni334F (5'-ACTCCTACGGGAGGCAGCAGT-3') [20] and Uni514R (5'-ATTACCGCGGCTGCTGGC-3') [21] were used. Reaction conditions included an initial hold at 95°C for 20 sec, followed by 40 cycles of 4 sec at 95°C and 25 sec at 65.5°C. For amplification of total yeast, the primers YEASTF (5'-GAGTCGAGTTGTTTGGGAATGC-3') and YEASTR (5'-TCTCTTTCCAAAGTTCTTTTCATCTT-3'), producing a 124-bp fragment, were used [22]. Reaction conditions involved an initial step at 95°C for 10 min, followed by 40 cycles of 15 sec at 95°C, 1 min at 60°C, and 30 sec at 72°C. Cell concentration was calculated by comparing sample threshold values (C_T) to a standard curve of *S. cerevisiae* or *Escherichia coli* genomic DNA extracted from known cell concentrations. All reactions were performed in triplicate in optical-grade 96-well plates on an ABI Prism 7500 Fast Quantitative PCR System (Applied Biosystems). The instrument automatically calculated cycle threshold (C_T), efficiency (E), and confidence intervals. Melt curve analysis was performed after thermal cycle program completion for both assays to assess the specificities of the amplicons.

Culture-dependent Identification

Fresh samples were serially diluted in 1% saline buffer and plated on Mann-Ragosa-Sharp Agar (MRS) supplemented with 25 mg/L cycloheximide (for the selective detection of lactic acid bacteria) and Wallerstein Differential Agar (WLD) supplemented with 1% (wt/wt) Sodium Bicarbonate and 25 mg/L cyclohexamide (for the differential detection of bacteria). All plates were incubated aerobically at 25°C for 3–7 days and then colonies were counted and recorded. Morphologically distinct colonies growing on differential agar were isolated by restreaking three consecutive times on plate count agar. Two isolates were obtained for each colony morphotype. Single colonies of pure isolates growing on plate count agar were removed and suspended in 20 µL GeneReleaser (BioVentures, Inc., Murfreesboro, TN). This suspension was microwaved at medium power for 10 min prior to addition of PCR reagents without mixing the contents of the tube. PCR amplification was performed in 50-µL reactions containing 25 µL 2× Promega GoTaq Green Master Mix (Promega), 1 mM $MgCl_2$, and 0.2 µL of each primer. Bacterial 16S rDNA gene amplification was performed using the forward primer 27F (5'-AGAGTTTGATCCTGGCTCAG-3') and reverse primer 1492R (5'-GGTTACCTTGTTACGACTT-3') [8]. The thermocycler program consisted of a denaturation at 94°C for 10 min; followed by 35 cycles of a denaturation at 94°C for 1 min, annealing at 50°C for 1 min, and an extension at 72°C for 1.5 min, and a final extension step of 72°C for 10 min. All positive amplicons were purified using QIAquick PCR Purification Kit (Qiagen), following the manufacturer›s instructions, and submitted to the UC Davis College of Biological Sciences Sequencing Facility for sequencing using the same primers. Sequence construction was performed using 4 peaks (www. mekentosj.com/science/4peaks) and alignment using NCBI BLAST.

RESULTS

All batches were brewed in the Spring and Winter of 2008, 2009, and 2010 in an independent craft brewery located in the Northeastern United States. Following wort production, boiled wort was pumped into the brewery's stainless steel coolship, located in a separate portion of the facility from the brewhouse and cellar, and allowed to cool overnight under circulating air. The following morning, once cooled to approximately 22°C, the wort was pumped into barrels and left to ferment at room temperature. Fermentation typically began within one week (as indicated by gas formation). Samples were collected starting in the coolship and continuing throughout the fermentation at regular intervals (Figure 1A). In addition to tracking bulk microbial composition using qPCR (Figure 1B), pH and °Plato were measured to track the rate of fermentation (Figure 1C).

Quantitative PCR of ACA fermentations

Quantitative PCR (qPCR) was used to quantify total yeasts and bacteria during ACA fermentation (Figure 1B). As TRFLP and Illumina sequencing provide relative community structure data but no indication of absolute abundance, this step was necessary to compare differences among time points. Total bacterial load was initially high in all batches, with an average of 6.5×10^7 cells/mL ($\pm 1.2 \times 10^6$) at wk 0 and growing steadily to 1.4×10^8 cells/mL ($\pm 7.9 \times 10^6$) at wk 2 (Figure 1B). Counts then steadily declined in all samples after wk 2 (coordinately with the decrease in pH and °Plato; Figure 1C), reaching 2.1×10^3 cells/mL ($\pm 2.0 \times 10^2$) at wk 60. From here, the average populations gradually regenerated, reaching 2.3×10^4 cells/mL ($\pm 1.1 \times 10^3$) at wk 137; average populations were 4.8×10^3 cells/mL ($\pm 9.4 \times 10^1$) at wk 148 in bottled, refermenting ACA.

The average total yeast population quickly grew from 7.2×10^3 cells/mL ($\pm 3.6 \times 10^3$) cells/mL at wk 0, peaking at 2.6×10^6 cells/mL ($\pm 4.1 \times 10^5$) at wk 4. Populations remained near this level through wk 9 but quickly dropped to 2.2×10^4 cells/mL ($\pm 4.0 \times 10^3$) at wk 11 (following ~80% attenuation of soluble solids as °Plato); average total yeast populations reached a minimum of 3.2×10^3 cells/mL ($\pm 2.0 \times 10^2$) at wk 36 before regrowing into the 10^4 range from wk 60 onward. After bottle refermentation on fruit, the average total yeast population reached 1.0×10^8 cells/mL ($\pm 1.4 \times 10^7$) in all bottles.

Yeast community structure

In lambic, a diverse range of yeasts are present, but *B. bruxellensis* is the most dominant yeast and exerts a significant impact on the aroma [1]. Therefore, ITS-TRFLP [5] was used to describe ACA yeast diversity over time and to determine whether ACA fermentation displays a similar yeast community to lambic. Initially, all fermentations displayed relatively high yeast diversity, but the dominant population was already 60–80% (relative abundance) *S. cerevisiae* from wk 0 (Figure 2A).

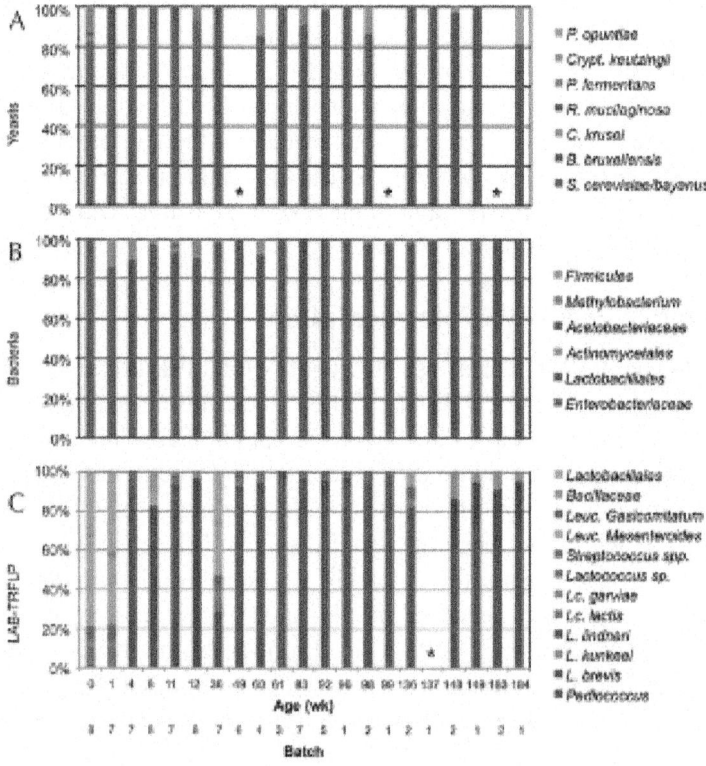

Figure 2: TRFLP analysis of ACA fermentation succession across complete timescale. Normalized relative OTU peak areas for multiple ACA fermentations observed over a 3-year period by ITS-TRFLP (yeasts, *Panel A*), 16S-TRFLP (bacteria, *Panel B*), and LAB-TRFLP (lactic acid bacteria, *Panel C*). All samples were tested in duplicate, when possible, and each bar represents averaged duplicates for a single time point, single barrel. *y*-axes indicate relative OTU abundance. *Sample was not amplifiable with these specific primers. doi:10.1371/journal.pone.0035507.g002

Other yeasts detected at wk 0 in different samples include: *Candida krusei*, *Pichia fermentans/kluyveri*, *Cryptococcus keutzingii*, and *Rhodotorula mucilaginosa*. By wk 1, all batches demonstrated a shift to ~60% *S. cerevisiae* and ~40% *R. mucilaginosa*. By wk 4, all fermentations were homogeneously *S. cerevisiae* until wk 11, when *B. bruxellensis*first appeared in some of the fermentations (average relative abundance 7%). As of wk 36, all fermentations were dominated by *B. bruxellensis* with an occasional trace of *S. cerevisiae*(including at wk 148, in a bottled, refermenting beer). *B. bruxellensis* remained the dominant yeast (64–100% relative abundance) until the end of fermentation and after bottling, with minor populations of *P. opuntiae*, *P. fermentans/kluyveri*, *C. keutzingii*, and *C. krusei* cropping up intermittently.

Bacterial community structure

TRFLP was used to track bacterial community structure in ACA over 3 yr of fermentation. In lambic, *Pediococcus* and enterobacteria play prominent parts in acidifying the beer and producing fatty acids related to characteristic aroma development, respectively [1]. Thus, total bacterial community structure was queried using 16S-TRFLP in order to assess what bacteria are present throughout ACA fermentation and followed by LAB-TRFLP [9] to further dissect the substructure of *Lactobacillales*. TRFLP indicated that the early fermentation (wk 0–4) of ACA was consistently dominated by enterobacteria (Figure 2B). Plating identified the primary enterobacteria involved to be predominantly *Klebsiella oxytoca* and *Enterobacter agglomerans*but also *Enterobacter ludwigii*, *Enterobacter cloacae*, *Enterobacter mori*, *Klebsiella pneumonia*, and *Serratia ureilytica* at different times over the first 12 wk (Table 1). None were cultured after this time point, though traces of enterobacteria were detected at later time points in some batches by molecular methods.

Table 1: Taxonomic Assignments of Bacteria Isolated from ACA. doi:10.1371/journal.pone.0035507.t001

Closest Match	Accession #	Total Score	Max ID	Colony Morphotype	Weeks Detected*
Paenibacillus provencensis strain 4401170	EF212893.1	1916	95%	Sm., creamy, translucent, slightly glossy	1
Enterobacter ludwigii isolate PSB2	HQ242715.1	1977	97%	Med., creamy, rd., glossy, slimy	1–4
Enterobacter mori strain R3-3 16S	GQ406569.1	1977	97%	Med-lg., creamy, opaque, matte, smooth	1–4
Serratia ureilytica isolate PSB22	HQ242735.1	2097	98%	Sm., wh., opaque, rd.	1–4
Pectobacterium carotovorum subsp. carotovorum strain EccB-5 16S	FJ527484.1	2049	97%	Med., rd., creamy, glossy	1–8
Klebsiella pneumoniae strain 27F 16S	GU327663.1	2056	97%	Med., hazy, copious slimy yellow coating	1–8
Enterobacter cloacae	EF120473.1	1982	96%	Tiny, hazy, translucent, glossy	1–8
Enterobacter hormaechei strain M.D.NA5-9	JF690889.1	1995	96%	Med., rd., creamy, glossy	1–8
Enterobacter agglomerans strain A84	AF130948.2	1991	96%	Lg., snotty, opaque, yellow	1–12
Klebsiella oxytoca strain SHD-1	GU361112.1	2032	98%	Med., rd., creamy, glossy	1–12
Lactobacillus brevis strain b4	FJ227317.1	2073	98%	Med., opaque qhite, rough margin	8–148
Acetobacter fabarum strain NM118-1	HM218478.1	2043	98%	Med., rd., translucent orange, spreading	8–148
Acetobacter lovaniensis strain KS1	FJ157228.1	2074	98%	Med., hazy/translucent, glossy, smooth	1–148

By wk 4, *Lactobacillales* succeeded as the dominant population in all fermentations, their relative population steadily growing from 50–70% (wk 4–12) to >90% (wk 12) for the remainder of the fermentation (Figure 2B). The only LAB cultured was *Lactobacillus brevis* (Table 1), leading to the initial assumption that this was the dominant LAB in all batches. However, LAB-TRFLP was used to resolve populations of LAB more deeply, revealing a more intricate tapestry of LAB involved throughout the fermentation, and larger differences across batches than observed using 16S-TRFLP (Figure 2C). A rich diversity was observed in the first 2 wk, dominated by *Leuconostoc* spp. and with measurable contingents of *Lactococcus lactis,Lactococcus garviae, Streptococcus* sp., *Lactobacillus delbreuckii, Lactobacillus curvatus, L. brevis,* and *Lactobacillus kunkeei.* By wk 4, the LAB population was predominantly*Pediococcus* (>80%), with minor detection of *Lactobacillus fermentum, L. brevis, L. kunkeei,* and *Lc. lactis.* In the very late fermentation of several batches, *Lactobacillus lindneri* emerged as a prevalent minority population. As *Bacilli* are also well covered by the LAB-TRFLP protocol used [9], small populations of *Bacillales* (typically <5%, except for batch eight) were detected in different batches throughout the fermentation, as was *Paenibacillus* in the first 1–2 wk of some batches. A bacterium identified as *Paenibacillus provenciensis* was isolated at wk 1 from batch eight (Table 1), but was not detected in any batches thereafter (by culturing or TRFLP).

In order to reveal relationships among samples based on community structure and to select representatives for BAS, samples were grouped via UPGMA hierarchical clustering based on Euclidean distance of MspI 16S-TRF profiles (Figure 3). Three distinct clusters form, dividing early (wk 0–1), middle (wk 2–12), and late fermentation (wk 36–184) samples, with the early- and middle-fermentation clusters each clearly bifurcating into Winter (batch eight) and Spring (batch seven) groups.

Therefore, 16 samples were selected from this set for Illumina sequencing, representing all defined clusters (at least 3 selected per cluster, with two independent samples representing each time point), all batches, and the complete time course of the fermentation, with the early fermentation highly sampled from batches seven and eight to further probe batch-to-batch variation.

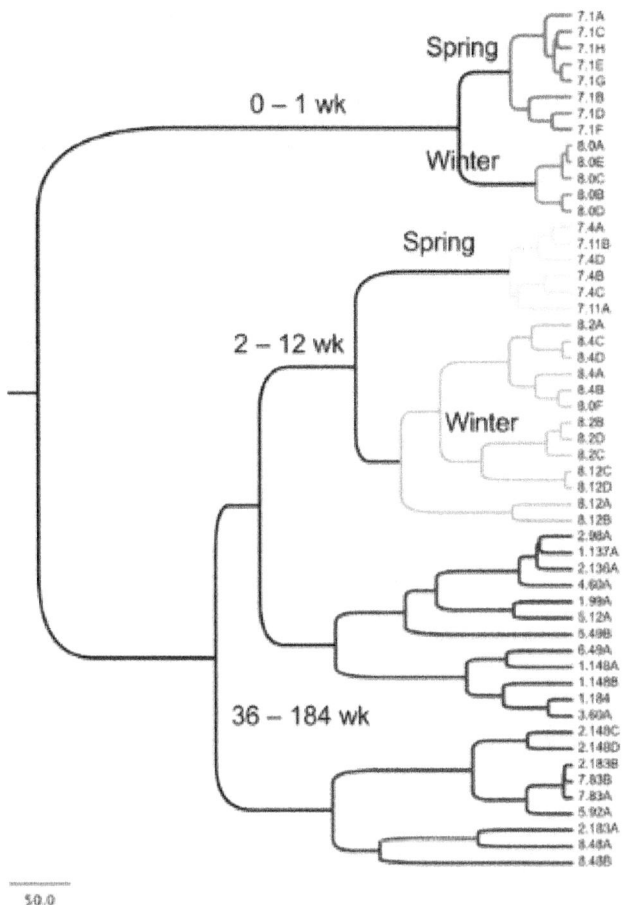

Figure 3: UPGMA hierarchical clustering demonstrates age-based grouping of samples. Hierarchical relationship among samples based on Euclidean distance of 16S-TRFLP OTU abundance profiles derived from the MspI restriction digest. Node labels (and associated colors) indicate age- and batch-based groups. Tip labels indicate batch number.wk. doi:10.1371/journal.pone.0035507.g003

Barcoded Amplicon Sequencing

As TRFLP is based on restriction mapping rather than true sequence data, BAS of universal 16S rDNA amplicons was applied to a representative subset of ACA to confirm taxonomic classifications, to identify low-abundance populations, and to perform phylogeny-based beta-diversity comparisons. Bacterial community profiles recovered by BAS revealed similar structures at the order level and genus level, compared to those observed by 16S-TRFLP and LAB-TRFLP, respectively (Figure 4). These data demonstrate the same general fermentation profile, initially dominated by *Enterobacteriaceae* and quickly succeeded by *Lactobacillales*, though by this method enterobacteria were detected at higher levels longer into the fermentation. Unfortunately, BAS could not differentiate the majority of *Enterobacteriaceae*beyond the family level (same level as TRFLP), so we must still rely on the culturing data to provide species identification of enterobacteria present. As no enterobacteria could be cultured beyond wk 12, it is unclear whether this population is experiencing its own internal species succession, or whether this is nonviable or non-active artifact from the group observed in the early fermentation.

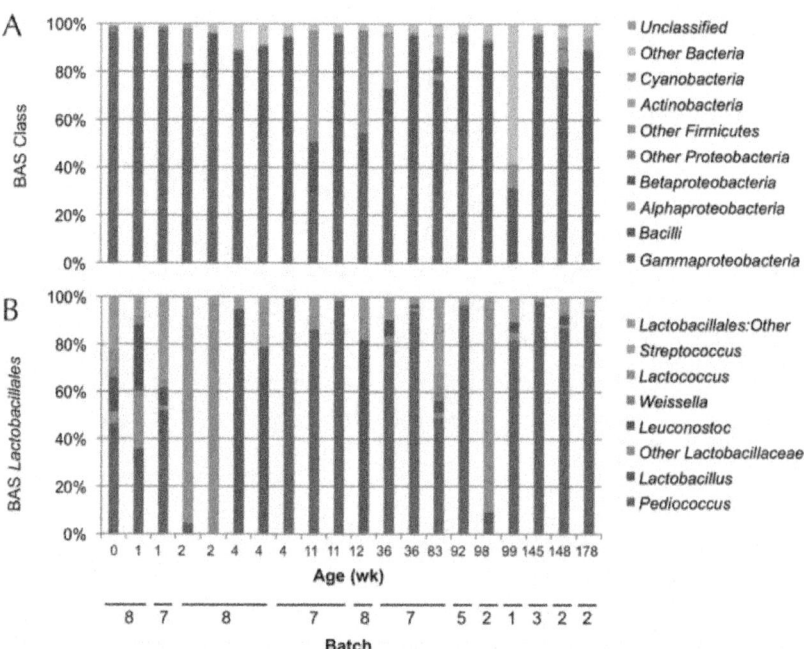

Figure 4: Bacterial taxa abundance measured using BAS 5′ sequences. *Panel A*: Class-level taxon relative abundance per sample as a percentage of total sequences. *Panel B*: Genus-level relative abundance of *Lactobacillales*. All bars represent a single sample from a single batch, and most time points are represented by at least two independent samples (where available), as presented. *y*-axes indicate relative OTU abundance. doi:10.1371/journal.pone.0035507.g004

At the genus level, the LAB community structure uncovered by BAS was highly similar to that observed by LAB-TRFLP, with a diverse early mixture of *Leuconostoc spp.*, *Lactococcus*,*Streptococcus*, and *Lactobacilli*, quickly dominated by *Pediococcus* within the first few weeks. BAS data suggest, however, that *Pediococcus* was abundant at wk 0. These data also indicate a large population of "other *Lactobacillaceae*" dominant at wk 2 and 98; this OTU (at this and other timepoints) most likely represents an amalgam of *Pediococcus* and the minor LAB observed at other time points grouped into one OTU due to sequencing error resulting in shorter sequences following quality filtration, not a distinct population. Notably, BAS was unable to differentiate LAB below the species level, so we must likewise rely on LAB-TRFLP to show species-level changes.

Two major advantages of BAS over TRFLP are that OTU identification is actually sequence-based and that sequencing is much more sensitive than TRFLP (which relies on detection of fluorescent labels), facilitating the identification of minor populations (such as the unidentified Firmicutes detected by TRFLP). One of the most abundant minor populations was *Alphaproteobacteria* (as much as 7%), including *Acetobacter* and *Brevundimonas*; the *Betaproteobacteria Ralstonia* and *Comamonadaceae* were also detected in low abundance, as were *Actinomycetales*, corroborating TRFLP data. While enterobacteria certainly represent a major population, emerging minor populations of other *Gammaproteobacteria* were also detected in wk 99–173, including *Acinetobacter* (as high as 1%, at wk 148), *Halomonas*, and *Pseudomonas*. These bacteria were detected after rigorous OTU filtration was applied, thus all represent >0.01% of total sequence abundance. A complete list of OTUs detected above this threshold, as well as relative abundance by sample, is presented in Table S2.

Batch-to-batch Consistency

Most batches exhibited consistent microbial transitions involving the same bacterial and yeast taxa. One exception was batch eight, for which LAB-TRFLP revealed batch-to-batch inconsistency not observed with 16S-TRFLP. Figure 5 compares batches seven (representative of the typical profile observed across batches) and eight using 16S-TRFLP (A,E), Class-level BAS (B,F), LAB-TRFLP (C,G), and genus-level BAS data for *Lactobacillales* (D,H). Class-level structure (Illumina) and resolution to family level (16S-TRFLP) suggest very close correspondence of these batches, with an initial dominance of enterobacteria quickly succeeded by *Lactobacillales*, the only dramatic observable differences being in minor OTUs (Figure 5A,B,E,F). However, LAB-TRFLP demonstrates a rapid divergence: whereas wk 0 (batch 8) and wk 1 (batch 7) showed very similar profiles (primary *Leuconostoc spp.*, as described above), by wk 2 (batch 8) and wk 4 (batch 7)—and through the remainder of the observation period—batch 7 was dominated by *Pediococcus* (also the dominant LAB in batches 1–6) and batch 8 was dominated by *L. brevis*. Beyond *Lactobacillales*, however, this batch demonstrated typical profiles for all other taxa, as shown by BAS data (increased levels of *Proteobacteria* correspond to *Acetobacteraceae*; Figure 5).

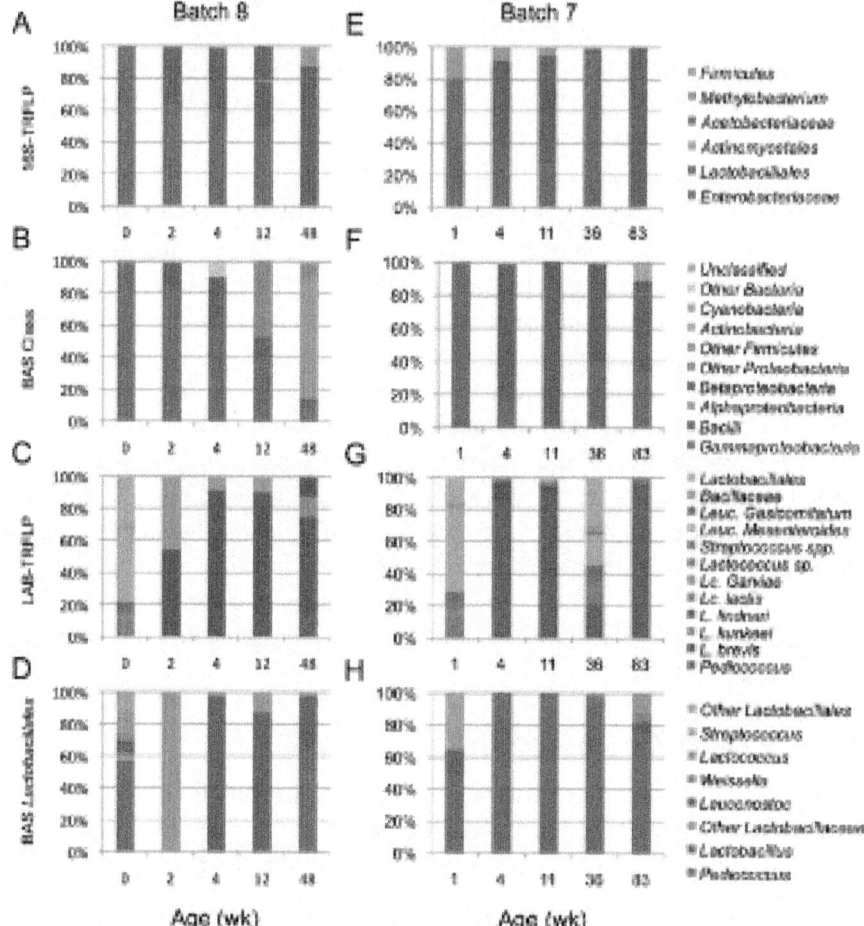

Figure 5: Parallel ACA batch comparison. Comparison of universal bacterial community structure (A,B,E,F) and *Lactobacillales*(C,D,G,H) in Spring batch 7 (*right*) and Winter batch 8 (*left*). *Panel A/E*: Relative abundance of 16S-TRFLP bacterial OTUs. *Panel B/F*: Class-level abundance per sample as percentage of total BAS sequences. *Panel C/G*: Relative abundance of LAB-TRFLP OTUs. *Panel D/H*: Genus-level relative abundance of *Lactobacillales* BAS sequences. All TRFLP samples are averages of duplicate samples. *y*-axes indicate relative OTU abundance. doi:10.1371/journal.pone.0035507.g005

Beta-diversity

Principal coordinates analysis (PCoA) was used to visualize relationships among samples and OTUs (as loadings) in a two- or three-dimensional space, in order to reveal underlying trends in these time-based data. Principal coordinates (PC) were calculated from both TRFLP and BAS data for two separate purposes: non-phylogenetic (Bray-Curtis) beta-diversity analysis using TRFLP to compare diversity across the complete dataset (as Illumina only represents a subsample), and weighted UniFrac clustering based on BAS of a representative subset to determine whether microbial community differences among samples are phylogenetically significant. Non-phylogenetic, count-based beta-diversity measures (such as Bray-Curtis) assume equidistant similarity among OTUs (representing a "star phylogeny") [23], so true phylogenetic measures are extremely useful (when available) to determine whether communities are not only structurally distinct but also represent significant evolutionary diversity [23].

PCs of Bray-Curtis dissimilarity scores derived from raw TRFLP data (prior to taxonomy assignment/grouping) were calculated to compare OTU diversity among samples at different time points. PCoA of both 16S-TRFLP (Figure 6A) and ITS-TRFLP (Figure 6B) data reveal time-based clustering of samples. For both bacterial and yeast communities, the progression from early to late-fermentation is strongly correlated with the first PC (explaining 43.5% and 60.7% variance, respectively), whereas the second PC primarily explains variance among late-fermentation samples (accounting for 19.8% and 10.4% of the variance, respectively). PCoA of Bray-Curtis scores calculated from LAB-TRFLP data demonstrates clustering related to batch but not to age (except at the individual-batch level, data not shown), due to batch-to-batch variation in dominant LAB. PCoA of weighted UniFrac distance (Figure 6C) reveals a similar relationship among samples, with a time-based progression strongly correlated with the first PC (57%). PCoA loadings (as OTUs), represented by grey bubbles, are plotted to demonstrate which OTUs best describe sample variance along PCs. The first PC is positively correlated with *Lactobacillales* and *Acetobacteraceae* and negatively with *Enterobacteriaceae*, explaining the majority of phylogenetic variance observed among samples. The second PC explains much less variance (22%), probably because the minor OTUs accounting for spread among Bray-Curtis dissimilarity scores for late-fermentation samples are phylogenetically related to the temporally dominant microbiota, to the effect that UniFrac concentrates and focuses sample variation around the core taxonomic shift from *Enterobacteriaceae* to*Lactobacillales*. Greater deviation from the age-based trend is seen with UniFrac clustering, as samples displaying a bloom of *Acetobaceraceae* correlate more positively with the first PC, removing from the trend and compressing the variance otherwise explained by the core taxonomic shift.

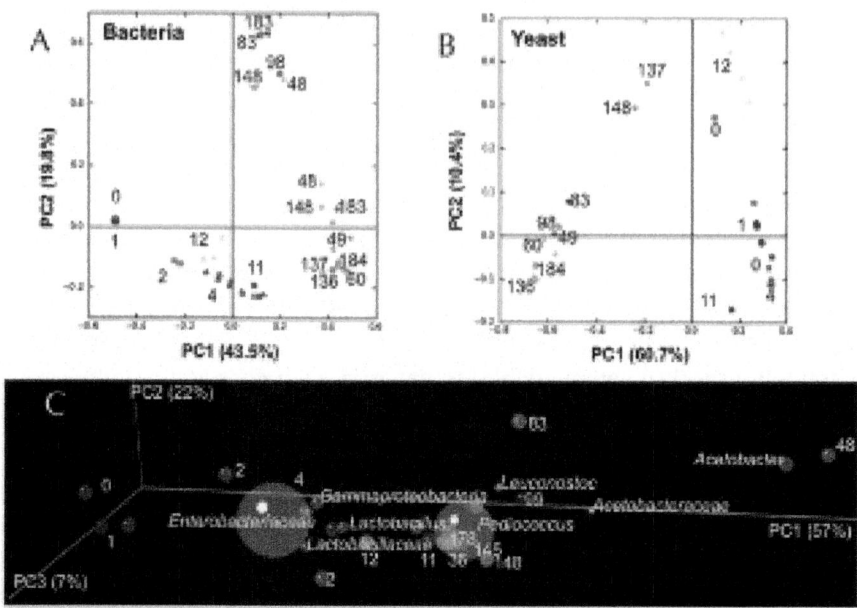

Figure 6: Principal coordinates analysis of ACA microbial succession. PCoA of Bray-Curtis dissimilarity scores derived from 16S-TRFLP of universal bacterial communities (*Panel A*) and ITS-TRFLP of yeasts (*Panel B*). Samples are colored by age (wk), as indicated by adjacent number. Sample distance is a function of shared OTU similarity. *Panel C*: 3-dimensional PCoA of weighted UniFrac distance of BAS of ACA samples. Samples are colored by age (wk), as indicated by adjacent number, their distance within the 3-dimensional space being a function of the phylogenetic similarity and abundance of their constituent taxa. Grey bubbles represent correlation of loadings (as taxonomic groups) along the same coordinates; their placement explains how much variance along each PC is explained by these taxa, with size as a function of relative abundance. doi:10.1371/journal.pone.0035507.g006

DISCUSSION

Culture-independent, molecular techniques for community profiling have not been used previously to study mixed-microbial beer fermentations, including lambic-style beers. The microbial ecology of lambic, the Belgian predecessor of ACA, has only been characterized using traditional, culture-dependent techniques [1], [24] paired with extensive chemical analyses [3], [4], [25], setting the foundations of current knowledge on spontaneous beer fermentation phenomena. Therefore, this study utilizing TRFLP and BAS for community profiling of ACA fermentations was useful not only for characterizing the complete microbiota of this previously uncharacterized fermentation, but also for updating our current knowledge of spontaneous beer fermentation ecology.

Viewed at higher taxanomic levels (e.g., judging from 16S-TRFLP alone), the microbial succession of ACA fermentation appears to closely parallel that previously observed in lambic[1], [24]: a brief, early bloom of enterobacteria before LAB and *Saccharomyces* conduct the main fermentation, and finally domination by *Brettanomyces* during the long maturation period. However, lower-level taxa reveal subtle difference between these fermentations. A different set of enterobacteria are involved, with ACA primarily dominated by *Klebsiella* spp., as well as*Enterobacter* spp., *P. carotovorum*, and *S. ureilytica*, as compared to lambic, dominated by*Enterobacter*, *Hafnia*, *Escherichia*, and *Citrobacter* [24]. Whereas *Pediococcus* spp. were the only LAB cultured during lambic fermentation [1], ACA involves several lactobacilli as well as*Leuconostoc*, *Lactococcus*, and a range of other LAB (though pediococci dominated most batches). Finally, BAS uncovered several minor populations of *Alphaproteobacteria*,*Gammaproteobacteria*, and *Actinomycetales*, most of which have never before been detected in beer, most likely as molecular community profiling methods have never been used to analyze beer fermentations. Likewise, ACA fermentation involves a diverse set of minor yeasts, including *Rhodotorula*, *Cryptococcus*, *Pichia*, and *Candida*, many of which have not been described previously in beer, let alone lambic.

This core microbial succession in ACA closely follows a pH/nutrient gradient, with the initial explosive growth of *Enterobacteriaceae* curtailed by a rapid pH drop in the first 8 wk, decreasing overall diversity and selecting primarily for *Lactobacillaceae* for the remainder of fermentation. *Saccharomyces* also follow the decline of enterobacteria, possibly due to relief from competition and acclimation to inhibitory factors such as carboxylic acids produced by enterobacteria and oxidative yeasts [26], [27]. Between wk 2–9, the majority of sugars (as °Plato) were consumed, reaching ~80% attenuation (roughly the amount of *Saccharomyces*-fermentable sugars contained in an all-malt wort). During this time *Saccharomyces* were the only yeast detected by ITS-TRFLP and total yeast populations were 100-fold those at both wk 0 and 11. The rapid drop in yeast counts and disappearance of *Saccharomyces* quickly follows this point, most illustrative of *Saccharomyces* flocculation following full attenuation, as is typical for beer fermentation. This niche is in turn filled by *Brettanomyces*, which gradually superattenuates the beer (seen as °Plato reduction beyond 20%) during the remainder of the fermentation. This gradient can also be observed via clustering of samples based on UniFrac distance or Bray-Curtis dissimilarity, with the most dramatic shifts in correlation along the first PC (i.e., diversity turnover) occurring at these stages (pH drop and full attenuation).

The core microbial profile observed in ACA appears highly conserved at the taxonomic depth achieved here, in spite of the fact that no inoculum is ever introduced to ACA wort. This includes the rare microbiota detected by BAS, the domination of *L. brevis* in one batch detected by LAB-TRFLP being the only considerable deviation. While reused barrels are an obvious source of microbial carryover, barrels for ACA production are sourced from diverse locations, including spirits production, so cannot be identified as the sole source, and even unused barrels led to replication of the microbial profile (data not shown). Enterobacteria are known regular contaminants of wort in typical brewhouse operations [26] and their ubiquity would make them difficult to eliminate from ACA when conducting spontaneous fermentation. Likewise, LAB and yeasts are common brewery microbiota, so could be introduced by surface exposure during transfer or on airborne particulates. However, the fact that the same lower taxa and succession dynamics are observed suggests the involvement of a stable brewhouse microbiota, as well as the general selective pressure exerted during fermentation (pH drop, nutrient sequestration).

This stable population may hint at the existence of "microbial terroir," or the establishment of stable, site-specific microbiota potentially impacting product quality criteria of beverage fermentations. Chances are species of *Saccharomyces* and *Brettanomyces* become enriched and established within the brewery environment, as shown in lambic breweries in Belgium [2], resulting in a semi-consistent inoculation as these yeasts are introduced from the brewery environment during coolship exposure, much in the same way that spontaneously fermented wine most likely becomes inoculated by *Saccharomyces* spp. enriched on winery equipment surfaces [27], [28], [29], [30]. In all of these spontaneously fermented beverages, such enrichment is likely a saving grace, ensuring a certain degree of fermentation consistency as competently fermentative, competitive yeasts dominate early in the fermentation, exerting selective pressure for ethanol- and pH-tolerant microbes. This is particularly pertinent to spontaneously fermented beers, as the wort is sterilized by boiling and thus would rely entirely on environmentally introduced and barrel-associated microorganisms to conduct the fermentation. As the ACA studied in this work is a seasonal product of a full-scale-production craft brewery, unlike the dedicated lambic breweries of Belgium, the house strains of *S. cerevisiae* are likely heavily enriched in this environment, resulting in the detected dominance of this yeast in these fermentations from wk 0, already exerting enough pressure to select for fairly consistent community profiles batch-to-batch. This concept of microbial terroir must be followed up in future studies comparing ACA microbial succession at the strain level at multiple sites to fully assess the sources, consistency, and quality impact of ACA microbiota.

Surprisingly, BAS did not dramatically increase the taxonomic depth beyond that achieved by TRFLP for the dominant microbiota. Neither 16S-TRFLP nor deep sequencing could delve below family-level assignment of enterobacteria, and while sequencing did identify genera of*Lactobacillales* in many cases, LAB-TRFLP could identify most populations to species with greater reliability than sequencing, so was used to improve resolution in both cases. This may be partially explained by two reasons: 1) sequencing-error-truncated reads resulted in large gaps of OTUs only identifiable as *Lactobacillaceae*; and 2) at the time of writing, paired-end reads are currently not supported in QIIME, so paired-end reads are analyzed and interpreted as separate, shorter single-end reads. How much taxonomic acumen would be added by the extra 100 nt of a full, concatenated V4 read (~250 nt) is yet to be determined and is likely regionally dependent. What BAS did improve is detection and identification of the rare microbiota of ACA, the detection limits (fluorescent peak-to-noise ratios) and taxonomic assignment (multiple restriction fragment comparison) of TRFLP being inadequate to identify most OTUs representing <1% of the community. However, BAS is still limited for deep exploration of rare taxa due to the lack of a codified quality filtration strategy [15], and OTUs currently require rigorous filtration to confidently differentiate real organisms from erroneous OTUs. Thus, some minor OTUs detected in ACA by TRFLP, such as *Methylobacterium* and some *Lactobacillales*, were detected below 0.01% relative sequence abundance and consequently removed, leading to an underestimate of true sample diversity. Advancements in sequencing technology continue to improve gigabase-per-run recovery—increasing multiplexing capacity and decreasing cost per sample—and error rates—permitting greater read length and post-filtration recovery of longer, high-quality sequences—providing hope that these limitations will be resolved in the near future. These improvements and reductions in cost per-sample are bringing NGS technologies within grasp of more researchers, promising larger-scale, higher-resolution, multivariate studies of fermented food systems utilizing the rare-microbiome exploration and phylogeny-based comparisons provided by these new tools.

ACKNOWLEDGMENTS

The authors thank the brewery and brewers involved in this study for their generous contributions of beer samples.

AUTHOR CONTRIBUTIONS

Conceived and designed the experiments: DAM NAB CWB. Performed the experiments: NAB. Analyzed the data: DAM NAB. Contributed reagents/ materials/analysis tools: DAM CWB. Wrote the paper: NAB DAM CWB.

REFERENCES

1.   Vanoevelen D, Spaepen M, Timmermans P, Verachtert H (1977) Microbial aspects of spontaneous fermentation in production of lambic and gueuze. J Inst Brew 83: 356–360.

2.   Verachtert H, Shanta Kumara HMC, Dawoud E (1990) Yeasts in Mixed Cultures. In: Verachtert H, De Mot R, editors. Yeast Biotechnology and Biocatalysis. New York: Marcel Dekker.

3.   Vanoevelen D, Delescaille F, Verachtert H (1976) Synthesis of aroma compounds during spontantous fermentation of lambic and gueuze. J Inst Brew 82: 322–326.

4.   Spaepen M, Vanoevelen D, Verachtert H (1978) Fatty acids and esters produced during spontaneous fermentation of lambic and gueuze. J Inst Brew 84: 278–282.

5.   Bokulich NA, Hwang CF, Liu S, Boundy-Mills K, Mills DA (2012) Profiling the yeast communities of wine using terminal restriction fragment length polymorphism. Am J Enol Vitic 63: doi:10.5344/ajev.2011.11077.

6.   White T, Burns T, Lee S, Taylor J (1990) Amplification and Direct Sequencing of Fungal Ribosomal RNA Genes for Phylogenetics. In: Innis MA, Gelfand DH, Sninsky JJ, White TJ, editors. PCR Protocols: A Guide to Methods and Applications. San Diego: Academic Press. pp. 315–322.

7.   Nadkarni MA, Martin FE, Jacques NA, Hunter N (2002) Determination of bacterial load by real-time PCR using a broad-range (universal) probe and primers set. Microbiology 148: 257–266.

8.   Lane DJ (1991) 16S/23S rRNA Sequencing. In: Stackebrandt E, Goodfellow M, editors. Nucleic Acid Techniques in Bacterial Systematics. New York: John Wiley & Sons. pp. 115–175.

9.   Bokulich NA, Mills DA (2012) Differentiation of Mixed Lactic Acid Bacteria Communities In Beverage Fermentations Using Targeted Terminal Restriction Fragment Length Polymorphism. Food Microbiol. doi:10.1016/j.fm.2012.1002.1007.

10. Abdo Z, Schuette UME, Bent SJ, Williams CJ, Forney LJ, et al. (2006)

Statistical methods for characterizing diversity of microbial communities by analysis of terminal restriction fragment length polymorphisms of 16S rRNA genes. Env Microbiol 8: 929–938.

11. Shyu C, Soule SJ, Bent SJ, Forster JA, Forney LJ (2007) MiCA: A Web-Based Tool for the Analysis of Microbial Communities Based on Terminal-Restriction Fragment Length Polymorphisms of 16S and 18S rRNA Genes. J Microbial Ecol 53: 562–570.

12. Cole JR, Chai B, Farris RJ, Wang Q, Kulam-Syed-Mohideen AS, et al. (2007) The ribosomal database project (RDP-II): introducing myRDP space and quality controlled public data. Nucleic Acids Res 35: D169–D172.

13. Cole JR, Wang Q, Cardenas E, Fish J, Chai B, et al. (2009) The Ribosomal Database Project: improved alignments and new tools for rRNA analysis. Nucleic Acids Res 37: D141–D145.

14. Caporaso JG, Kuczynski J, Stombaugh J, Bittinger K, Bushman FD, et al. (2010) Qiime allows analysis of high-throughput community sequence data. Nat Methods 7: 335–336.

15. Caporaso JG, Lauber CL, Walters WA, Berg-Lyons D, Lozupone CA, et al. (2011) Global patterns of 16S rRNA diversity at a depth of millions of sequences per sample. PNAS 108: 4516–4522.

16. Hamady M, Walker JJ, Harris JK, Gold NJ, Knight R (2008) Error-correcting barcoded primers for pyrosequencing hundreds of samples in multiplex. Nat Methods 5: 235–237.

17. Edgar RC (2010) Search and clustering orders of magnitude faster than BLAST. Bioinformatics 26: 2460–2461.

18. Wang Q, Garrity GM, Tiedje JM, Cole JR (2007) Naive Bayesian classifier for rapid assignment of rRNA sequences into the new bacterial taxonomy. Appl Environ Microbiol 73:

19. Lozupone CA, Knight R (2005) UniFrac: A new phylogenetic method for comparing microbial communities. Appl Environ Microbiol 71: 8228–8235.

20. Lafarge V, Ogier JC, Girard V, Maladen V, Leveau JY, et al. (2004) Raw cow milk bacterial population shifts attributable to refrigeration. Appl Environ Microbiol 70: 5644–5650.

21. Muyzer G, Dewaal EC, Uitterlinden AG (1993) Profiling of complex microbial populations by denaturing gradient gel-electrophoresis analysis of polymerase chain reaction-amplified genes coding for 16S ribosomal RNA. Appl Environ Microbiol 59: 695–700.

22. Hierro N, Esteve-Zarzoso B, Gonzalez A, Mas A, Guillamon JM (2006) Real-time quantitative PCR (QPCR) and reverse transcription-QPCR for detection and enumeration of total yeasts in wine. Appl Environ Microbiol 72: 7148–7155.

23. Hamady M, Knight R (2009) Microbial community profiling for human microbiome projects: Tools, techniques, and challenges. Genome Res 19: 1141–1152.

24. Martens H, Dawoud E, Verachtert H (1991) Wort enterobacteria and other microbial populations involved during the first month of lambic fermentation. J Inst Brew 97: 435–439.

25. Martens H, Dawoud E, Verachtert H (1992) Synthesis of aroma compounds by wort enterobacteria during the first stage of lambic fermentation. J Inst Brew 98: 421–425.

26. Priest FG, Cowbourn MA, Hough JS (1974) Wort Enterobacteria - Review. J Inst Brew 80: 342–356.

27. Fleet GH (2003) Yeast interactions and wine flavour. Int J Food Microbiol 86: 11–22.

28. Blanco P, Orriols I, Losanda A (2011) Survival of commercial yeasts in the winery environment and their prevalence during spontaneous fermentations. J Indust Microbiol Biotechnol 38: 235–239.

29. Ciani M, Mannazzu I, Marinangeli P, Clementi F, Martini A (2004) Contribution of winery-resident Saccharomyces cerevisiae strains to spontaneous grape must fermentations. Antonie van Leeuwenhoek 85: 159–164.

30. Santamaria P, Lopez R, Lopez E, Garijo P, Gutierrez AR (2008) Permanence of yeast inocula in the winery ecosystem and presence in spontaneous fermentations. Eur Food Res Technol 227: 1563–1567.

Chapter 3

FERMENTATION OF SOY MILK VIA LACTOBACILLUS PLANTARUM IMPROVES DYSREGULATED LIPID METABOLISM IN RATS ON A HIGH CHOLESTEROL DIET

Yunhye Kim., Sun Yoon., Sun Bok Lee. Hye Won Han, Hayoun Oh, Wu Joo Lee, Seung-Min Lee

Department of Food and Nutrition, College of Human Ecology, Yonsei University, Seoul, South Korea

ABSTRACT

We aimed to investigate whether *in vitro* fermentation of soy with *L. plantarum* could promote its beneficial effects on lipids at the molecular and physiological levels. Rats were fed an AIN76A diet containing 50% sucrose (w/w) (CTRL), a modified AIN76A diet supplemented with 1% (w/w) cholesterol (CHOL), or a CHOL diet where 20% cascin was replaced with soy milk (SOY) or fermented soy milk (FSOY). Dietary isoflavone profiles, serum lipids, hepatic and fecal cholesterol, and tissue gene expression were examined. The FSOY diet had more aglycones than did the SOY diet. Both the SOY and FSOY groups had lower hepatic cholesterol and serum triglyceride (TG) than did the CHOL group. Only FSOY reduced hepatic TG and serum free fatty acids and increased serum HDL-CHOL and fecal cholesterol. Compared to CHOL, FSOY lowered levels of the nuclear forms of SREBP-1c and SREBP-2 and expression of their target genes, including FAS, SCD1, LDLR, and HMGCR. On the other hand, FSOY elevated adipose expression levels of genes involved in TG-rich lipoprotein uptake (ApoE, VLDLR, and Lrp1), fatty acid oxidation (PPARα, CPT1α, LCAD, CYP4A1, UCP2, and UCP3), HDL-biogenesis (ABCA1, ApoA1, and LXRα), and adiponectin signaling (AdipoQ, AdipoR1, and AdipoR2), as well as levels of phosphorylated AMPK and ACC. SOY conferred a similar expression profile in both liver and adipose tissues but failed to reach statistical significance in many of the genes tested, unlike FSOY. Our data indicate that fermentation may be a way to enhance the beneficial effects of soy on lipid metabolism, in part via promoting a reduction of SREBP-dependent cholesterol and TG synthesis in the liver, and enhancing adiponectin signaling and PPARα-induced expression of genes involved in TG-rich lipoprotein clearance, fatty acid oxidation, and reverse cholesterol transport in adipose tissues.

INTRODUCTION

Moderate intake of dietary cholesterol has been recommended to avoid undesirable health conditions. High serum cholesterol and/or triglyceride (TG) levels are well-known risk factors for the development of disorders of lipid regulation, including cardiovascular disease [1]. For instance, people with atherosclerosis, a hallmark of cardiovascular disorders, often exhibit hypercholesterolemia, hypertriacylglyceridemia, and/or low serum HDL-CHOL [2]. Dietary cholesterol, along with fatty acids, has been shown to be a significant contributor to blood cholesterol levels [3]. The negative impact of high dietary cholesterol on health appears to be lessened by other food components such as isoflavones or polyphenols [4]. In contrast, a cholesterol-rich diet is more harmful when the diet contains high saturated fatty acids and high sucrose [5], a common feature of diets of Western society.

Previous meta-analyses and animal studies have suggested that soy is a dietary component that can reduce risk factors for diseases associated with dysregulated cholesterol metabolism[4], [6], [7]. Soy proteins rich in isoflavones have been suggested to have protective effects against disturbed serum lipid profiles in animals [6], [8]. Despite considerable data, there have also been conflicting reports, either demonstrating that soy protein isolates lacking isoflavones have a significant effect on blood lipids [9] or alternately demonstrating the necessity of soy isoflavones for beneficial effects [4], [8]. It is possible that the effects of separate components of soy may be synergistic, producing greater effects in concert than they do as isolated components [7]. Additionally, differences in the bioavailability of soy constituents could influence the efficacy of soy products. The majority of naturally occurring soy isoflavones exists in the form of conjugated glucosides that have undergone modifications such as malonylation and acetylation. *In vitro* fermentation by probiotics has been shown to increase the bioavailability of isoflavones by removing glycosyl moieties and transforming them into aglycone structures [10]. Soy isoflavone aglycones are absorbed more easily by the intestine [11]. Therefore, fermentation may increase the efficacy of soy products by enhancing the intestinal bioavailability of soy constituents.

Serum lipid metabolism is tightly regulated, in part through modulation of hepatic gene expression [12]. At the molecular level, the sterol regulatory element binding protein (SREBP)-1 transcribes genes required for fatty acid synthesis, thus increasing TG, while SREBP-2 governs cholesterol metabolism by regulating the expression of its target genes, including low density lipoprotein receptor (LDLR) and 3-hydroxy-3-methylglutaryl-CoA reductase (HMGCR) [12]. LDLR is a receptor for LDL-CHOL, and HMGCR is a rate-limiting enzyme in cholesterol biosynthesis; their expression ultimately increases intracellular cholesterol levels[12]. In addition to SREBP-mediated transcriptional regulation, other transcription factors, such as liver X receptor alpha (LXRα) [13], and/or various regulatory steps at the post-transcriptional and/or post-translational level are also important for regulation of lipid metabolism [14], [15].

Although there have been numerous studies on soy protein or soy constituents, the effects of fermented whole soy products on the dysregulation of lipid metabolism and gene expression resulting from a high cholesterol diet have not been studied at the molecular level [16], [17]. The objectives of the current study were to examine the efficacy of soy milk fermentation in facilitating its beneficial effects on serum lipids, and to investigate novel molecular mechanisms of whole soy milk as a serum lipid-based health-improving food source. To achieve this, we first investigated soy isoflavone profiles in both SOY and FSOY and then evaluated the effects of SOY and FSOY on serum lipids and hepatic and fecal lipid content in rats fed a high cholesterol diet. Subsequent evaluation of their effects on gene expression in liver and adipose tissues was also performed. Molecular analyses of the effects of SOY and FSOY may allow us to further define the impact of fermentation on lipid metabolism, thereby providing an opportunity to explore the use of fermentation by lactic acid bacteria to maximize the beneficial effects of soy products.

MATERIALS AND METHODS

Animals

Thirty-two five-week-old male Sprague Dawley rats (Koatech, Pyungtek, Korea) were divided into four experimental groups (CTRL, CHOL, SOY, and FSOY) of eight animals each. Each group was fed an experimental diet *ad libitum* for six weeks (CTRL: AIN76A diet; CHOL: 1% high cholesterol-containing AIN76A diet; SOY: modified AIN76A diet in which 20% casein and 40% corn oil were replaced with freeze-dried SOY; FSOY: modified AIN76A diet in which 20% casein and 40% corn oil were replaced with freeze-dried FSOY) (Table 1). All animals were fed experimental diets daily and housed in a pathogen-free environment with controlled temperature and humidity (18–24°C, 50–60%, respectively). Rats were fasted for 12 h prior to sacrifice and were euthanized with diethyl ether. Tissues were dissected, snap-frozen in liquid nitrogen, and then stored at −80°C until use. For analysis of blood lipids, blood samples were collected through the abdominal inferior vena cava. Serum was obtained by centrifugation at 3,000 rpm for 20 min and stored in a freezer at −80°C until analysis. Feces from each group were collected for three days prior to sacrifice. All experimental procedures were reviewed and approved by the Committee on Animal Experimentation and Ethics of Yonsei University (YLARC Permit #: 2010-0044).

Table 1: Diet compositions (g ingredient/kg diet) in the experimental groups. doi:10.1371/journal.pone.0088231.t001

Ingredients	Diet Groups			
	CTRL[1]	CHOL[2]	SOY[3]	FSOY[4]
Sucrose	499.99	499.99	461.59	461.59
Corn starch	150	150	138.3	138.3
Casein	200	200	160	160
Soy milk/Fermented soy milk	–	–	124.2	124.2
Corn oil	50	50	28.8	28.8
Mineral mix[5]	35	35	35	35
Vitamin mix[6]	10	10	10	10
Cellulose	50	50	50	50
DL-methionine	3	3	3	3
Choline Bitartrate	2	2	2	2
Ethoxyquin	0.01	0.01	0.01	0.01
Cholesterol	–	10	10	10
Cholic acid	–	5	5	5
Total	1000	1015	1027.9	1027.9

Preparation of Diets Containing Freeze-dried SOY and FSOY

SOY was produced by Yonsei Milk (Asan, Korea). Soy milk, derived from whole soy, contained 5 g of carbohydrate, 7 g of protein, 3.5 g of lipids, 2 g of dietary fiber, and 150 mg of sodium per kg. To produce FSOY, 0.1% (w/v) *Lactobacillus plantarum* KCTC10782BP (Cell Biotech International Korea, Seoul, Korea) was used. After the addition of 20 g of Galacto-oligosaccharide (Samyang, Seoul, Korea), fermentation proceeded at 25°C until the acidity reached pH 5.1. After being filtered, SOY and FSOY were freeze-dried (ILShinBioBase, Dongduchen, Korea).

Determination of Isoflavone Content

In brief, 1 g of freeze-dried SOY and FSOY diets were mixed with 25% methanol (1:4, v/v) containing 400 µg of caffeine as an internal standard at 25°C for 1 h with sonication and then centrifuged to remove the insoluble fraction. Each supernatant was filtered through a 0.45 µm GHP column (PALL Life Sciences, Port Washington, NY, USA). The filtrate was analyzed using ultraperformance liquid chromatography (UPLC) (Waters, Palo Alto, CA, USA) with a cyano column (2.1 mm × 50 mm, 1.8 µm, HSS Cyano; Waters). The mobile phases were water containing 0.1% formic acid and acetonitrile containing 0.1% formic acid. The temperature of the column was 30°C and chromatograms were recorded at 260 nm. Daidzein, daidzin, genistein, and genistin were identified based on their retention times after comparison with corresponding standards purchased from Sigma Aldrich (St. Louis, MO, USA).

Determination of Isoflavone Derivatives by Mass Spectrometry

LC-MS identification was performed using UPLC (Acquity UPLC™; Waters, Milford, MA, USA) and LTQ Orbitrap (Thermo Fisher Scientific, Waltham, MA, USA), equipped with ESI. Data analysis was performed with MassLynx™ 4.0 data system software (Waters).

Determination of Hepatic and Fecal Lipids, Blood Lipid Profiles, and Fecal Bile Acid

Lipids from liver and feces were extracted using a modified Folch method [18]. The isolated lipid fraction was used to measure TOTAL-CHOL and TG concentrations using commercially available kits (Asan Pharm, Seoul, Korea). Serum TOTAL-CHOL, HDL-CHOL, and TG levels were also measured with commercially available kits (Asan Pharm, Seoul, Korea). All determinants were obtained by absorbance at 550 nm using a microplate reader (Molecular Device, Sunnyvale, CA, USA). LDL-CHOL was calculated using the Friedewald formula [19](LDL-CHOL (mg/dl)=TOTAL-CHOL – HDL-CHOL – TG/5). Atherogenic indices were obtained by the Haglund method (Atherogenic index=(TOTAL-CHOL – HDL-CHOL)/HDL-CHOL) [20]. Bile acid was extracted using the DeWael method [21]. Extracted fecal bile acid was measured using a kit from Bioquant (Nashville, TN, USA). Serum glucose levels were measured using the Fuji DRI-CHEM 4000i analyzer (Fuji Film Co., Tokyo, Japan).

RNA Extraction and Quantitative RT-PCR

Total RNA was extracted from tissues using Trizol according to the manufacturer's protocol (Invitrogen, Carlsbad, CA, USA). Total RNA (1 μg) was used to synthesize cDNA with ImProm-II reverse transcriptase (Promega, Madison, WI, USA). Real-time PCR was performed on a CFX96 sequence detection system (Bio-Rad, Hercules, CA, USA) using EvaGreen qPCR Mix Plus (Solis BioDyne, Estonia). Expression levels were normalized to the amount of 18S rRNA. Relative mRNA levels were then calculated by the difference in C_t values among animal groups, expressed as the fold change.

Western Blot Analysis

Tissues were homogenized in RIPA buffer and then centrifuged at $11,000 \times g$ for 10 min at 4°C. Equal amounts of protein lysates were loaded on an SDS-polyacrylamide gel. Anti-SREBP-2, anti-SREBP-1c, and anti-adiponectin antibodies from Santa Cruz Biotechnology (Santa Cruz, CA, USA), anti-p-AMP-activated protein kinase (AMPK) and anti-p-Acetyl CoA carboxylase (ACC) antibodies from Cell Signaling Technology (Beverly, MA, USA), anti-LDLR antibody from BioVision (Milpitas, CA, USA), and anti-GAPDH antibody from Signalway Antibody (College Park, MD, USA) were used to detect the respective proteins.

Statistical Analyses

SPSS was used to perform statistical analyses (Statistical Package for the Social Sciences; SPSS Inc., Chicago, IL, USA). Student's t-test and one-way analysis of variance (ANOVA) followed by Duncan's multiple comparisons test were used to determine statistically significant differences among the experimental groups. A P value <0.05 was the criterion for statistical significance.

RESULTS

FSOY had a Higher Proportion of unmodified Soy Isoflavones and their Aglycones than SOY

Soy isoflavone composition was compared between SOY and FSOY in terms of percentages (Table 2). Fermentation lowered the amounts of both malonyl and acetyl groups in isoflavones. Interestingly, equol was below the detection limit in SOY but represented 0.03% of the total isoflavones in FSOY. In addition, soy isoflavone aglycones, including genistein, daidzein, and glycitein, were increased in FSOY compared to levels in SOY. These results indicate that fermentation increased the proportion of soy isoflavone aglycones as well as equol in FSOY and diminished soy isoflavone glycones conjugated with malonyl and acetyl residues relative to those in SOY.

Table 2: Percentage of soy isoflavones in SOY and FSOY[1].doi:10.1371/journal.
pone.0088231.t002

	SOY	FSOY	p^2
Malonyl daidzin	9.27±0.06	0.29±0.01	<0.05
Acetyl daidzin	1.07±0.02	0.43±0.00	<0.05
Daidzin	22.42±0.07	33.14±0.26	<0.05
Daidzein	0.10±0.01	0.72±0.02	<0.05
Equol	0.00±0.00	0.03±0.00	<0.05
Malonyl glycitin	0.8±0.01	0.00±0.00	<0.05
Acetyl glycitin	0.77±0.00	0.00±0.00	<0.05
Glycitin	2.93±0.01	3.73±0.04	<0.05
Glycitein	0.02±0.00	0.09±0.00	<0.05
Malonyl genistin	19.20±0.12	0.33±0.01	<0.05
Acetyl genistin	2.49±0.01	0.89±0.01	<0.05
Genistin	39.85±0.13	58.72±0.30	<0.05
Genistein	1.08±0.01	1.64±0.01	<0.05
Total	100±0.00	100±0.00	

FSOY Lowered Hepatic Lipids and Serum TG and FFAs, and Elevated HDL-CHOL

When fed *ad libitum*, the CHOL group showed a modest decrease in food intake compared to the other groups (Fig. 1A and B). The CHOL, SOY, and FSOY groups demonstrated similar levels of serum TOTAL-CHOL and LDL-CHOL as the CTRL group (Fig. 1C and D), suggesting that SOY and FSOY failed to lower serum TOTAL-CHOL or LDL-CHOL. FSOY significantly restored the CHOL-driven reduction of serum HDL-CHOL, which was not observed with SOY (Fig. 1E). This led to a noticeable reduction of atherogenic index values in the FSOY group compared to the CHOL group (Fig. 1F). Compared to the CHOL diet, both SOY and FSOY lowered hepatic TOTAL-CHOL, although only FSOY increased fecal TOTAL-CHOL, which was not accompanied by a change in fecal bile acid content (Fig. 1I, K, and L). On the other hand, only FSOY significantly lowered CHOL-elevated hepatic TG levels and serum FFAs (Fig. 1J and H). CHOL did not affect serum TG levels, but SOY and FSOY greatly decreased serum TG, to below that of the CTRL group (Fig. 1G). Overall, our data suggest that, despite the higher dietary cholesterol intake due to *ad libitum* feeding, SOY and FSOY did not worsen serum or liver lipid profiles relative to CHOL. On the contrary, SOY and FSOY decreased serum

TG and CHOL-elevated hepatic TOTAL-CHOL. Moreover FSOY reduced hepatic TG and serum FFAs, and increased the levels of serum HDL-CHOL.

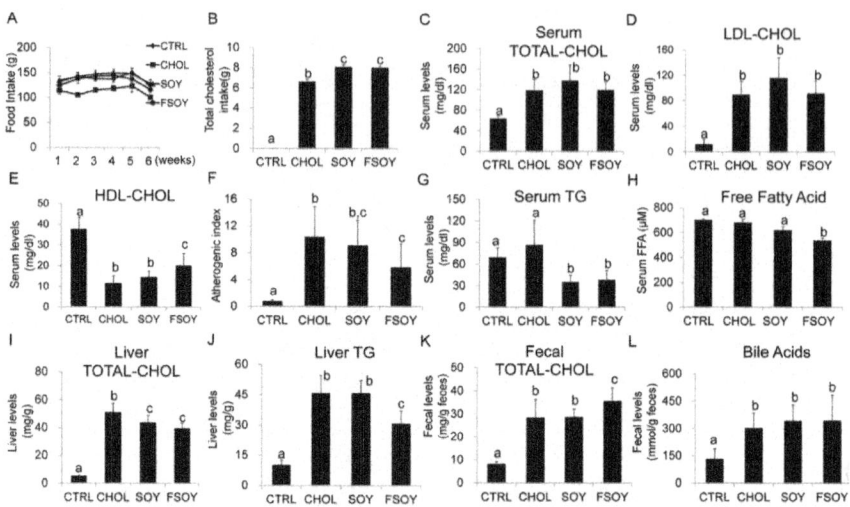

Figure 1: Food intake and lipid profiles of rats fed SOY or FSOY as opposed to CHOL. Food intake (A), total cholesterol intake (B), serum levels of TOTAL-CHOL (C), LDL-CHOL (D), and HDL-CHOL (E), atherogenic index (F), serum levels of TG (G) and free fatty acids (H), hepatic TOTAL-CHOL (I), hepatic TG (J), fecal TOTAL-CHOL (K), and fecal bile acid (L) in each animal group. The results are expressed as means ± SD of eight animal tissues in each group. Values not sharing the same letter were significantly different ($P<0.05$) between all groups by ANOVA. doi:10.1371/journal.pone.0088231.g001

FSOY Downregulated Expression of Genes Involved in Lipid Metabolism in the Liver

Next, we analyzed whether a significant reduction of hepatic cholesterol and TG by FSOY is associated with downregulation of genes involved in lipid metabolism. As for cholesterol metabolism, only FSOY showed a significant reduction in SREBP-2 mRNA and protein compared to levels in the CHOL group (Fig. 2A–C), which was accompanied by a reduction in HMGCR mRNA and LDLR protein (Fig. 2D and E). However, the reduction of LDLR mRNA in the FSOY group was not great enough to reach statistical difference between FSOY and CHOL (data not shown). On the other hand, CHOL increased hepatic transcripts of SREBP-1c, fatty acid synthase (FAS), and stearoyl CoA desaturase (SCD1) relative to CTRL (Fig. 2F, I, and J). FSOY downregulated the mature and nuclear forms of SREBP-1c protein, and its target genes, including FAS and SCD-1, compared to levels in the CHOL group (Fig. 2G–J). SOY and FSOY did not lower the CHOL-elevated SREBP-1c mRNA (Fig. 2F), These data suggest that FSOY might lower hepatic cholesterol and TG in the liver in part through downregulating SREBP-dependent gene expression.

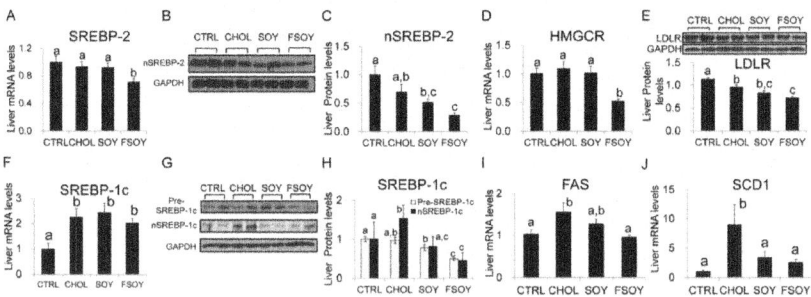

Figure 2: Hepatic expression levels of genes involved in cholesterol and TG metabolism in rats on SOY or FSOY diet compared with CHOL diet. Expression levels of SREBP-2 mRNA (A), nuclear form of SREBP-2 protein (B and C), HMGCR mRNA (D), LDLR protein (E), SREBP-1c mRNA (F), premature and nuclear forms of SREBP proteins (pre-SREBP and nSREBP)(G and H), FAS mRNA (I), and SCD1 mRNA (J). Representative band images indicating protein levels for SREBP-2 (B), SREBP-1c (G), and LDLR (E) are presented. Bands were quantified, normalized, and presented as a graph (C, D, and F). The results are expressed as means ± SE of eight animal tissue samples. Values not sharing the same letter were significantly different ($P<0.05$) between all groups by ANOVA. doi:10.1371/journal.pone.0088231.g002

FSOY Increased Expression of Genes Involved in TG-rich Lipoprotein Clearance and Fatty Acid Oxidation

To gain insight into the molecular mechanisms of FSOY-lowered serum TG levels, adipose gene expression levels were compared among the groups. ApoE mRNA was significantly increased by both SOY and FSOY (Fig. 3A). However, only FSOY demonstrated a significant increase in mRNA levels of very low density lipoprotein receptor (VLDLR) and low density lipoprotein receptor-related protein 1 (Lrp1) relative to levels in CHOL (Fig. 3B and C). Lipoprotein lipase (LPL) mRNA levels were not affected by any diet (data not shown). On the other hand PPARα, a master transcription factor for fatty acid oxidation, and its target genes, including mitochondrial uncoupling protein 2 (UCP2), UCP3, acyl-CoA oxidase 1, palmitoyl (ACOX1), long-chain specific acyl-CoA dehydrogenase (LCAD), and cytochrome P450 family 4 subfamily a polypeptide 2 (CYP4A2) were significantly upregulated by FSOY, relative to CHOL (Fig. 3 F, H–J). SOY also upregulated some PPAR target genes, such as UCP3, ACOX1, and LCAD (Fig. 3 E, H, and I). Only SOY significantly statistically increased transcript levels of carnitine palmitoyltransferase 1 alpha (CPT1α) compared to levels in the CHOL group (Fig. 3G). These results indicate that CHOL-suppressed fatty acid oxidation genes were partly upregulated by SOY, but that FSOY had a greater effect, on more genes. Only FSOY significantly upregulated the ApoE, VLDLR, and Lrp1 genes, which are involved in TG-rich lipoprotein uptake.

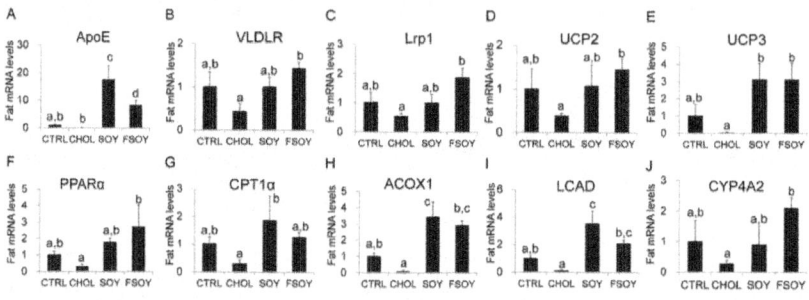

Figure 3: Adipose expression levels of genes related to lipoprotein uptake and fatty acid oxidation in SOY and FSOY groups compared to CTRL and CHOL groups.

mRNA levels of ApoE (A), VLDLR (B), Lrp1 (C), UCP2 (D), UCP3 (E), PPARα (F), CPT1α (G) ACOX1 (H), LCAD (I) and CYP4A1 (J) in rats fed experimental diets. The results are expressed as means ± SE of eight animal tissues. Values not sharing the same letter were significantly different ($P<0.05$) between all groups by ANOVA. doi:10.1371/journal.pone.0088231.g003

FSOY upregulated Genes Involved in Reverse Cholesterol Transport

ApoA1 and ATP-binding cassette, sub-family A, member 1 (ABCA1) genes play crucial roles in the formation of HDL-CHOL. FSOY significantly increased transcript levels of ApoA1 and ABCA1 relative to levels in the CHOL group (Fig. 4A and B). LXRα, a transcription factor for ABCA1, was also significantly elevated only by FSOY (Fig. 4C). mRNA levels of peroxisome proliferator activated receptor gamma (PPARγ), a transcription factor for LXRα, did not show any statistical difference among the groups (Fig. 4D). These results indicate that FSOY increased adipose expression levels of genes for reverse cholesterol transport, probably leading to FSOY elevation of serum HDL-CHOL.

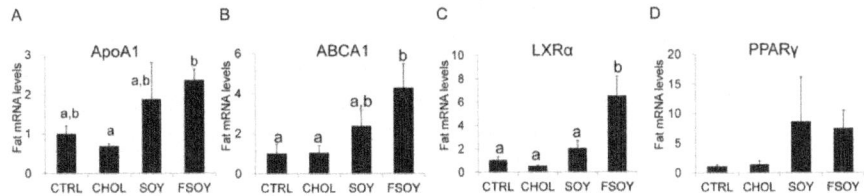

Figure 4: The effects of SOY or FSOY on the mRNA levels of genes involved in reverse cholesterol transport in rats on a high cholesterol diet. Adipose mRNAs of ApoA1 (A), ABCA1 (B), LXRα (C), and PPARγ (D) in rats fed experimental diets. The results are expressed as means ± SE of eight animal tissues. Values not sharing the same letter were significantly different ($P<0.05$) between all groups by ANOVA. doi:10.1371/journal.pone.0088231.g004

FSOY Increased Expression of Adiponectin Gene and AMPK Signals

Compared to CHOL, FSOY increased mRNA and protein levels of AdipoQ and transcript levels of the AdipoR1 and -R2 genes in adipose tissue, whereas SOY resulted in a significant increase only in AdipoQ protein and AdipoR1 mRNA (Fig. 5A–D). Phosphorylation of AMPK, a target of adiponectin signaling, was elevated in the liver and adipose tissues by both SOY and FSOY compared to CHOL, but with greater statistical significance by FSOY than by SOY (Fig. 5E and F). ACC, an enzyme involved in fatty acid synthesis, is inactivated by AMPK-mediated phosphorylation [22]. FSOY increased the levels of p-ACC in the liver and adipose tissues (Fig. 5C and D). Our data indicate that FSOY is likely to increase adiponectin signals as well as AMPK activation.

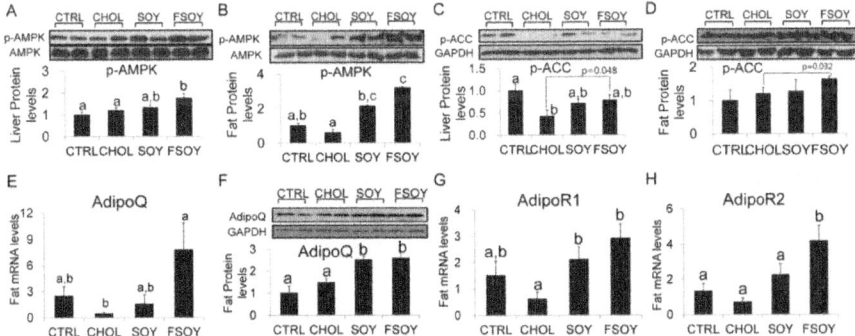

Figure 5: The effects of SOY or FSOY on mRNA levels of adiponectin and adiponectin receptors, and on AMPK activation. Hepatic and adipose levels of phosphorylated AMPK (A and B) and phosphorylated ACC (C and D), fat adi-

ponectin mRNA (E) and protein (F), and mRNA levels of adiponectin receptor 1 (G) and adiponectin receptor 2 (H). The results are expressed as means ± SE of eight animal tissues. Values not sharing the same letter were significantly different ($P<0.05$) between all groups by ANOVA, and Student's t-test was performed to compare the hepatic and adipose protein levels of p-ACC between the CHOL and FSOY groups. doi:10.1371/journal.pone.0088231.g005

DISCUSSION

In the present study we provided evidence demonstrating that beneficial effects of soy milk on lipid metabolism in rats on a high cholesterol and high sucrose diet were further augmented by fermentation of soy milk. *L. plantarum* fermentation of soy milk resulted in modification of the isoflavones in soy milk, increasing deconjugated and aglycone forms of isoflavones. As expected from a previous report showing that *L. plantarum* has strong β-glucosidase activity[23], FSOY not only had a greater proportion of aglycones, but also glycones without malonyl and acetyl groups than did SOY. These chemical changes in the soy isoflavone components of FSOY are likely to contribute to the specific effects of FSOY by increasing the amount of bioavailable soy isoflavones compared to those found in SOY. In fact, a human study demonstrated that *in vivo* bioavailability of isoflavones is enhanced by fermentation, compared to untreated soy milk [17]. The fermentation-promoted beneficial effects of soy milk on lipid metabolism were demonstrated by FSOY-specific reduction of hepatic TG, serum FFAs, and atherogenic indices, and an FSOY-specific increase in serum HDL-CHOL in addition to a decrease in hepatic cholesterol and serum TG as shown in SOY, compared to CHOL.

In order to understand the molecular mechanisms that mediate these beneficial effects of fermented soy milk, we evaluated the expression levels of hepatic and adipose genes related to lipid metabolism. Not only SREBP-2, a master transcription factor governing increases in cellular cholesterol levels, but also the SREBP-2 target genes, HMGCR and LDLR, were significantly downregulated only by FSOY. FSOY-specific significant downregulation of SREBP-2 and its target genes would contribute to the greater reduction of hepatic cholesterol levels relative to that conferred by SOY. Our data regarding FSOY effects on SREBP-2 expression levels are not consistent with a previous report showing that a soy protein diet increases SREBP-2 expression levels [24]. This discrepancy might have arisen from the fact that we used whole soy milk containing many other soy components, including fatty acids and/ or isoflavones. In addition, SREBP-1c, a key transcription factor involved in the synthesis of fatty acids, and its target genes, FAS and SCD1, were significantly downregulated by FSOY. SREBP-1c increases the transcription of FAS and SCD-1, resulting in an increase in the synthesis of TG [12]. CHOL induced SREBP-1c gene expression and processing, resulting in an increase in the nuclear form of SREBP-1c protein, which is consistent with the marked elevation of hepatic TG in CHOL. Thus FSOY-induced downregulation of SREBP-1c may lower CHOL-increased hepatic TG. SOY demonstrated similar effects in its regulation of SREBP-1c, but they were not strong enough to lower hepatic TG relative to that in CHOL with statistical significance. In addition, FSOY increased phosphorylated AMPK levels in the liver. AMPK phosphorylates and inactivates ACC and HMGCR [25], thus the AMPK activation observed in FSOY may further decrease synthesis of fatty acids and cholesterol by inhibiting the catalytic enzyme activities of ACC and HMGCR, respectively. Taken together, the downregulation of expression and processing of SREBP genes and of the expression of their target genes, as well as of AMPK activation, appear to contribute to the reduction of hepatic lipids in the FSOY group.

Hepatic cholesterol levels could also be affected by decreased dietary cholesterol absorption. FSOY increased fecal cholesterol levels without lowering fecal bile acid contents, suggesting that FSOY might confer decreased dietary cholesterol absorption instead of increasing biliary cholesterol excretion. According to our unpublished data, *L. plantarum* bound to cholesterol lowers its solubility in *in vitro* experiments, suggesting that fermented products containing lactic acids could reduce the bioavailability of dietary cholesterol for intestinal absorption by lowering the solubility of cholesterol in the intestine.

Upregulation of ApoA1, ABCA1, and liver X receptor α (LXRα) was detected only in the FSOY group. ApoA1 is an apolipoprotein found in nascent HDL-CHOL, and the ABCA1 transporter incorporates cholesterol and phospholipids into lipid-poor HDL particles [26]. FSOY-specific upregulation of ApoA1 and ABCA1 genes may result in an increase in ApoA1-dependent cholesterol efflux via ABCA1 in adipose tissue in HDL-CHOL biogenesis. A significant elevation of liver X receptor α (LXRα) mRNA found in the FSOY group could mediate ABCA1 upregulation, given that the nuclear receptors LXRα and/or LXRβ, and retinoid X receptor RXR, activate ABCA1 expression [27]. On the other hand, we did not observe significant changes in serum LDL-CHOL even though it has been reported in a meta-analysis that the consumption of soy isoflavones lowers serum LDL-CHOL [4]. It is possible that the replacement of 20% of dietary protein with SOY or FSOY might not have been sufficient to elicit a serum LDL-CHOL-lowering effect.

FSOY-driven serum TG reduction may be related to an increase in the uptake and catabolism of TG-rich lipoproteins in peripheral tissues. The transcript levels of ApoE, an apolipoprotein found in TG-rich lipoproteins, VLDLR, a receptor for apoE-containing VLDL, and Lrp1, an apoE-receptor protein, were increased in the FSOY group. Lack of VLDLR has been demonstrated to elevate TG content in circulating serum [28]. Lrp1 mRNA levels have been shown to be positively associated with levels of serum TG clearance [29]. FSOY may promote the oxidation of fatty acids taken up from TG-rich lipoproteins and remove them by dissipating heat through the respiratory uncoupling process. In our study, FSOY upregulated fatty acid oxidation-related genes, including PPARα and its target genes, such as ACOX1, LCAD, and CYP4A1, which encode products involve in various types of fatty acid oxidation processes such as peroxisomal, mitochondrial, and microsomal oxidation, respectively, as well as UCP2 and UCP3, which dissipate energy as heat [30]. The roles of UCP proteins in the control of serum TG were demonstrated by a marked reduction of blood TG in transgenic mice overexpressing adipose tissue-specific Ucp1 [31]. Thus FSOY-induced increases in fatty acid oxidation and the respiratory uncoupling process in adipose tissue could lead to a significant clearance of serum TG. At the same time the FSOY effects on adipose fatty acid oxidation may also be enhanced by increased activation of the adiponectin signaling pathway. PPARα activity is facilitated by adiponectin signaling [32]. FSOY increased the expression of AdipoQ, AdipoR1, and AdipoR2, which could enhance adiponectin signaling and promote PPARα activity. Furthermore, FSOY-induced phosphorylation of AMPK and ACC may have promoted fatty acid oxidation in the FSOY group. ACC is an enzyme that generates malonyl-CoA, which inhibits fatty acid influx into mitochondria for β-oxidation [22]. Activated AMPK phosphorylates ACC, resulting in a decrease in the production of malonyl-CoA, thereby permitting an influx of fatty acids into mitochondria for β-oxidation [22]. Additionally, FSOY-induced reduction of serum FFAs may lower the production of TG rich-VLDL by the liver, given that an elevation of serum FFA levels increases the influx of FFA into the liver for VLDL production [33]. Overall, FSOY-induced upregulation of genes involved in TG-rich lipoprotein uptake and fatty acid oxidation, as well as AMPK activation, may increase serum TG clearance.

In the present study we provide mechanistic evidence supporting the beneficial effects of soy milk in terms of regulation of expression of genes in lipid metabolism (Fig. 6). In addition, we suggest that fermentation might be an effective way to increase the beneficial effects of soy on the dysregulation of lipid metabolism induced by a high cholesterol diet.

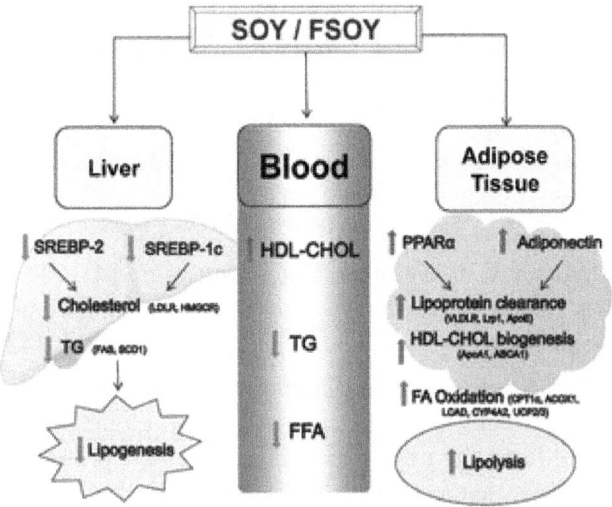

Figure 6: Proposed model of the beneficial effects of fermented soy on lipid profiles and related gene expression in the liver and adipose tissues of rats on a high cholesterol diet.doi:10.1371/journal.pone.0088231.g006

ACKNOWLEDGMENTS

We thank Seung Yun Rim for technical assistance.

AUTHOR CONTRIBUTIONS

Conceived and designed the experiments: YHK SBL SY SML. Performed the experiments: YHK SBL HWH HYO WJL SML. Analyzed the data: YHK SBL HWH SY SML. Contributed reagents/materials/analysis tools: SY SML. Wrote the paper: YHK SBL SML.

REFERENCES

1. Khoo KL, Tan H, Liew YM, Deslypere JP, Janus E (2003) Lipids and coronary heart disease in Asia. Atherosclerosis 169: 1–10. doi: 10.1016/s0021-9150(03)00009-1

2. Grundy SM (1998) Hypertriglyceridemia, atherogenic dyslipidemia, and the metabolic syndrome. Am J Cardiol 81: 18B–25B. doi: 10.1016/s0002-9149(98)00033-2

3. Spady DK, Woollett LA, Dietschy JM (1993) Regulation of plasma LDL-cholesterol levels by dietary cholesterol and fatty acids. Annu Rev Nutr 13: 355–381. doi: 10.1146/annurev.nu.13.070193.002035

4. Taku K, Umegaki K, Sato Y, Taki Y, Endoh K, et al. (2007) Soy isoflavones lower serum total and LDL cholesterol in humans: a meta-analysis of 11 randomized controlled trials. Am J Clin Nutrit 85: 1148–1156.

5. Yasutake K, Nakamuta M, Shima Y, Ohyama A, Masuda K, et al. (2009) Nutritional investigation of non-obese patients with non-alcoholic fatty liver disease: the significance of dietary cholesterol. Scand J Gastroenterol 44: 471–477. doi: 10.1080/00365520802588133

6. Lin Y, Meijer GW, Vermeer MA, Trautwein EA (2004) Soy protein enhances the cholesterol-lowering effect of plant sterol esters in cholesterol-fed hamsters. The Journal of nutrition 134: 143–148.

7. Hoie LH, Morgenstern EC, Gruenwald J, Graubaum HJ, Busch R, et al. (2005) A double-blind placebo-controlled clinical trial compares the cholesterol-lowering effects of two different soy protein preparations in hypercholesterolemic subjects. European journal of nutrition 44: 65–71. doi: 10.1007/s00394-004-0492-0

8. Anthony MS, Clarkson TB, Hughes CL Jr, Morgan TM, Burke GL (1996) Soybean isoflavones improve cardiovascular risk factors without affecting the reproductive system of peripubertal rhesus monkeys. J Nutr 126: 43–50.

9. Fukui K, Tachibana N, Wanezaki S, Tsuzaki S, Takamatsu K, et al. (2002) Isoflavone-free soy protein prepared by column chromatography reduces plasma cholesterol in rats. Journal of agricultural and food chemistry 50: 5717–5721. doi: 10.1021/jf025642f

10. Chien HL, Huang HY, Chou CC (2006) Transformation of isoflavone phytoestrogens during the fermentation of soymilk with lactic acid bacteria and bifidobacteria. Food Microbiol 23: 772–778. doi: 10.1016/j.fm.2006.01.002

11. Izumi T, Piskula MK, Osawa S, Obata A, Tobe K, et al. (2000) Soy isoflavone aglycones are absorbed faster and in higher amounts than their glucosides in humans. J Nutr 130: 1695–1699.

12. Horton JD, Goldstein JL, Brown MS (2002) SREBPs: activators of the complete program of cholesterol and fatty acid synthesis in the liver. J Clin Invest 109: 1125–1131. doi: 10.1172/jci0215593

13. Joseph SB, Laffitte BA, Patel PH, Watson MA, Matsukuma KE, et al. (2002) Direct and indirect mechanisms for regulation of fatty acid synthase gene expression by liver X receptors. J Biol Chem 277: 11019–11025. doi: 10.1074/jbc.m111041200

14. Field FJ, Born E, Murthy S, Mathur SN (2001) Regulation of sterol regulatory element-binding proteins by cholesterol flux in CaCo-2 cells.

J Lipid Res 42: 1687–1698.

15. Chambers CM, Ness GC (1997) Translational regulation of hepatic HMG-CoA reductase by dietary cholesterol. Biochem Biophys Res Commun 232: 278–281. doi: 10.1006/bbrc.1997.6288

16. Ascencio C, Torres N, Isoard-Acosta F, Gomez-Perez FJ, Hernandez-Pando R, et al. (2004) Soy protein affects serum insulin and hepatic SREBP-1 mRNA and reduces fatty liver in rats. J Nutr 134: 522–529.

17. .Kano M, Takayanagi T, Harada K, Sawada S, Ishikawa F (2006) Bioavailability of isoflavones after ingestion of soy beverages in healthy adults. J Nutr 136: 2291–2296.

18. .Folch J, Lees M, Sloane Stanley GH (1957) A simple method for the isolation and purification of total lipids from animal tissues. J Biol Chem 224: 497–509.

19. Friedewald WT, Levy RI, Fredrickson DS (1972) Estimation of the concentration of low-density lipoprotein cholesterol in plasma, without use of the preparative ultracentrifuge. Clin Chem 18: 499–502.

20. Haglund O, Luostarinen R, Wallin R, Wibell L, Saldeen T (1991) The effects of fish oil on triglycerides, cholesterol, fibrinogen and malondialdehyde in humans supplemented with vitamin E. J Nutr. 121: 165–169.

21. de Wael J, Raaymakers CE, Endeman HJ (1977) Simplified quantitative determination of total fecal bile acids. Clin Chim Acta 79: 465–470. doi: 10.1016/0009-8981(77)90443-0

22. Srivastava RA, Pinkosky SL, Filippov S, Hanselman JC, Cramer CT, et al. (2012) AMP-activated protein kinase: an emerging drug target to regulate imbalances in lipid and carbohydrate metabolism to treat cardio-metabolic diseases. J Lipid Res 53: 2490–2514. doi: 10.1194/jlr.r025882

23. Di Cagno R, Mazzacane F, Rizzello CG, Vincentini O, Silano M, et al. (2010) Synthesis of isoflavone aglycones and equol in soy milks fermented by food-related lactic acid bacteria and their effect on human intestinal Caco-2 cells. J Agric Food Chem 58: 10338–10346. doi: 10.1021/jf101513r

24. Torre-Villalvazo I, Tovar AR, Ramos-Barragan VE, Cerbon-Cervantes MA, Torres N (2008) Soy protein ameliorates metabolic abnormalities in liver and adipose tissue of rats fed a high fat diet. J Nutr 138: 462–468.

25. Carling D, Clarke PR, Zammit VA, Hardie DG (1989) Purification and characterization of the AMP-activated protein kinase. Copurification of acetyl-CoA carboxylase kinase and 3-hydroxy-3-methylglutaryl-CoA

reductase kinase activities. Eur J Biochem 186: 129–136. doi: 10.1111/j.1432-1033.1989.tb15186.x

26. Lawn RM, Wade DP, Garvin MR, Wang X, Schwartz K, et al. (1999) The Tangier disease gene product ABC1 controls the cellular apolipoprotein-mediated lipid removal pathway. J Clin Invest 104: R25–31. doi: 10.1172/jci8119

27. .Schwartz K, Lawn RM, Wade DP (2000) ABC1 gene expression and ApoA-I-mediated cholesterol efflux are regulated by LXR. Biochem Biophys Res Commun 274: 794–802. doi: 10.1006/bbrc.2000.3243

28. Tacken PJ, Teusink B, Jong MC, Harats D, Havekes LM, et al. (2000) LDL receptor deficiency unmasks altered VLDL triglyceride metabolism in VLDL receptor transgenic and knockout mice. J Lipid Res 41: 2055–2062. doi: 10.1016/s0021-9150(00)81351-9

29. Espirito Santo SM, Rensen PC, Goudriaan JR, Bensadoun A, Bovenschen N, et al. (2005) Triglyceride-rich lipoprotein metabolism in unique VLDL receptor, LDL receptor, and LRP triple-deficient mice. J Lipid Res 46: 1097–1102. doi: 10.1194/jlr.c500007-jlr200

30. Rakhshandehroo M, Knoch B, Muller M, Kersten S (2010) Peroxisome proliferator-activated receptor alpha target genes. PPAR Res 2010.

31. Rossmeisl M, Kovar J, Syrovy I, Flachs P, Bobkova D, et al. (2005) Triglyceride-lowering effect of respiratory uncoupling in white adipose tissue. Obes Res 13: 835–844. doi: 10.1038/oby.2005.96

32. Kadowaki T, Yamauchi T (2005) Adiponectin and adiponectin receptors. Endocr Rev 26: 439–451. doi: 10.1210/er.2005-0005

33. Lewis GF (1997) Fatty acid regulation of very low density lipoprotein production. Curr Opin Lipidol 8: 146–153. doi: 10.1097/00041433-199706000-00004

Chapter 4

PHENOTYPIC LANDSCAPE OF SACCHAROMYCES CEREVISIAE DURING WINE FERMENTATION: EVIDENCE FOR ORIGINDEPENDENT METABOLIC TRAITS

Carole Camarasa[1,2,3], Isabelle Sanchez[1,2,3,] Pascale Brial[1,2,3,] Fre´de´ ric Bigey[1,2,3,] Sylvie Dequin[1,2,3]

[1] INRA, UMR1083, Montpellier, France,

[2] SupAgro, UMR1083, Montpellier, France,

[3] Universite´ Montpellier 1, UMR1083, Montpellier, France

ABSTRACT

The species *Saccharomyces cerevisiae* includes natural strains, clinical isolates, and a large number of strains used in human activities. The aim of this work was to investigate how the adaptation to a broad range of ecological niches may have selectively shaped the yeast metabolic network to generate specific phenotypes. Using 72 *S. cerevisiae* strains collected from various sources, we provide, for the first time, a population-scale picture of the fermentative metabolic traits found in the *S. cerevisiae* species under wine making conditions. Considerable phenotypic variation was found suggesting that this yeast employs diverse metabolic strategies to face environmental constraints. Several groups of strains can be distinguished from the entire population on the basis of specific traits. Strains accustomed to growing in the presence of high sugar concentrations, such as wine yeasts and strains obtained from fruits, were able to achieve fermentation, whereas natural yeasts isolated from "poor-sugar" environments, such as oak trees or plants, were not. Commercial wine yeasts clearly appeared as a subset of vineyard isolates, and were mainly differentiated by their fermentative performances as well as their low acetate production. Overall, the emergence of the origin-dependent properties of the strains provides evidence for a phenotypic evolution driven by environmental constraints and/or human selection within *S. cerevisiae*.

INTRODUCTION

Despite the extensive diversity of *S. cerevisiae*, most work on this model organism has been carried out using only a handful of domesticated laboratory strains. Since the discovery and identification of yeast as a fermentative microorganism in the 19th century, a large number of *S. cerevisiae* strains have been isolated from diverse sources all over the world, corresponding to extremely different living environments. These include natural habitats of yeast in fruits, soil, cacti and the bark of oak trees, as well as the facultative infections of immunocompromised patients. However, *S. cerevisiae* is found most often in connection with human activities, which include baking, brewing, winemaking and fermented beverage production (sake, palm wine). Indeed, this yeast has been exploited by man for millennia for the fermentation and preservation of beverages and food [1], [2].

Recent advances in genomic tools allow the genetic diversity of *S. cerevisiae* to be assessed in unprecedented detail. The overall genetic variation between strains has been estimated to be between 0.1 and 0.5%, based on approaches using multilocus sequence typing, multilocus microsatellite analysis, genome sequencing and whole-genome tiling arrays [3], [4], [5], [6],[7]. Specific and large-scale genome sequencing projects have resulted in a massive amount of genomic data for *S. cerevisiae* [8], [9], [10], [11]. Phylogenetic analysis of strains from a broad-range of ecological niches, revealed that *S. cerevisiae* originated in a wild habitat, probably the bark of oak trees, and that a subset of strains specialized for fermentation were emerged from subsequent selection and cultivation [4]. In addition, domestication events, rather than geography, substantially impacted the genetic structure of the *S. cerevisiae*population [7], [8], [12], [13], [14], [15], [16]. These domestication events were followed by human-associated dissemination of these yeasts throughout the world.

To date, the phenotypic variation of yeast populations originating from diverse environments has been only partially characterized. Several studies have focused on identifying the genetic bases for specific physiological traits, such as high-temperature growth [17], [18], ethanol resistance [19], sporulation efficiency [20],[21], drug responses [22], [23] and morphology [24]. These studies generally concerned growth determinations for a limited number of laboratory or vineyard strains or clinical isolates. Recently, extensive phenotypic variation in the mitotic proliferation ability of strains, was reported following high-throughput stress resistance analysis or adaptation to diverse environments (carbon and nitrogen sources, presence of toxins, nutrient limitations) for collections of *S. cerevisiae* strains [8], [25] [26], [27].

The variability between strains for phenotypes other than growth, particularly for metabolic traits such as glycolytic flux, metabolite yields, or the ability to use various substrates, has been poorly investigated despite their considerable industrial interest. In connection with this, eight strains with diverse genetic backgrounds (laboratory strains, vineyard and clinical isolates) were reported to be highly variable for a simple phenotypic trait, namely the utilization of di/tripeptides as nitrogen source [28]. Similarly, a population of 19 strains assembled from five different habitats (industry, forest, laboratory, clinic, fruit) exhibited an important variability both in life-history traits of yeast growth and in metabolic traits (glycolytic rate and ethanol production) [29]. More recently, the diversity between 9 food-processing strains (brewery, enology, and distillery) has been analyzed regarding their growth and metabolic behaviors in three industrial processes [30].

In view of the limited information available, the natural genetic resources of *S. cerevisiae* and the phenotypic variations between strains, and particularly those related to metabolism, bear further systematic exploration. The first aim of this study was to assess the extensive diversity of *S. cerevisiae* yeast strains by investigating a large panel of yeasts with respect to their specific phenotypic traits. A special attention was paid to phenotypes that have been directly the target of human selection for industrial purposes such as fermentation performance and kinetics, production of acetate, glycerol and aromatic compounds. Strains from diverse sources and environments (clinical, industrial, laboratory and wild isolates) were grown under the conditions of wine fermentation. These conditions are characterized by high sugar and ethanol concentrations, high acidity, low nitrogen availability, and anaerobiosis. Since wine fermentation can be regarded as an extreme environment expected to highlight variations between strains, it constitutes a model system for studying yeast phenotypic diversity.

The wide variety of the environments from which these strains were collected represent an array of conditions and stressors which likely contributed to the emergence and divergence of different phenotypes. These arose as the organisms developed distinct strategies to face the selective pressures of their living environments. In addition, fermentation yeasts typically have been specialized for a particular industrial process (e.g., baking, brewing, winemaking, etc.) through human manipulation. Consequently, both environmental and human selective pressures may have resulted in specific properties being shared by strains which live in similar habitats. We also analyzed the resulting phenotypic dataset to determine whether some traits were specific to a particular ecological niche. This allowed us to investigate the relationships between yeasts and their environments and to assess whether the evolution of certain phenotypes was driven by environmental and/or human factors.

RESULTS

Strain phenotypes under extreme fermentation condition

To investigate the phenotypic diversity among the *S. cerevisiae* strains, we characterized the fermentation performances of 72 strains obtained from widely different environments and sources (Table 1). The strains included the reference strain, S288C, and other lab strains (8), natural strains (19), clinical isolates (13), yeasts used in fermentative processes (10), yeasts found in vineyards (8) or in commercial winemaking (9), and several baker's yeast (5) strains. Anaerobic fermentations were carried out in synthetic MS medium containing a high glucose concentration and limited amounts of nitrogen and lipids. Although anaerobiosis was not imposed, it occurred rapidly and spontaneously due to the design of fermentors and the large amount of CO_2 produced. Under these conditions, yeast proliferation was rapidly limited by the low nitrogen concentration in the medium, so that most of the sugar was consumed by resting cells during the stationary phase. Throughout the fermentation process, the CO_2 production rate increased rapidly as the number of cells increased, then progressively decreased during the stationary phase. To simplify the analysis of the complex fermentation rate profiles, five variables were extracted from each fermentation curve. These included the total amount of released CO_2 (CO_{2F}), and four kinetics variables: the maximal fermentation rate (V_{max}), the time at which 50% and 75% of sugars were consumed, as estimated by 55 and 80 g/L of CO_2 produced (T_{50} and T_{75}, respectively) and the fermentation rate at T_{50} (V_{50}). In addition, the phenotypic description of each strain consisted of two growth features (dry weight, population size) and 11 metabolic variables (glycerol, acetate, succinate and eight volatile organoleptic compounds), measured at T_{75}. This resulted in a data set of 18 variables for each of the 72 strains.

Table 1: Collection of *S. cerevisiae* strains from diverse environments and geographical locations. doi:10.1371/journal.pone.0025147.t001

Environment	Strain	Geographical origin	Comments	Collection
Baker				
	CLIB 324	Saigon, Vietnam	Baker strain	Washington
	CLIB 215	New Zealand	Baker strain	Washington
	YS2	Australia	Baker strain	Sanger
	YS4	Netherlands	Baker strain	Sanger
	YS9	Singapore	Baker strain, Le Saffre	Sanger
Clinical				
	YJM280	USA	Peritoneal fluid	Washington
	YJM320	USA	Blood	Washington
	YJM326	USA	Patient	Washington
	YJM421	USA	Ascites fluid	Washington
	YJM653	USA	Broncho-alveolar lavage	Washington
	273614N	Newcastle UK	Fecal isolate	Sanger
	322134S	Newcastle UK	Throat-sputum isolate	Sanger
	378604X	Newcastle UK	Sputum isolate	Sanger
	YJM428	USA	Paracentesis fluid	Washington
	YJM451	Europe	Patient	Washington
	YJM975	Bergamo, Italy	Vaginal isolate	Sanger
	YJM978	Bergamo, Italy	Vaginal isolate	Sanger
	YJM981	Bergamo, Italy	Vaginal isolate	Sanger
Fermentation processes				
Beer				
	CLIB 382		Beer	Washington
	NCYC361	Ireland	Beer spoilage strain from wort	Sanger
Palm wine				
	Y12	Ivory Coast, Africa	Palm wine,	Washington
	DBVPG1853	Ethiopia	White Tecc	Sanger
	DBVPG6044	West Africa	Bili wine	Sanger
	NCYC110	Nigeria, West Africa	Ginger beer from Z. officinale	Sanger
	PW5	Africa	Raphia palm wine	Washington
Sake				
	K11	Japan	Shochu sake strain Awamori-1	Sanger
	UC5	Japan	Sene sake	Washington
	Y9	Japan	Ragi (similar to sake wine)	Sanger
Laboratory				
	FL 100	France	Crossing from D2339-17 and S1786	Washington
	CEN.PK	Germany		INSAT*
	ENY.WA-1A p			
	S288c	California, USA	Rotting fig	Sanger
	SK1	USA	Soil	Sanger
	W303	USA		Sanger
	W303 p	USA		
	Y55	France	Wine	Sanger
Natural				
Bertam palm				
	UWOPS03-461.4	Malaysia	Nectar, Bertam palm	Sanger
	UWOPS05-217.3	Malaysia	Nectar, Bertam palm	Sanger
	UWOPS05-227.2	Malaysia	Trigona, Bertam palm	Sanger
Cactus				
	UWOPS83-787.3	Bahamas	Fruit, Opuntia stricta	Sanger

Environment	Strain	Geographical origin	Comments	Collection
	UWOPS87-2421	Hawaii	Cladode, *Opuntia megacantha*	Sanger
Fruit				
	Y10	Phillipines	Coconut	Washington
	CBS 7960	Sao Paulo, Brazil	Produces ethanol from cane-sugar syrup.	Washington
	DBVPG6040	Netherlands	Fermenting fruit juice	Sanger
	DBVPG6765	Indonesia	Lici fruit	Sanger
Oak				
	NC-02	North Carolina, USA	Oak tree exudates	Washington
	T7	Missouri, USA	Oak tree exudates	Washington
	YPS1009	New Jersey, USA	Oak tree exudates	Washington
	YPS128	Pennsylvania, USA	Oak tree exudates	Sanger
	YPS163	Pennsylvania, USA	Oak tree exudates	Washington
	YPS606	Pennsylvania, USA	Oak tree exudates	Sanger
Soil				
	DBVPG1373	Netherlands	Soil	Sanger
	DBVPG1788	Finland	Soil	Sanger
	I14	Italy	Soil sample	Washington
	IL-01	Illinois, USA	Soil sample	Washington
Vineyard				
	YJM269	Portugal	Blauer Portugieser grapes	Washington
	BC187	Napa Valley, USA	Barrel fermentation, haploid derivative UCD2120	Sanger
	DBVPG1106	Australia	Grapes	Sanger
	L-1374	Chile	Wine	Sanger
	L-1528	Chile	Wine	Sanger
	M22	Italy	Vineyard isolate	Washington
	YIIc17_E5	France	Sauternes wine	Sanger
	RM 11	California, USA	Haploid derivative Bb32	Washington
Wine commercial				
	59-A	France	Meiotic spore of strain EC1118	
	V5p	France	Meiotic spore of strain CIVC8130	
	T73	Spain	Red wine (Monastrel)	Lalvin
	71B	Germany	Vineyard	Lalvin
	EC1118	France	Champagne fermentation	Lalvin
	L2226	France	Vineyard (Côte du Rhone)	Enoferm
	WE372	South Africa	Wine (Cape Town)	Anchor
	K1M	France	Grapes	Lalvin
	VL1	France		Laffort

We tested the reproducibility of our phenotypic analysis by fermenting several strains at least in duplicate. For each of these strains, the fermentation profiles were almost identical) and could be considered as a fingerprint of the strain's performance under standardized culture conditions. Moreover, we detected no substantial variation between the independent determinations of the fermentation kinetics, growth and metabolic variables and the intra-class correlation coefficients ranged between 86% and 99%, with a mean value of 95%. . This reproducibility analysis indicated the feasibility of our approach for assessing phenotypic diversity in *S. cerevisiae*.

Phenotypic variations among *S. cerevisiae* strains

The fermentation profiles of the 72 strains varied substantially from each other, reflecting their diverse fermentative performances (Figure 1). Many of the 72 strains were able to complete the fermentation of 240 g/L glucose. However, 45% of them exhibited a stuck profile and stopped fermenting before glucose was exhausted (i.e., the residual glucose concentration was above 10 g/L and CO_2 production was below 105 g/L). Great variations in kinetics variables were observed between the strains. For example, V_{max} was between 0.4 and 2.1 g/L/h and T_{75} was between 64 and 444 h. Due to the broad diversity in the origins of the strains, the value ranges for the measured variables, especially for V_{max}, were considerably greater than what was previously reported for commercial wine yeasts [31], [32] or strains from industry (distillery, wine, bakery) [30]. Nevertheless, for most strains, V_{max}, T_{50}, and to a lesser extent, T_{75} were very similar and only a few individuals exhibited extreme behaviors. The values for the V_{50} variable, which described the activity of the yeast during the latter stages of fermentation, were more dispersed than other variables, and were predominantly between 0.2 and 1.2 g/L/h. This observation suggested that considerable diversity exists in the abilities of yeast to face the multiple stresses conditions present at the end of fermentation (e.g., ethanol toxicity, nitrogen and micro-nutrient starvation).

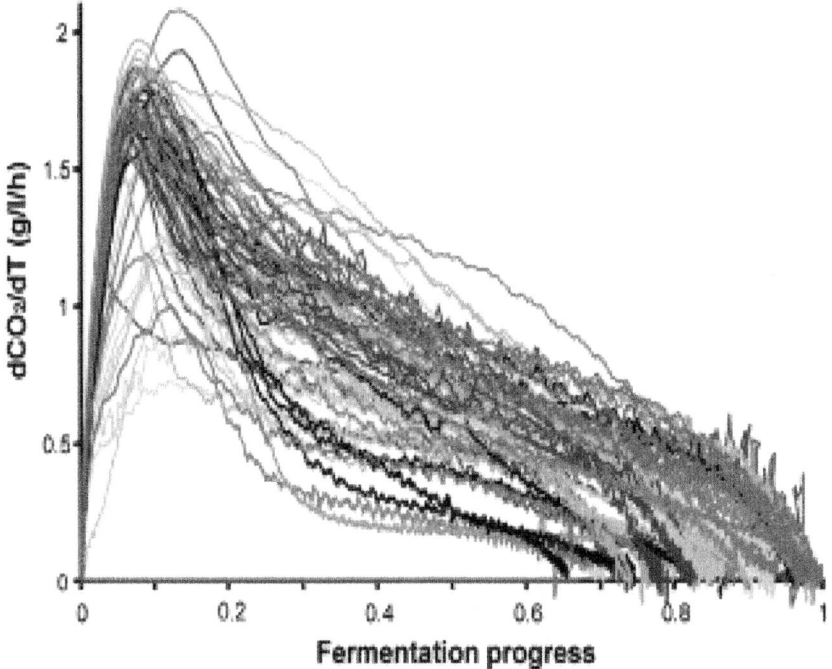

Figure 1: Comparison of the fermentation profiles for 72 *S. cerevisiae* strains from diverse geographical locales and environments. Fermentations were carried out in synthetic medium containing 240 g/L glucose, 200 mg/L nitrogen, pH 3.5. The fermentation profiles are presented as the CO_2 production rate vs. the fermentation progress, which corresponded to the ratio of consumed glucose to initial glucose. The lines are

colored according to the origin of the strains: vineyard (purple), soil (grey), sake (light grey), palm wine (pink), oak (brown), laboratory (yellow), fruit (red), wine commercial (dark green), Bertam-palm (blue), baker (black), beer (dark grey), cactus (light blue). doi:10.1371/journal.pone.0025147.g001

Considerable differences between strains, which were between 2- and 15-fold, were also found in the formation of biomass (dry weight) and in the synthesis of fermentation by-products (except ethanol). A large part of the variables was symmetrically distributed about the mean (Figure 2), although a few outliers existed for strains displaying extreme production levels (such as the synthesis of 12 g/L glycerol by CBS7960 and NCYC361). Positive skewed or reverse J-shaped distributions were observed for the production of acetate esters and some ethyl esters derivatives, reflecting the null or limited production of these compounds by the majority of the studied strains. Finally, in contrast to the great differences found in the formation of other metabolites, the conversion of glucose to ethanol remained almost constant among the whole population as usually viewed, with a mean value of 0.46 ± 0.01 g ethanol/g glucose.

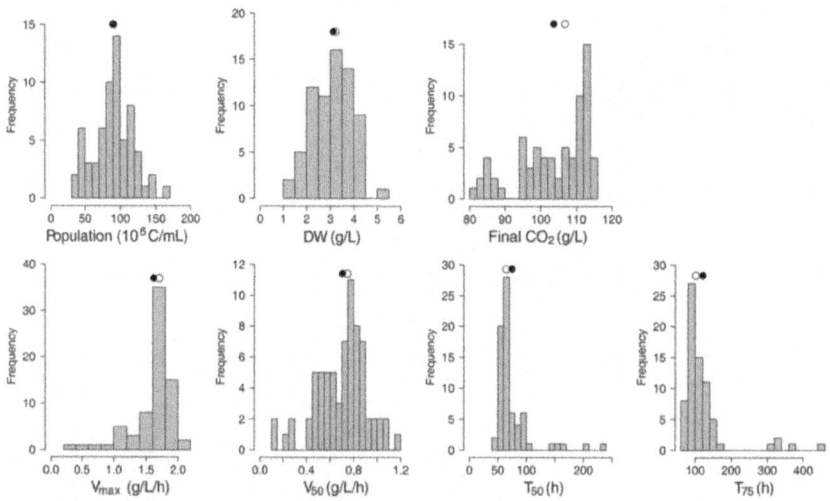

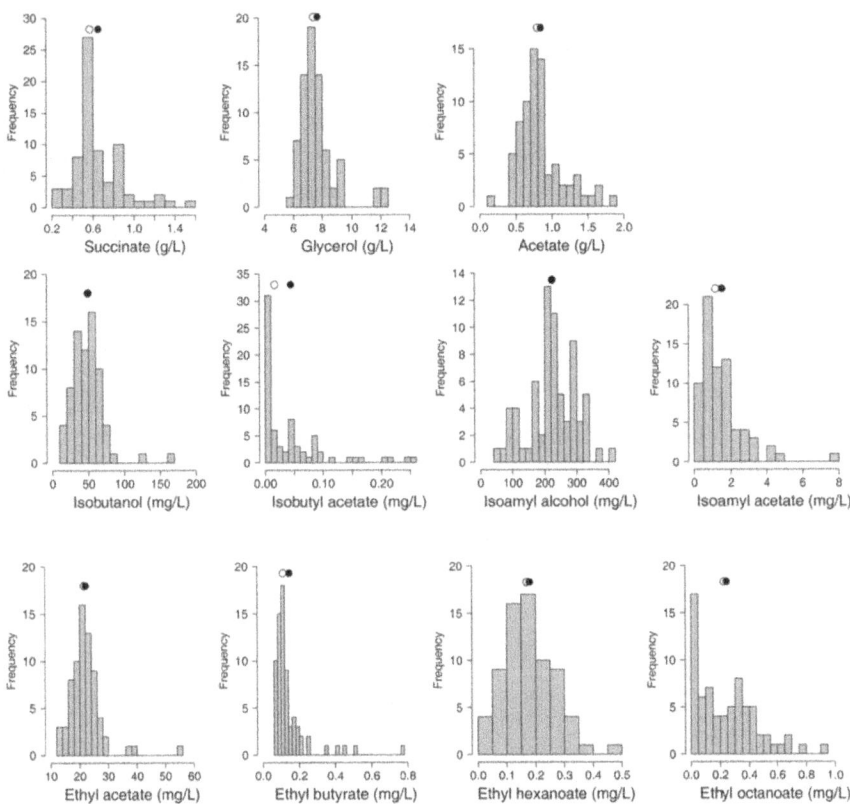

Figure 2: Distribution frequency of the phenotypic variables in the total population (72 strains). Closed circle: mean value; open circle: median value. Kinetics variables (CO_{2F}, V_{max}, V_{50}, T_{50} and T_{75}) were determined from the fermentation curves. Growth (cell number and dry weight) and metabolic variables (glycerol, acetate, succinate, two higher alcohols, two acetate esters and four ethyl esters) were measured when 75% of the sugars were consumed (180 g/L). doi:10.1371/journal.pone.0025147.g002

Relationships between the phenotypic traits

Considering the size of the population, we used the Pearson's Product Moment method to perform a correlation analysis between all the phenotypic traits (Tables S2, S3). Most of the variables varied independently within the population. However, as expected, T_{75} and T_{50} were strongly correlated (r=0.92, p<0.001) with each other and both of them were correlated negatively with V_{50} (r=−0.81, p<0.001 and r=−0.78, p<0.001, respectively, Figures 3A & B). The maximal fermentation rate V_{max} and the kinetics variables characterizing the last steps of the fermentation process (V_{50} and T_{75}) were moderately but significantly correlated ($r=0.48$, $p<0.001$ and $r=-0.52$, $p<0.001$, respectively). This strong correlation significance indicated that V_{max}, which measures activity at the beginning of the process (growth phase), impacted at least partially the behavior of the strain during the stationary phase (end of fermentation). However, the fairly weak correlation coefficients, close to 0.5, suggested a partial decoupling between the first and the last parts of fermentation, likely due to the contribution of other parameters in the control of the stationary phase, as the tolerance of the strains to inhibitory compounds (ethanol). A correlation ($r=0.71$, $p<0.001$) was found between the biomass production and the final amount of CO_2 released, indicating that most poorly growing strains exhibited stuck profiles (Figure 3C). This is consistent with previous reports citing the inability of commercial yeasts to complete wine fermentation when nutrient limitations affect their growth [33], [34], [35], [36]. Regarding the metabolic variables, substantial correlations were found between the productions of isoamyl acetate and either its isoamyl alcohol precursor ($r=0.49$, $p<0.001$) or isobutyl acetate ($r=0.51$, $p<0.001$), which is the other main acetate ester produced by yeast during alcoholic fermentation (Figure 3D).

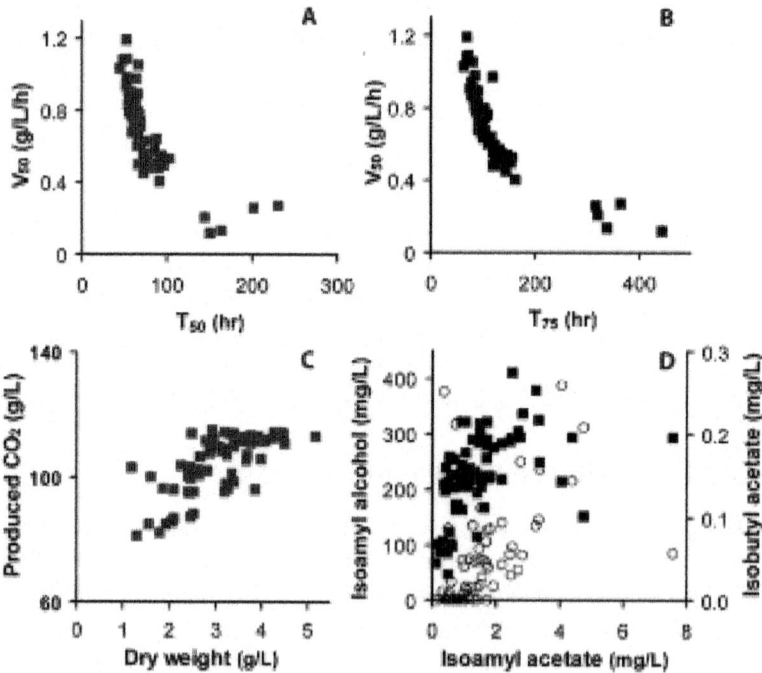

Figure 3: Relationships between the phenotypic variables within the total population of strains. Correlations were found between: the fermentative activity at mid-fermen-

tation V_{50} and the time necessary to consume 50% or 75% of sugars, T_{50} (A) and T_{75} (B), respectively; the final CO_2 release and the dry weight (C); and the production of isoamyl acetate and the productions of isoamyl alcohol (■) and isobutyl acetate (○) (D). doi:10.1371/journal.pone.0025147.g003

Metabolic traits discriminate strains from different origins

To identify potential relationships between strain origin and fermentation phenotype, the data set was first analyzed regarding seven ecological niches: baker, clinical, fermentation processes, laboratory, and vineyard, natural and wine commercial (Figure S2, Table S4). To obtain a general overview of the data, a principal component analysis [PCA] and a linear discriminant analysis [LDA] (with origin as factor) were first performed using all the traits and all the strains. Only 38% of the variation was explained by the two first components of the global PCA and the LDA analysis did not allow to discriminate the origin of strains, due to the complexity of the dataset (Figure S3). Consequently, an exploratory study was performed in order to select the variables that exhibited a significant global effect among the seven groups of strains (Table 2, Table S5). Univariate analyses of variance (ANOVA) without multiplicity adjustment identified several variables relevant to discriminating the strains on the basis of their origin (p-value<0.05): dry weight, population size, CO_{2F}, T_{75}, T_{50}, glycerol, acetate and ethyl butyrate. For the other descriptive variables, variations were mainly attributed to the large intra-group variability. The analysis of the most selective variables allowed us to identify specific traits common to all the strains from the same ecological niche, for three groups: wine commercial, baker and laboratory. Indeed, a noteworthy characteristic of laboratory strains was their high level of ethyl butyrate synthesis compared to strains from other habitats. In addition, we found that these yeasts produced little biomass, fermented sugars slowly, produced high amounts of acetate and low amounts of isoamyl acetate. Conversely, commercial wine strains were able to completely and rapidly ferment the available sugars, while producing high biomass and little acetate. Finally, we found that bakery yeasts were characterized by low acetate, succinate and glycerol productions but, contrary to wine commercial strains, exhibited poor growth and fermentative performances.

Table 2: Univariate analysis of the variance for the phenotypic variables. doi:10.1371/journal.pone.0025147.t002

Phenotypic variable	p-value 7 groups	p-value 13 groups
Cell number	0.035	0.15
Dry weight	0.004	0.002
V_{max}	0.06	0.004
V_{50}	0.09	0.005
T_{75}	0.02	0.0006
T_{50}	0.01	<0.0001
CO_{2F}	0.02	0.0005
Glycerol	0.03	0.001
Acetate	0.001	0.04
Succinate	0.65	0.46
Isobutanol	0.12	0.22
Isobutyl acetate	0.34	0.33
Isoamyl alcohol	0.34	0.46
Isoamyl acetate	0.06	0.16
Ethyl acetate	0.08	0.08
Ethyl butyrate	<0.0001	0.0004
Ethyl hexanoate	0.090	0.007
Ethyl octanoate	0.71	0.14

For the other origins (vineyard, clinical, nature, fermentation processes), we found a large intra-group variability for all the phenotypic traits. This may be explained by the intrinsic diversity within each class. The clinical strains were isolated from human infections. Since these yeasts are generally considered to originate from other environments [7], [37], substantial phenotypic variability can be expected in this group. The nature, vineyard and fermentation processes groups consisted of strains from habitats with a strong heterogeneity regarding the living conditions. The nature group also consisted of strains from different environments, including sugar-rich and sugar-poor ones, which may have likely affected their cell physiology. Consequently, we redefined these categories (Table 1) by separating the nature isolates into fruit, cactus, Bertram palm, oak and soil subgroups and the fermentation processes group into beer, sake, and palm wine. In this way, a total of 13 categories were established and analyzed as described above. This reclassification substantially decreased intra-group variability for most of the phenotypic traits (Figure S2). The two most significant variables for discriminating the strains among the 7 groups, namely ethyl butyrate and acetate, as well as the kinetic variables (T_{75}, T_{50}, V_{50}, V_{max} and CO_{2F}), the dry weight and the production of ethyl hexanoate were identified as contributed substantially to the variance between the 13 habitats (Table 2). Palm wine strains consumed sugar at high rates throughout the fermentation course, resulting in short fermentation times. These strains were specifically differentiated by their low succinate production and, to a lesser extent, by their high acetate and isoamyl acetate production. Yeasts used for sake and beer fermentation exhibited low fermentation rates (V_{50}) and long fermentation times. Beer strains were distinguished from sake strains and all other strains, by their low biomass production and by their low levels of aromatic compound synthesis, particularly of the ethyl ester derivatives (ethyl hexanoate). Regarding the nature group, all the strains isolated from soil and fruits were able to complete the fermentation of sugar, unlike the strains derived from cactus, oak and palm habitats. These strains were further discriminated by biomass production, which was high for cactus and fruit strains and low for oak and palm strains. Furthermore, the profile of ethyl ester synthesis during fermentation varied greatly among these groups. Oak and Bertam-palm strains produced low levels of ethyl acetate compared to the other strains, whereas cactus and Bertam-palm strains produced high levels of ethyl octanoate. Ethyl hexanoate production was high for the soil and oak strains and low for the Bertam-palm strains.

Phenotypic differentiation of commercial wine from vineyard yeasts

We compared the specific phenotypic properties of the wine commercial strains to those of yeasts from other environments, particularly the vineyard strains. The wine commercial strains, as well as strains originating from fruits and the majority of those from the soil (3/4) and vineyard (6/8) groups, were able to completely ferment 240 g/L glucose (Figure 4A). Whereas all the wine commercial strains had short fermentation times (<270 h), some individuals in the three other groups of strains having good fermentative capacities, including the vineyard set, displayed prolonged fermentation profiles (Figure 4B). More generally, the variability of wine commercial strains was lower than that of the vineyard group (Figure S2,Table S4), and substantially lower standard deviation values than those of the vineyard group were observed (Table 3). In addition, the wine commercial strains exhibited extreme values for some specific phenotypic traits compared to those of vineyard strains. These included production of acetate (0.6±0.1 g/L versus 0.9±0.4 g/L, respectively), of isoamyl alcohol (and its acetate ester derivative) (242±64 g/L versus 194±87 mg/L, respectively) and V_{50} (0.8 g/L·h±0.1 versus 0.7 g/L·h±0.2, respectively). All together, these results showed that wine commercial strains constituted a minimally diverse subset of yeasts from vineyard, in agreement with the selection of these strains for their technological traits by man.

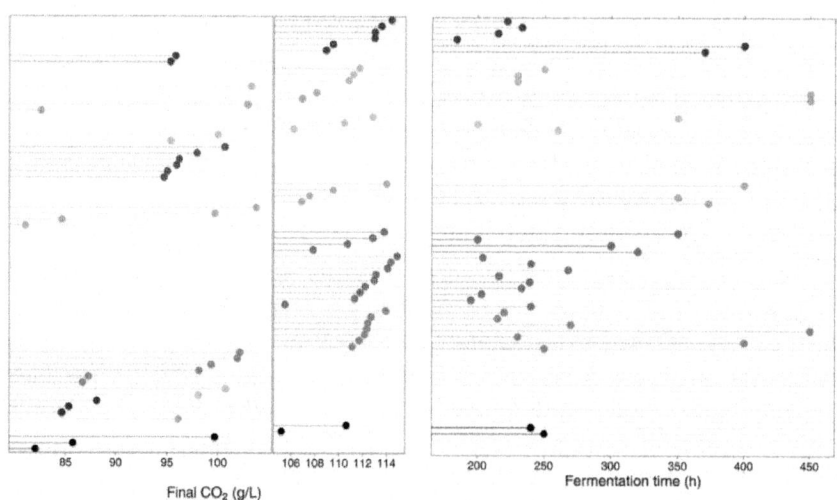

Final CO₂ (g/L)   Fermentation time (h)

Figure 4: Abilities of the 72 strains from different environments to efficiently ferment a high concentration of sugar.

Strains were considered able to achieve fermentation of 240 g/L of sugar when their production of CO_2 at the end of fermentation (A) was higher than 105 g/L. These strains were further discriminated by the fermentation time (B). Vineyard: purple symbols; soil: grey symbols; sake: green symbols; palm: pink symbols, oak: brown symbols; laboratory: yellow symbols; fruit: red symbols; wine commercial: dark green symbols; clinical: tan symbols, cactus: blue symbols; Bertam –palm wine: dark blue symbols; beer: beige symbols ; baker: black symbols. doi:10.1371/journal.pone.0025147.g004

Table 3: Comparison of the phenotypic variables between strains isolated from the wine commercial and vineyard strain groups. doi:10.1371/journal.pone.0025147.t003

Phenotypic variable	Wine commercial		Vineyard	
	Mean	S.D.	Mean	S.D.
Cell number, 10^6 c/mL	119	27	92	22
Dry weight, g/L	3.6	0.5	3.8	0.9
V_{max}, g/L/h	1.7	0.2	1.6	0.3
CO_2 produced g/L	112	3	108	8
T_{75} hr	88	10	109	27
T_{50} hr	61	5	67	16
V_{50} g/L·h	0.8	0.1	0.7	0.2
Succinate g/L	0.62	0.14	0.63	0.29
Glycerol g/L	7.0	0.4	7.1	0.7
Acetate g/L	0.6	0.1	0.9	0.4
Isobutanol g/L	52	7.8	70	48
Isobutyl acetate g/L	0.04	0.04	0.08	0.06
Isoamyl alcohol g/L	242	64	194	87
Isoamyl acetate g/L	1.9	1.2	1.3	0.7
Ethyl acetate g/L	24	3	23	3
Ethyl butyrate g/L	0.18	0.13	0.13	0.04
Ethyl hexanoate g/L	0.2	0.1	0.2	0.1
Ethyl octanoate g/L	0.2	0.2	0.3	0.3

Effects of strains origin on phenotypic profiles

Finally, we carried out a comprehensive assessment of the relationships between their habitats (qualitative variable) and their quantitative phenotypic traits. A population of 57 strains from 10 different groups was initially considered for this analysis: laboratory, baker, wine commercial, sake, palm wine, vineyard, oak, soil, fruit, and Bertam palm. Clinical strains were excluded from the analysis because of their high intragroup variability, as well as *beer* and *cactus* groups, with only two individuals each. A linear discriminant analysis (LDA), was applied to the most discriminant phenotypic traits: dry weight, T_{75}, CO_{2F}, acetate, and ethyl butyrate. This analysis explained 45% and 22% of the intergroup variance in the first two discriminant axes, respectively (Figure 5, Figure S4).

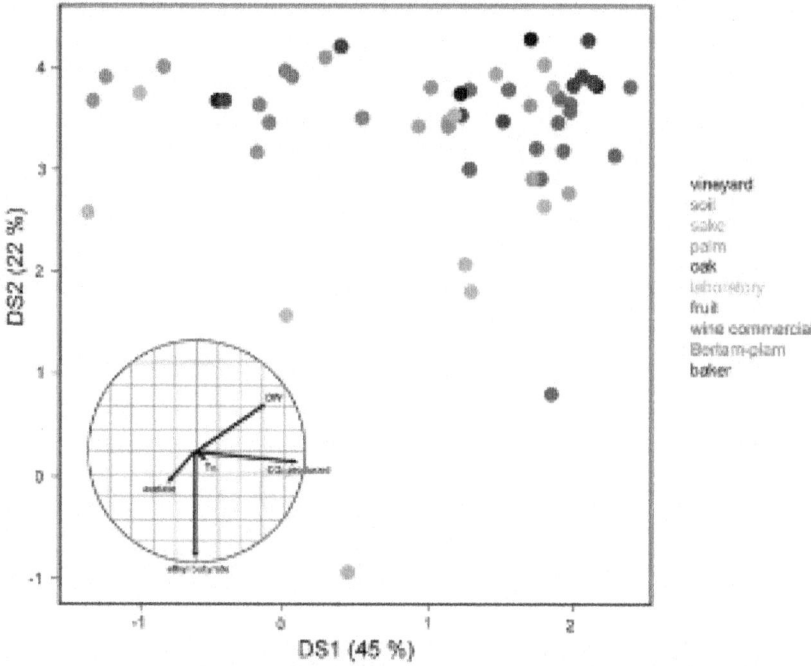

Figure 5: Linear discriminant analysis (LDA) of the population based on five discriminating phenotypic traits. A linear discriminant analysis was applied to the most discriminating variables: dry weight, T_{75}, CO_{2F}, acetate and ethyl butyrate measured for 53 strains representing 10 different groups. Clinical isolates were not included, due to the large phenotypic variability observed among the strains in this group. *Beer* and *cactus* groups, with only two strains each, were removed for this analysis. Groups of origin include vineyard (purple), soil (grey), sake (green), palm wine (pink), oak (brown), laboratory (yellow), fruit (red), wine commercial (dark green), Bertam-palm (dark blue), baker (black). doi:10.1371/journal.pone.0025147.g005

The LDA first showed that the laboratory strains clearly separated from other individuals according to their high production of ethyl butyrate and, to a lesser extent, that of acetate. Moreover, the yeasts from rich-sugar environment, composed of the wine commercial (8/9), fruit (3/4) and vineyard (5/8) strains, were close to each other in the LDA representation on the basis of their high levels of CO_2 production, which reflected their good fermentative capacities, high dry weights, and low levels of acetate production. Surprisingly, these strains displayed phenotypic similarities and clustered together with some sake, baker and soil yeasts. The univariate analysis showed that some of these groups of strains were discriminated on the basis of other specific variables (Figure S2), which were not included in the global LDA analysis.

Finally, two groups consisting of oak isolates on one hand and of strains from Bertam palm on the other hand, emerged due to their low dry weight and defective fermentation abilities. Together, these observations show that these few phenotypic traits, measured under these extreme fermentation conditions, allowed some strains to be discriminated based on their origin. However, this did not include yeasts from the soil and palm wine groups, which were not clearly distinguishable.

DISCUSSION

Wine Fermentation Phenotypes Reflect a Wide Diversity in Yeast Response to a Stressful Environment

The large data set of 18 kinetic, metabolic and growth characters determined during glucose fermentation for 72 *S. cerevisiae* strains (1296 distinct measurements) provides a detailed picture of the extent of metabolic diversity within this species, which has been poorly explored until now. Wine fermentation conditions represent a combination of various stresses (osmotic, ethanol, acidic, nutrient limitation) that accentuate the metabolic differences between strains. All variables except ethanol production varied independently within the population for most strains, and to a large extent (between 2- and 15-fold, depending on the variable). This substantial variability highlighted the disparate strategies used by the strains to cope with the numerous environmental stresses found in alcoholic fermentation and their different levels of ability to adapt to this extreme environment. Moreover, there was little or no correlation between the vast majority of variables, including metabolic ones, either within the entire population as a whole or within groups of strains separated according to their ability to complete fermentation or to their environmental origin. This suggests great inter-strain diversity in the metabolic strategies they used to deal with these unfavorable conditions, which derives from a substantial flexibility of the *S. cerevisiae* metabolic network.

A key factor that differentiated the strains was their ability to complete fermentation: 45% of the strains were unable to entirely consume the available sugars (240 g/L) and exhibited stuck fermentation profiles. Two main reasons for these problematic fermentations were identified in the case of commercial wine yeasts growing on grape juices. First, a nutrient limitation (e.g., nitrogen or lipids) may result in a low fermentation rate as a consequence of inefficient growth[33], [38], [39]. Second, the toxicity of fermentation by-products (e.g., ethanol and fatty acids) during the latter stages of fermentation may inhibit sugar transport and alter cellular membrane integrity, leading to reduced metabolic activity and viability [40], [41], [42], [43]. Consistent with the first hypothesis, we found that fermentation efficiency was related to biomass production. Furthermore, for all strains unable to complete the fermentation, the fermentation rate became asymptotic toward the end of the culture, which has been reportedly due to loss of viability [34]. This supports the main role of biomass production as the factor governing the fermentation course under stressful conditions. However, contrary to previous observations from wine commercial strains [34], [38] or from industrial environments [30], we found no correlation between V_{max} and biomass or population size. This is likely due to the use of highly diverse *S. cerevisiae* strains from a broader range of environments.

Adaptation to Environment Results in Emergence of Specific Metabolic Traits

Shared phenotypes among strains from similar environments has been reported for yeasts collected from oaks, which exhibit freeze-thaw resistance crucial for survival in wintry environments, and for vineyard isolates, which have low sensitivities to copper sulfate, an anti-microbial agent widely used in European fields [5], [25], [44]. Similarly, the profile of resistance of sake-producing yeasts to various stresses was consistent with their specialized metabolism for growing under the defined conditions of sake fermentation [25]. Our study revealed additional specific traits that characterize strains originating from the same ecological niche. Examples include the low level production of fermentation by-products (e.g., glycerol, acetate and succinate) by baker's yeasts or the very high production levels of ethyl butyrate and limited biomass formation exhibited by laboratory strains. Surprisingly, in addition to wine commercial strains, most of the strains in the vineyard, fruit and soil groups also displayed good fermentative properties, whereas strains from oak, plant and brewery environments exhibited most of the stuck fermentation profiles.

Recently, an exhaustive mapping of the mitotic proliferation traits of *S. cerevisiae* growing under a wide-range of environments, reported a strong effect of population genetic history on trait variations within this species, suggesting that the relationships between ecological niche and phenotypes may be fortuitous or due to a common influence from a shared genetic lineage[27]. Our analysis, based on a population-scale phenotyping of *S. cerevisiae* restricted to only wine fermentation conditions but measuring a large number of growth, kinetics and metabolic characters allows to reveal links between source environments and specific phenotypes of some groups of strains. These relationships between the origin and the properties of some strain groups likely reflect phenotypes that evolved in response to environmental constraints. The stresses and conditions of particular habitats may have shaped the metabolism and physiology of these strains, resulting in adaptations and the emergence of environment-specific traits. Two different *S. cerevisiae* life-history strategies (grasshopper ant) have been previously defined on the basis of specific growth characteristics (rate, cell size, final population size) of strains related to the resources available in their original environment [29],[45].Similarly, the higher fermentative capacities of strains in the fruit, vineyard and wine commercial groups, may have arisen through selection in response to the prevalence of high sugar concentrations in these environments. Adaptations to osmotic stress and to the toxicity of fermentations products, such as the ethanol and fatty acids generated in the presence of abundant sugar, allow strains found in these environments to have selective advantages which contribute to their prevalence and to their capacity to efficiently ferment large amounts of sugar. Conversely, we found that yeasts isolated from "poor-sugar" environments (e.g. oaks and other plants) do not exhibit these efficient fermentation features. In the same way, laboratory strains, which have been propaged for many generations on rich media, optimal for growth [46], most likely lost their capacity to thrive in harsh environmental conditions. This might explain why they grew poorly and exhibited stuck profiles during alcoholic fermentation.

Human selection is another factor that has contributed to the environment-specific properties of strains used in industrial processes [13]. The physiologic and metabolic features common to all the wine commercial strains include low acetate production, substantial biomass production, aroma production (in the form of isoamyl alcohol and ester-acetate) and the fast and efficient fermentation of high sugar concentrations. These traits differentiate the wine commercial strains from strains found in other environments, including the vineyard strains. These phenotypes are the consequences of human selection since wine commercial strains have been intentionally picked out from vineyard environments and exploited for winemaking due to their advantageous kinetic and metabolic characteristics. Similarly, the low production of acetate, glycerol and succinate by-products exhibited by baker strains likely reflects the human selection of strains for their high CO_2 production rates, which are needed for bread-making [10], [47] and are detrimental to by-product formation. The clade of laboratory strains, from which most *S. cerevisiae* knowledge has been acquired, was significantly differentiated from the other strain groups for many phenotypic traits, including low biomass formation, poor fermentation performances and synthesis of specific metabolites. This may be explained by the fact that most of the commonly used laboratory strains were derived from the S288C genetic background [7], [48], [49]. However, our analysis included two strains with S288C independent genetic background, SK1 and Y55. The divergence of all the laboratory strains with the rest of *S. cerevisiae* population may also reflect their long-term domestication under optimal growth conditions which likely repressed some protective and adaptive mechanisms essential for survival in natural environments [46].

In contrast to the other groups of strains, the phenotypes of the clinical strains were broadly distributed compared to those of the entire population. This may be explained by the opportunistic colonization of human tissues by *S. cerevisiae* strains which normally inhabit different environments, and thus differed considerably in their physiologic and metabolic traits. Consistent with our phenotypic observations, the clinical isolates were also highly diverse genetically and did not form a coherent group [7], [25], [37].

Overall, the low variability of laboratory strains compared to the total population highlights the need to continue to study the genomic and phenotypic diversity of *S. cerevisiae*. This research will provide new insights on the relationships and interactions between *S. cerevisiae* and its highly varied environments, and on the molecular mechanisms involved as yeasts adapt to their habitats. Ultimately, this will allow better use of this species ample natural genetic resources.

Genomic analyses of panels of *S. cerevisiae* strains identified distinct subgroups based on the identification of SNPs, which demonstrated that the *S. cerevisiae* population structure at least partly reflected the numerous different environments from which yeasts were isolated [7], [8]. Interestingly, it was recently reported that the genetic subgroup Wine/European could be differentiated from other lineages, namely Malaysian, West African and North American, based on the phenotypes of strains grown in different environments and in the presence of different drugs [8] [27], indicating that trait variations in yeasts reflect for an important part the genetic structure of *S. cerevisiae* population. Accordingly, the genetic polymorphism may be a contributor to the particular abilities of strains to adapt to their environments and consequently, in the emergence of environment-specific phenotypes, as those observed in this study. Recently, quantitative genetic studies of segregating populations from crosses between strains from divergent lineages, were described as a powerful tool for investigating the genetic determinants of polygenic phenotypes [26], [50]. These approaches may be further developed, using parental strains selected from *S. cerevisiae* population on the basis of our phenotypic database, to identify the genetic architecture of particular physiologic and metabolic traits, including those of technological interest.

MATERIALS AND METHODS

Yeast Strains

Seventy-two *S. cerevisiae* strains, all prototrophic except *S. cerevisiae* W303-1A (*MATα leu2-3, 112 ura3-1 trp1-1 his3-11, 15 ade2-1*), collected from ecologically and geographically diverse environments (Table 1) were characterized in this study. Many of the strains came from Washington University (22) and Sanger Institute (36), whereas others were obtained from several different companies or laboratories. The genome sequences of most of these strains are available. According to their origin and/or existing classifications in the Sanger Institute and Washington University databases, our strains were first classified into seven major groups: *baker, clinical, fermentation processes, laboratory, vineyard, natural* and *wine commercial*. The *fermentation processes* and *natural* clades were further separated into three (*beer, palm wine* and *sake*) and five (*oak, Bertam palm, soil, fruit* and *cactus*) sub-groups, respectively. For each strain, an aliquot of a reference stock, conserved at −80°C, was transferred to a YPD agar plate (1% Bacto yeast extract, 2% bactopeptone, 2% glucose, 1.5% agar) 48 h before fermentation.

Fermentation Conditions

Initial cultures in YPD medium were grown in 50 mL flasks at 28°C, with shaking, (150 rpm) for 12 h. These cultures were used to inoculate secondary cultures at a density of $1×10^6$ cells/mL. Fermentations were carried out in synthetic MS medium, which contained 240 g/L glucose, 6 g/L malic acid, 6 g/L citric acid and 200 mg/L nitrogen in the form of amino acids (148 mg N/L) and NH_4Cl (52 mg N/L), at pH 3.5 (5). Ergosterol (1.875 mg/L), oleic acid (0.625 mg/L) and Tween 80 (0.05 g/L) were provided as anaerobic growth factors. Fermentations took place in 1.1 liter fermentors equipped with fermentation locks to maintain anaerobiosis, at 28°C, with continuous magnetic stirring (500 rpm). The CO_2 release was followed by automatic measurement of fermentor weight loss every 20 minutes. The rate of CO_2 production (dCO_2/dt, where t is time) was calculated by polynomial smoothing of the last ten values of CO_2 production. The frequent acquisition of CO_2 release values and highly precise bioreactor weighing (±10 mg) allowed accurate CO_2 production rates to be calculated, with good repeatability and a small variation coefficient: $(dCO_2/dt)_{max}$=0.8% [31]. For data analysis, five variables were determined from the entire fermentation rate curve. (1) The total amount of CO_2 released, allowed us to estimate the fermentative capacity of the yeasts and to identify the strains unable to completely ferment the available glucose (240 g/L) ("stuck" profiles). (2 & 3) The times required to ferment 50% (T_{50})

and 75% (T_{75}) sugars were recorded because some strains displayed "stuck" fermentation profiles and were not able to complete the fermentation process. (4 & 5) Finally, the maximal CO_2 production rate (V_{max}) and the rate at mid-fermentation (V_{50}) reflected yeast activity at the beginning of the process and during the stationary phase.

Analytic Methods

Cells were counted using an electronic particle counter (Multisizer 3 Coulter Counter, Beckman Coulter) fitted with a probe with a 100-µm aperture. The dry weight of the yeast was measured by filtering 50 mL of culture though a 0.45 µm-pore Millipore nitrocellulose filter, which was washed twice with 50 mL distilled water and dried for 24 h at 105°C. These analyses were performed at T_{75} (when 75% sugar was consumed).

Glucose and fermentation products (acetate, succinate, glycerol and ethanol) were analyzed by high-pressure liquid chromatography (HPLC 1100, Agilent Technologies) on an HPX-87H Aminex column (Bio-Rad Laboratories Inc.). Dual detection was performed with a refractometer and a UV detector (Hewlett Packard).

The concentrations of volatile compounds were assayed by gas chromatography using an Agilent 6890 chromatograph equipped with an headspace injector and a ZB-WAX column (60 m×0.32 mm×0.5 µm) from Phenomenex Inc. The pressure was held constant (120 kPa) and temperature was progressively increased. The column temperature was: initially held at 38°C for 3 min, then increased to 65°C at a rate of 3°C/min, then increased to 160°C at a rate of 6°C/min, then held at 160°C for 5 min, then increased to 230°C at a rate of 8°C/min, and finally held at 230°C for 5 min. The compounds were detected using a flame ionization detection (FID) system. The final ethanol concentration of the samples was adjusted to 11%, to standardize transfer between the liquid and the headspace.

Statistical Analyses

Statistical analyses were performed using the R software, version 2.9.2.

For each trait, usual descriptive statistics were calculated by strain origin with mean, standard deviation and coefficient of variation.

Reproducibility of measurements was evaluated from data of fermentations achieved at least in duplicate with the same strains (n=21 strains) using the intra-class correlation coefficient[51].

Pairwise correlations between variables were calculated using Pearson's Product Moment correlation coefficients (r). Since 153 multiple correlations were computed, p-values were corrected for multiple testing using Benjamini-Hochberg methods [52] by means of R's multtest package [53]. As some distributions were not gaussian, the robustness of the Pearson's coefficients was studied comparing this approach with the Spearman's rank correlations. Both approaches were consistent and only Pearson's correlations were presented.

For a global analysis of phenotypic diversity within *S. cerevisiae* population, a principal component analysis [PCA] and a linear discriminant analysis [LDA] (with origin as factor) were first performed considering the whole data set using ade4 package [54].

To analyse the diversity regarding the origin of the strains, the most discriminant variables according to strain origin were selected. Each variables was tested using a one-way ANOVA in order to keep the ones having a significant impact on strain origin (at a p-value threshold of 0.05), without any multiplicity adjustment. For each trait, normality of residual distributions and homogeneity of variance were studied using usual diagnostic graphics. As some traits failed in showing strict homogeneity of variance, a robustness analysis was done using a Kruskall-Wallis test (non-parametric test) to assess a global effect of the strain origin. As the results between the 2 analyses were consistent (Table S5), the one-way ANOVA has been considered as robust and only these results were systematically presented.

To reveal the structure of the population according to the origin of the 72 strains and the chosen phenotypes, a LDA was performed on the selected subset of phenotypic variables using the *discrimin* function of the ade4 package [54].

LDA is a supervised multivariate method that uses the class information to characterize the structure of the data by maximizing the ratio of "between-class" variance to "within-class" variance. The resulting combination was used as a linear classifier for dimensionality reduction, to separate the strains origin according to the studied phenotypes.

ACKNOWLEDGMENTS

We thank Jean-Luc Legras for his insightful comments and suggestions on the manuscript. The authors would also like to thank Virginie Angenieux for her contribution to the analysis of volatile compounds. We thank Gianni Liti and Justin Fay for providing the strains collections used in this study.

AUTHOR CONTRIBUTIONS

Conceived and designed the experiments: CC SD FB. Performed the experiments: PB IS CC. Analyzed the data: CC IS SD FB. Contributed reagents/materials/analysis tools: PB IS. Wrote the paper: CC SD.

REFERENCES

1. Cavalieri D, McGovern PE, Hartl DL, Mortimer R, Polsinelli M (2003) Evidence for S. cerevisiae fermentation in ancient wine. J Mol Evol 57: Suppl 1S226–232.

2. McGovern PE, Zhang J, Tang J, Zhang Z, Hall GR, et al. (2004) Fermented beverages of pre- and proto-historic China. Proc Natl Acad Sci U S A 101: 17593–17598.

3. Winzeler EA, Castillo-Davis CI, Oshiro G, Liang D, Richards DR, et al. (2003) Genetic diversity in yeast assessed with whole-genome oligonucleotide arrays. Genetics 163: 79–89.

4. Fay JC, Benavides JA (2005) Evidence for domesticated and wild populations of Saccharomyces cerevisiae. PLoS Genet 1: 66–71.

5. Aa E, Townsend JP, Adams RI, Nielsen KM, Taylor JW (2006) Population structure and gene evolution in Saccharomyces cerevisiae. FEMS Yeast Res 6: 702–715.

6. Schacherer J, Ruderfer DM, Gresham D, Dolinski K, Botstein D, et al. (2007) Genome-wide analysis of nucleotide-level variation in commonly used Saccharomyces cerevisiae strains. PLoS One 2: e322.

7. Schacherer J, Shapiro JA, Ruderfer DM, Kruglyak L (2009) Comprehensive polymorphism survey elucidates population structure of Saccharomyces cerevisiae. Nature 458: 342–345.

8. Liti G, Carter DM, Moses AM, Warringer J, Parts L, et al. (2009) Population genomics of domestic and wild yeasts. Nature 458: 337–341.

9. Wei W, McCusker JH, Hyman RW, Jones T, Ning Y, et al. (2007) Genome sequencing and comparative analysis of Saccharomyces cerevisiae strain YJM789. Proc Natl Acad Sci U S A 104: 12825–12830.

10. Argueso JL, Carazzolle MF, Mieczkowski PA, Duarte FM, Netto OV, et al. (2009) Genome structure of a Saccharomyces cerevisiae strain widely used in bioethanol production. Genome Res 19: 2258–2270.

11. Novo M, Bigey F, Beyne E, Galeote V, Gavory F, et al. (2009) Eukaryote-to-eukaryote gene transfer events revealed by the genome sequence of the wine yeast Saccharomyces cerevisiae EC1118. Proc Natl Acad Sci U S A 106: 16333–16338.

12. Azumi M, Goto-Yamamoto N (2001) AFLP analysis of type strains and laboratory and industrial strains of Saccharomyces sensu stricto and its application to phenetic clustering. Yeast 18: 1145–1154.

13. Diezmann S, Dietrich FS (2009) Saccharomyces cerevisiae: population divergence and resistance to oxidative stress in clinical, domesticated and

wild isolates. PLoS One 4: e5317.

14. Ezeronye OU, Legras JL (2009) Genetic analysis of Saccharomyces cerevisiae strains isolated from palm wine in eastern Nigeria. Comparison with other African strains. J Appl Microbiol 106: 1569–1578.

15. Hennequin C, Thierry A, Richard GF, Lecointre G, Nguyen HV, et al. (2001) Microsatellite typing as a new tool for identification of Saccharomyces cerevisiae strains. J Clin Microbiol 39: 551–559.

16. Legras JL, Merdinoglu D, Cornuet JM, Karst F (2007) Bread, beer and wine: Saccharomyces cerevisiae diversity reflects human history. Mol Ecol 16: 2091–2102.

17. Sinha H, Nicholson BP, Steinmetz LM, McCusker JH (2006) Complex genetic interactions in a quantitative trait locus. PLoS Genet 2: e13.

18. Steinmetz LM, Sinha H, Richards DR, Spiegelman JI, Oefner PJ, et al. (2002) Dissecting the architecture of a quantitative trait locus in yeast. Nature 416: 326–330.

19. Hu XH, Wang MH, Tan T, Li JR, Yang H, et al. (2007) Genetic dissection of ethanol tolerance in the budding yeast Saccharomyces cerevisiae. Genetics 175: 1479–1487.

20. Ben-Ari G, Zenvirth D, Sherman A, David L, Klutstein M, et al. (2006) Four linked genes participate in controlling sporulation efficiency in budding yeast. PLoS Genet 2: e195.

21. Magwene PM, Kayikci O, Granek JA, Reininga JM, Scholl Z, et al. (1987) Outcrossing, mitotic recombination, and life-history trade-offs shape genome evolution in Saccharomyces cerevisiae. Proc Natl Acad Sci U S A 108: 1987–1992.

22. Kim HS, Huh J, Fay JC (2009) Dissecting the pleiotropic consequences of a quantitative trait nucleotide. FEMS Yeast Res 9: 713–722.

23. Perlstein EO, Ruderfer DM, Roberts DC, Schreiber SL, Kruglyak L (2007) Genetic basis of individual differences in the response to small-molecule drugs in yeast. Nat Genet 39: 496–502.

24. Nogami S, Ohya Y, Yvert G (2007) Genetic complexity and quantitative trait loci mapping of yeast morphological traits. PLoS Genet 3: e31.

25. Kvitek DJ, Will JL, Gasch AP (2008) Variations in stress sensitivity and genomic expression in diverse S. cerevisiae isolates. PLoS Genet 4: e1000223.

26. Cubillos FA, Billi E, Zorgo E, Parts L, Fargier P, et al. (2011) Assessing the complex architecture of polygenic traits in diverged yeast populations. Mol Ecol 20: 1401–1413.

27. Warringer J, Zorgo E, Cubillos FA, Zia A, Gjuvsland A, et al. (2011) Trait variation in yeast is defined by population history. PLoS Genet 7: e1002111.

28. Homann OR, Cai H, Becker JM, Lindquist SL (2005) Harnessing natural diversity to probe metabolic pathways. PLoS Genet 1: e80.

29. Spor A, Nidelet T, Simon J, Bourgais A, de Vienne D, et al. (2009) Niche-driven evolution of metabolic and life-history strategies in natural and domesticated populations of Saccharomyces cerevisiae. BMC Evol Biol 9: 296.

30. Albertin W, Marullo P, Aigle M, Dillmann C, de Vienne D, et al. (2011) Population size drives industrial Saccharomyces cerevisiae alcoholic fermentation and is under genetic control. Appl Environ Microbiol 77: 2772–2784.

31. Bely M, Sablayrolles JM, Barre P (1990) Description of Alcoholic Fermentation Kinetics: Its Variability and Significance. Am J Enol Vitic 41: 319–324.

32. Marullo P, Mansour C, Dufour M, Albertin W, Sicard D, et al. (2009) Genetic improvement of thermo-tolerance in wine Saccharomyces cerevisiae strains by a backcross approach. FEMS Yeast Res 9: 1148–1160.

33. Bisson L (1999) Stuck and sluggish fermentations. Am J Enol Vitic 50: 107–117.

34. Blateyron L, Sablayrolles JM (2001) Stuck and slow fermentations in enology: statistical study of causes and effectiveness of combined additions of oxygen and diammonium phosphate. J Biosci Bioeng 91: 184–189.

35. Sablayrolles JM, Dubois C, Manginot C, Roustan JL, Barre P (1996) Effectiveness of combined ammoniacal nitrogen and oxygen additions for completion of sluggish and stuck wine fermentation. J Ferm Bioeng 82: 377–381.

36. Varela C, Pizarro F, Agosin E (2004) Biomass content governs fermentation rate in nitrogen-deficient wine musts. Appl Environ Microbiol 70: 3392–3400.

37. Muller LA, McCusker JH (2009) Microsatellite analysis of genetic diversity among clinical and nonclinical Saccharomyces cerevisiae isolates suggests heterozygote advantage in clinical environments. Mol Ecol 18: 2779–2786.

38. Manginot C, Roustan JL, Sablayrolles JM (1998) Nitrogen demand of

different yeast strains during alcoholic fermentation. Importance of the stationary phase. Enzyme Microb Technol 23: 511–517.

39. Santos J, Sousa MJ, Cardoso H, Inacio J, Silva S, et al. (2008) Ethanol tolerance of sugar transport, and the rectification of stuck wine fermentations. Microbiology 154: 422–430.

40. Alexandre H, Mathieu B, Charpentier C (1996) Alteration in membrane fluidity and lipid composition, and modulation of H(+)-ATPase activity in Saccharomyces cerevisiae caused by decanoic acid. Microbiology 142(Pt 3): 469–475.

41. Kim J, Alizadeh P, Harding T, Hefner-Gravink A, Klionsky DJ (1996) Disruption of the yeast ATH1 gene confers better survival after dehydration, freezing, and ethanol shock: potential commercial applications. Appl Environ Microbiol 62: 1563–1569.

42. Parrou JL, Teste MA, Francois J (1997) Effects of various types of stress on the metabolism of reserve carbohydrates in Saccharomyces cerevisiae: genetic evidence for a stress-induced recycling of glycogen and trehalose. Microbiology 143(Pt 6): 1891–1900.

43. Salmon JM (1989) Effect of Sugar Transport Inactivation in Saccharomyces cerevisiae on Sluggish and Stuck Enological Fermentations. Appl Environ Microbiol 55: 953–958.

44. Fay JC, McCullough HL, Sniegowski PD, Eisen MB (2004) Population genetic variation in gene expression is associated with phenotypic variation in Saccharomyces cerevisiae. Genome Biol 5: R26.

45. Spor A, Wang S, Dillmann C, de Vienne D, Sicard D (2008) "Ant" and "grasshopper" life-history strategies in Saccharomyces cerevisiae. PLoS One 3: e1579.

46. Palkova Z (2004) Multicellular microorganisms: laboratory versus nature. EMBO Rep 5: 470–476.

47. Codon A, Benitez T, Korhola M (1998) Chromosomal polymorphism and adaptation to specific industrial environments of Saccharomyces strains. Appl Environ Microbiol 49: 154–163.

48. Mortimer RK (2000) Evolution and variation of the yeast (Saccharomyces) genome. Genome Res 10: 403–409.

49. Mortimer RK, Polsinelli M (1999) On the origins of wine yeast. Res Microbiol 150: 199–204.

50. Kim HS, Fay JC (2007) Genetic variation in the cysteine biosynthesis pathway causes sensitivity to pharmacological compounds. Proc Natl Acad Sci U S A 104: 19387–19391.

51. Quan H, Shih W (1996) Assessing reproducibility by the within-subject coefficient of variation with random effects models. Biometrics 52: 1195–1203.

52. Benjamini Y, Hochberg Y (1995) Controlling the false discovery rate: a practical and powerful approach to multiple testing. J R Stat Soc Ser B 57: 289–300.

53. Gilbert H, Pollard K, van der Laan M, Dudoit S (2009) Resampling-Based Multiple Hypothesis Testing with Applications to Genomics: New Developments in the R/Bioconductor Package multtest. University of California, Berkeley, CA.

54. **54.**Dray S, Dufour AB (2007) The ade4 package: implementing the duality diagram for ecologists. J Stat Software 22: 1–20

Chapter 5

IDENTIFICATION OF POENI-1 AND RELATED PLASMIDS IN OENOCOCCUS OENI STRAINS PERFORMING THE MALOLACTIC FERMENTATION IN WINE

Marion Favier[1,2], Eric Bilhe` re[1] , Aline Lonvaud-Funel[1] , Virginie Moine[3] , Patrick M. Lucas[1]

[1] University of Bordeaux, ISVV, Unit of Enology EA 4577, Villenave d'Ornon, France,

[2] SARCO, research subsidiary of the Laffort group, BP 40, Bordeaux, France,

[3] Laffort, BP 17, Bordeaux, France

ABSTRACT

Plasmids in lactic acid bacteria occasionally confer adaptive advantages improving the growth and behaviour of their host cells. They are often associated to starter cultures used in the food industry and could be a signature of their superiority. *Oenococcus oeni* is the main lactic acid bacteria species encountered in wine. It performs the malolactic fermentation that occurs in most wines after alcoholic fermentation and contributes to their quality and stability. Industrial *O. oeni* starters may be used to better control malolactic fermentation. Starters are selected empirically by virtue of their fermentation kinetics and capacity to survive in wine. This study was initiated with the aim to determine whether *O. oeni* contains plasmids of technological interest. Screening of 11 starters and 33 laboratory strains revealed two closely related plasmids, named pOENI-1 (18.3-kb) and pOENI-1v2 (21.9-kb). Sequence analyses indicate that they use the theta mode of replication, carry genes of maintenance and replication and two genes possibly involved in wine adaptation encoding a predicted sulphite exporter (*tauE*) and a NADH:flavin oxidoreductase of the old yellow enzyme family (*oye*). Interestingly, pOENI-1 and pOENI-1v2 were detected only in four strains, but this included three industrial starters. PCR screenings also revealed that *tauE* is present in six of the 11 starters, being probably inserted in the chromosome of some strains. Microvinification assays performed using strains with and without plasmids did not disclose significant differences of survival in wine or fermentation kinetics. However, analyses of 95 wines at different phases of winemaking showed that strains carrying the plasmids or the genes *tauE* and *oye* were predominant during spontaneous malolactic fermentation. Taken together, the results revealed a family of related plasmids associated with industrial starters and indigenous strains performing spontaneous malolactic fermentation that possibly contribute to the technological performance of strains in wine.

INTRODUCTION

Lactic acid bacteria (LAB) contribute to winemaking during the malolactic fermentation (MLF). MLF usually takes place after the yeast-driven alcoholic fermentation (AF) and lasts a few days to several months depending on wine composition, temperature and LAB population [1],[2]. MLF mainly consists in the conversion of the strong dicarboxylic L-malate into the softer L-lactate and CO_2. It is beneficial in that it reduces the acidity of wine, improves its taste and aromas and contributes to its microbiological stability [3], [4]. MLF generally occurs when the LAB population exceeds 10^6 cells.ml^{-1}. *Oenococcus oeni* is often the only species detected in wine during MLF and thus considered as the best-adapted species [5]. At the intraspecies level, *O. oeni* strains differ considerably in terms of capacity to survive in wine and to conduct MLF. They are more or less tolerant to wine acidity (pH 2.9–4.0), alcohol (11–17%), phenols and sulfites [6], [7]. They also perform MLF more or less rapidly and have a beneficial or detrimental impact on wine aromas [8], [9]. To better control the onset, duration and aromatic impact of MLF, winemakers can make use of industrial *O. oeni* strains. A few dozens of malolactic starters are available to date. They are natural strains selected on the basis of their tolerance to wine stressors, kinetics of MLF, aromas production and safety regarding undesirable metabolisms such as the production of biogenic amines, bitterness or ropiness[10].

The molecular mechanisms at the origin of *O. oeni* survival and growth in wine, and differences existing between strains are still poorly understood. Diverse genes possibly involved in wine adaptation were described in the past decades. This includes genes related to general stress response, membrane composition and fluidity, pH homeostasis, multidrug resistance, or response to oxidative stress and DNA damage [11]–[18]. Comparative genomic analyses also revealed a number of genes that were statistically more often present in *O. oeni* strains of technological interest [19], [20], [21]. However, all the genes identified to date do not satisfactorily explain differences of survival in wine and kinetics of MLF observed amongst *O. oeni* strains [22].

Until now, little attention was paid to the plasmids of *O. oeni*. Plasmids are known as a source of phenotypic and genetic diversity in LAB and occasionally confer adaptive advantages to host strains. Besides maintenance and transfer mechanisms, they encode important traits such as secondary metabolisms, resistance to bacteriophages, antibiotics or heavy metals, and production of exopolysaccharides, bacteriocins and immunity proteins [23], [24]. Plasmids are often associated to starter cultures used in the food industry and could be a signature of their technological superiority and individuality [24]. In the dairy starter *Lactococcus lactis*, they confer phenotypes that reflect adaptation to the dairy environment, such as lactose catabolism, protease activity, peptide and amino acid uptake and bacteriophage resistance[25]. Six small cryptic plasmids of *O. oeni* were sequenced and described to date: pLo13 [26], p4028 [27], pOg32 [28], pRS1 [29], pRS2 and pRS3 [30]. They encode replication and mobilization proteins but do not carry any gene potentially involved in wine adaptation. Large plasmids were detected in a number of *O. oeni* strains but no sequence was reported to date[31]–[35]. A functional role has been assigned to the 22.5-kb plasmid pBL34 that seems to confer pesticide resistance to its *O. oeni* host cells [35].

This work was initiated with the aim to investigate whether plasmids may contribute to wine adaptation of *O. oeni* strains. Two large plasmids, named pOENI-1 (18.3 kb) and pOENI-1v2 (21.9 kb), were described for the first time in *O. oeni*. Their contribution to the technological properties of strains was investigated by analyzing their sequences, their distribution in the *O. oeni* species, their associated phenotypes and their frequency in wines at different steps of winemaking.

MATERIALS AND METHODS

Bacteria strains and culture conditions

O. oeni strains used in this study are listed in Table 1. They consist in 11 industrial starters, 14 strains of laboratory collections and 18 strains isolated for this study from red and white wines collected during spontaneous MLF of vintages 2008 and 2009. New isolates were deposited in the SARCO collection (SARCO Laboratory, Bordeaux, France). All the strains were stored at −80°C in the presence of 30% (v/v) glycerol and propagated under anaerobic conditions at 25°C in grape juice medium (GJ) containing 25% (v/v) commercial red grape juice, 0.5% (wt/v) yeast extract, 0.1% (v/v) Tween 80, pH 4.8.

Table 1: *O. oeni* strains used in this study. doi:10.1371/journal.pone.0049082.t001

Strain[a]	Collection	Origin
C1	IOEB	Commercial product Lactooenos 350 Preac, Laffort
C2	IOEB	Commercial product Lactooenos 450 Preac, Laffort
C3	IOEB	Commercial product Lactooenos B16, Laffort
C4	IOEB	Commercial product Vitilactic BL01, Martin Vialatte
C5	IOEB	Commercial product Viniflora Ciné, CHR Hansen
C6	IOEB	Commercial product Lalvin 31, Lallemand
C7	IOEB	Commercial product Oeno2, Lamothe-Abiet
C8	IOEB	Commercial product Lactoenos SB3, Laffort
C9	IOEB	Commercial product Vitilactic F, Matin Vialatte
C10	IOEB	Commercial product Lalvin VP41, Lallemand
PSU1	ATCC	Commercial starter, Red wine, California, 1977
IOEB 0026	IOEB	Red wine, France, 2000
IOEB 0501	IOEB	Red wine, France, 2005
IOEB 0608	IOEB	Red wine, France, 2006
IOEB 8419	IOEB	Red wine, France, 1984
IOEB 9115	IOEB	Red wine, France, 1991
IOEB 9304	IOEB	Cider, France, 1993
IOEB 89006	IOEB	Red wine, France, 1989
IOEB 89127	IOEB	Red wine, France, 1989
IOEB S268	IOEB-SARCO	Red wine, France, 2000
IOEB S384	IOEB-SARCO	White wine, France, 2002
IOEB S422	IOEB-SARCO	White wine, France, 2002
IOEB S343a	IOEB-SARCO	Red wine, France, 2002
IOEB S455	IOEB-SARCO	White wine, France, 2003
S4	SARCO	Red wine, France, 2009
S11	SARCO	Sparkling white wine, France, 2008
S12	SARCO	White wine, France, 2009
S13	SARCO	Red wine, France, 2009
S14	SARCO	Red wine, France, 2009
S15	SARCO	Red wine, France, 2009
S17	SARCO	Red wine, France, 2009
S18	SARCO	Red wine, France, 2009
S19	SARCO	Red wine, France, 2009
S20	SARCO	Red wine, France, 2009
S22	SARCO	White wine, France, 2009
S23	SARCO	White wine, England, 2009
S24	SARCO	Red wine, England, 2009
S25	SARCO	Red wine, France, 2009
S26	SARCO	Red wine, France, 2009
S27	SARCO	Red wine, France, 2009
S28	SARCO	Red wine, France, 2009
S29	SARCO	Red wine, France, 2009
ATCC BAA 1163	ATCC	Red wine, France, 1984

[a]IOEB: Institute of oenology of Bordeaux, S: SARCO, ATCC: American type culture collection.
doi:10.1371/journal.pone.0049082.t001

Strain typing

Strain typing was performed by NotI restriction of bacterial DNA followed by pulse field gel electrophoresis (PFGE) of restriction fragments as previously described [36]. DNA restriction patterns were compared in a dendrogram generated by the unweighted pair group method using arithmetic means (UPGMA) with the Dice coefficient of similarity and a tolerance limit of 2.3% in Bionumerics 5.1 software (Applied Maths, Kortrijk, Belgium). Multilocus sequence typing (MLST) was also performed for several strains according to the procedure described in [36]. MLST data was processed in a neighbor-joining tree constructed using MEGA4 [37].

Plasmid Sequencing and Analysis

Plasmid pOENI-1 was isolated from a 10-ml culture of commercial strain *O. oeni* C9 by the alkaline lysis method [38]. The plasmid DNA preparation was digested by EcoRI and PstI (New Englands Biolabs) to construct a library of inserts in the *E. coli* vector pBluescript II SK+ (Stratagene). Two inserts (1.3 and 2.5 kb) were sequenced. The rest of the plasmid was amplified in two PCR products (4.0 and 10.5 kb) obtained using primers designed in the first sequences and a high-fidelity DNA polymerase (iProof, Bio-Rad). The two PCR products were cloned in pGEM-T Easy (Promega), transferred in *E. coli* and sequenced. The sequences were assembled using Lasergene (DNA Star) in a single circular DNA molecule of 18,332-bp. The sequence of plasmid pOENI-1v2 was obtained by sequencing the genome of strain S11 by the 454 technology (single reads of 450-nt on average, GenoToul, Toulouse, France). Sequences assembled with Lasergene (DNA-Star) formed a circular DNA molecule of 21,926-bp (coverage >100X all along the plasmid sequence). Open reading frames (ORFs) were predicted with GeneMark [39] and Glimmer [40]. Gene annotation was performed manually using Blast and Interproscan analyses [41].

PCR-based detection of pOENI-1 in *O. oeni* strains

Bacterial genomic DNAs were extracted using the Wizard genomic DNA purification kit (Promega) according to the manufacturer's instructions. DNA preparations were used as template in PCR assays to detect the plasmid genes *repA*, *oye* and *tauE*, the chromosomal gene OEOE_0812 (locus tag of the *O. oeni* PSU-1 genome, NC_008528), and to confirm the integrity of plasmids using a combination of three overlapping PCRs that extend over the whole plasmid sequences. The primers used are listed in Table 2. PCR amplifications were performed in 20-μl mixtures containing 25 ng of template DNA, 0.25 μM of each primer and the Taq-&GO™ PCR mix (MP Biomedicals). The standard

PCR program was 95°C for 5 min; followed by 30 cycles of 95°C for 30 s, 50°C to 60°C for 30 s, 72°C for 30 s and a final step of 10 min at 72°C. PCR products were visualized under UV light exposure after electrophoresis in 1.2% (w/v) agarose gels and staining with ethidium bromide.

Table 2: Primers list. doi:10.1371/journal.pone.0049082.t002

Primer name	Forward sequence (5'–3')	Reverse sequence (5'–3')	Target	Product (bp)
Detection of plasmid genes				
8a/8b [a]	TAAGCAAACGGGGTCAACTC	TCAGGCCGAGGATCAATAAC	ORF 8	142
oye1/oye2	AGTAGTTATTCCGCCAATGA	ATGAATGGCTCCTTAGCATA	ORF 11	602
oyeQ1/oyeQ2 [a]	TAAGGGATTTGAAGGCCAACT	TTGAAGAATTGCTTTAGCACCA	ORF 11	106
repA1/repA2	ATCGGCTCGAATATTCTCTCAA	CGTATTCTCTAGCCGCTTGTTT	ORF 15	911
2repA/NC1	ATCACCTAGTAGACGAAGAG	GGTAGGCAGGTTCTAATC	ORF 15_ORF 1	6464
orf20a/orf10	AGTTAAGAACTATCGTAAGTCC	TTACTGGCCTCCTACTGAAC	ORF 20_ORF 10	9848
repA2/NC3	CGTATTCTCTAGCCGCTTGTTT	GCATTCGACTTTGCGGAATG	ORF 10_ORF 15	5199/8765 [b]
oye1/orf13-2	AGTAGTTATTCCGCCAATGA	TACAGCATACACTCACAGCA	ORF 11_ORF 13	2376/5942 [b]
orf20a/orf20b	AGTTAAGAACTATCGTAAGTCC	AACAGGATCATAGTACATCAC	ORF20	821
Detection of chromosomal genes				
O01/O02	GTGCCGCTTTTTTGGATATTA	AGCAATTTTATCTTTATAGCT	*mleA*	431
0812a/0812b	GATTATTACCAATTCGGCTG	ACGCCGGAAATAATGTAG	OEOE_0812	540
rpoB1/rpoB2 [a]	ATGGAACGTGTTGTCCGCGA	GGATTGGTTTGATCCATGAA	*rpoB*	148

[a] Primers used in qPCR assays.
[b] Product sizes obtained for pOENI-1 and pOENI-1v2.
doi:10.1371/journal.pone.0049082.t002

Determination of plasmid/*oye* gene copy number per cell

Copy number of plasmids pOENI-1, pOENI-1v2 and gene *oye* (ORF 11) per cell was determined by quantitative real-time PCR (qPCR) using a GoTaq® qPCR Master Mix (Promega) on a CFX96™ Real-Time Detection System (Bio-Rad). Amplification conditions were 95°C for 3 min, followed by 40 cycles of 95°C for 30 s, 60°C for 30 s and 72°C for 30 s and a final step of 70°C to 90°C with an increment of 0.5°C each 5 s. Two primer pairs (Table 2) were used to quantify the chromosomal gene *rpoB* and the plasmid gene *oye* in order to calculate their relative proportion. Serial decimal dilutions of *O. oeni* ATCC BAA 1163 genomic DNA were used to produce the standard curves. In this strain, both *oye* and *rpoB* are present on the chromosome at one copy per cell. Standard curve equations and coefficients of correlation calculated from three independent experiments were: $C_T = -3.33x + 35.11$, $R^2 = 0.996$ (*rpoB*), $C_T = -3.49x + 37.28$, $R^2 = 0.993$ (*oye*). Genomic DNA of all tested strains was extracted from bacterial colonies suspended in 200-µl sterile H_2O and heating for 10 min at 80°C prior cooling on ice. All determinations were done in triplicates.

Quantification of *oye/tauE* genes in wine

Determinations were performed by qPCR as described above, except that template DNAs were extracted from total microorganisms of 10-ml samples of must or wine by the method reported in [31]. Standard curves were produced using DNA extracted from decimal dilutions of *O. oeni*ATCC BAA 1163 or S24 (*rpoB* and *tauE* at one copy per cell) inoculated in sterile wine (10^7 to 10^2 cells.ml^{-1}). The corresponding equations indicate cycle threshold values for 1 ml of wine: ATCC BAA 1163, C_T =−3.61x +41.66, R^2=0.906 (*rpoB*), C_T =−3.64x +42.26, R^2=0.961 (*oye*); S24, C_T =−3.98x +39.38, R^2=0.975 (*rpoB*), C_T =−3.86x +39.14, R^2=0.979 (*tauE*). The tested samples were 95 red wines and musts collected at different stages of winemaking (must, alcoholic fermentation, MLF) in 86 wineries of Bordeaux's area. No industrial strain was employed to conduct MLF in these wines.

Plasmid Curing

O. oeni strains carrying pOENI-1 or pOENI-1v2 were cultivated in GJ medium for about 20 generations. Cultures were plated to analyze 30 colonies and determine the presence or absence of the plasmids. DNA templates were prepared by suspending each colony in 200-µl sterile H_2O and heating for 10 min at 80°C prior cooling on ice. Multiplex PCR were performed in order to detect simultaneously a chromosomal gene (PCR positive control, *mleA*) and a plasmid gene (ORF 20). PCRs were carried out in 20-µl reaction mixtures containing 1-µl of cell suspension, 0.25 µM of each primer and the Taq-&GO™ PCR mix (MP Biomedicals). Clones without plasmid were controlled by NotI-PFGE typing as described above and compared with parental strains.

Population dynamics during growth in wine and MLF kinetics

O. oeni strains carrying pOENI-1 or pOENI-1v2 and isogenic plasmid-less derivatives were used in two types of experiments: direct inoculation in wine for monitoring MLF kinetics and inoculation in grape juice for monitoring cell growth in wine and MLF kinetics. For the first experiment, cells were produced as freeze-dried industrial preparations (SARCO) and inoculated to 10^7cell.ml^{-1} in a red wine immediately after alcoholic fermentation. Wines were incubated at 20°C until completion of MLF. L-malic acid degradation was determined twice per week using an L-malic enzymatic kit (Roche Boehringer, R-biopharm). LAB populations were determined by plating on GJ medium once per week. For the second experiment, cells were propagated in GJ medium and inoculated to 10^3 cell.ml^{-1} in a commercial grape juice supplemented with glucose/fructose to 210 g.l^{-1}, L-malic acid to 4 g.l^{-1}, SO_2 to 20 mg.l^{-1} and adjusted to pH 3.6. Alcoholic fermentation was initiated by inoculating the industrial starter yeast F33 to 200 mg.l^{-1} according to manufacturer's instructions (Laffort, France). AF was monitored daily by weight-loss determinations. MLF and bacterial populations were determined three times per week as described above. Plasmids stability was investigated by PCR-analysis (see the paragraph "plasmid curing") of 30 to 50 colonies picked up on plates produced at the inoculation time, at the end of AF and the end of MLF.

Sequence Accession Numbers

The nucleotide sequences of pOENI-1 and pOENI-1v2 were submitted to GenBank and are available under the accession numbers JX416328 and JX416329, respectively.

RESULTS

Sequence analysis of pOENI-1

During a survey of *O. oeni* strains, we have detected a large plasmid in the industrial strain *O. oeni* C9. This plasmid, named pOENI-1, was analyzed to determine whether it contributes to the technological properties of *O. oeni* C9. Its complete sequence was obtained by sequencing diverse restriction fragments and PCR products. pOENI-1 is a circular DNA molecule of 18,332-bp in length. Its GC% is 40.8, compared to 38% in the *O. oeni* chromosome [19]. Sequence annotation revealed 18 complete ORFs and two truncated ORFs (ORFs 4 and 20) ranging from 210 to 1512-bp (Fig. 1A and Table 3). A function was ascribed to 15 of the 20 encoded proteins. The protein encoded by ORF 15 shares more than 70% sequence identity with replication initiator protein A (RepA) encountered in theta type plasmids plca36 of*Lactobacillus casei* Zhang [42], pLgLA39 of *Lactobacillus gasseri* LA39 [43] and pSF118-44 of*Lactobacillus salivarius* UCC118 [44]. The intergenic region located between ORF 14 and ORF 15 shows all the hallmarks of the theta-type replication origin [45], [46]. It is located upstream of *repA*, contains an AT-rich region (positions 12,832–12,915, 71% AT) and an 18-bp repetition present at 7 copies (atatatctgatatatcaa, positions 12,862–12,987). Therefore, it is likely that pOENI-1 is the first large theta-type plasmid described in *O. oeni*.

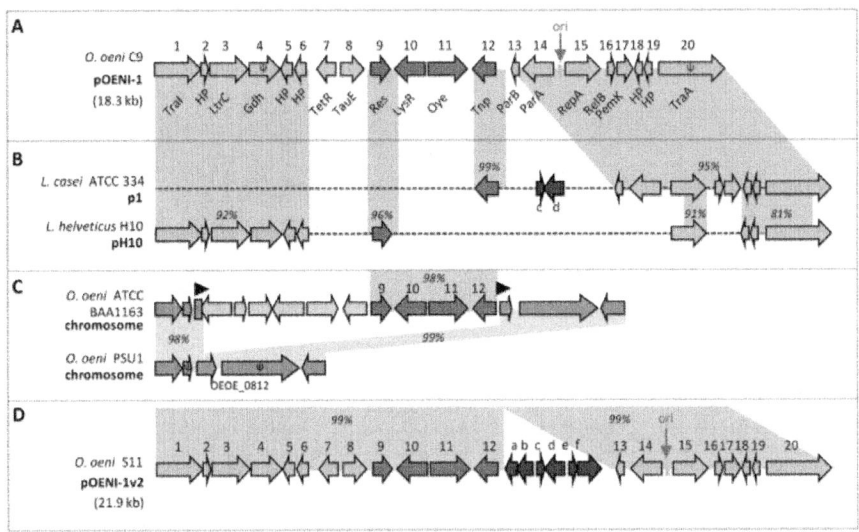

Figure 1: Genetic organization of pOENI-1 and comparison with related sequences. A. Genetic organization of plasmid pOENI-1. ORFs are represented by numbered arrows and identified by corresponding protein tags (see also Table 3). B. Sequence comparison of pOENI-1 and related plasmids p1 (CP000424) and pH10 (CP002430). ORFs "c, d" (purple arrows) share 99% similarity with ORFs of pOENI-1v2. C. Portions of chromosomes in *O. oeni* ATCC BAA 1163 and *O. oeni* PSU1. The gene OEOE_0812 in *O. oeni* PSU1 (green arrow) is disrupted in *O. oeni* ATCC BAA 1163 by an 10 genes insert comprising four genes conserved in pOENI-1 (red arrows) and six genes unrelated to pOENI-1 (pink arrows). The insert is bordered by an 8-bp repeated sequence (dark triangles). D. Genetic organization of pOENI-1v2. ORFs numbered from 1 to 20 share more than 99% nucleotide sequence similarity with corresponding ORFs in pOENI-1. ORFs shaded in purple are not detected in pOENI-1 and code for transposases (a, e, f,), hypothetical proteins (b, c) and a recombinase (d). Pseudogenes are symbolized by arrowheads containing the symbol ψ. Regions of sequence similarity are indicated in percentages and shaded in blue. ori: putative origin of replication. doi:10.1371/journal.pone.0049082.g001

Table 3: ORFs and predicted proteins of pOENI-1. doi:10.1371/journal. pone.0049082.t003

ORF	Position	%GC	Protein	Size (aa)	Predicted function	Best blast (organism, GenBank accession)	% identity
1	108-1619	44,0	TraI	503	DNA topoisomerase IA, TraI	L. pentosus, CCB84017	95
2	1742-1957	38,9	HP	71	Hypothetical protein	L. brevis, ZP_03940833	90
3	1961-3082	42,7	LtrC	373	LtrC-like protein	L. helveticus, ADX71206	96
4ª	3096-4053	36,9	Gdh	319	Glycerate dehydrogenase	Lc. kimchii, YP_003621246	70
5	4558-4172	42,1	HP	128	Hypothetical protein	P. claussenii, AEV96201	94
6	4871-4551	40,5	HP	106	Hypothetical protein	L. brevis, ZP_03940921	99
7	5865-5281	52,7	TetR	194	Transcriptional regulator, TetR	L.mali, ZP_09449471	99
8	6041-6787	53,3	TauE	248	Putative permease TauE	L. mali, ZP_09449472	99
9	7365-7952	41,8	Res	195	Resolvase	L. crispatus, YP_003601011	99
10	9124-8192	35,3	LysR	310	Transcriptional regulator, LysR	O. oeni, ZP_01543901	100
11	9267-10448	42,1	Oye	393	NADH: flavin oxidoreductase	O. oeni, ZP_01543900	100
12	11258-10575	45,5	Tnp	227	Transposase	O. oeni, ZP_01543898	100
13	11927-11649	41,6	ParB	92	Partition protein, ParB	L. casei, YP_794449	100
14	12831-11908	40,8	ParA	307	Partition protein, ParA	L. casei, YP_794448	99
15	13334-14449	39,5	RepA	371	Replication protein, RepA	P. claussenii, AEV96162	76
16	14709-14990	42,9	RelB	93	Addiction module antitoxin, RelB	L. hilgardii, ZP_03954201	100
17	14980-15486	35,9	PemK	168	Addiction module toxin, PemK	L. paracasei, ABA12818	99
18	15754-15476	34,8	HP	92	Hypothetical protein	L. brevis, YP_796419	97
19	15985-15776	30,0	HP	69	Hypothetical protein	L. mali, ZP_0944951	90
20ª	16256-18317	38,7	TraA	687	Putative nickase, TraA	L. pentosus, CCC15328	95

ªpseudogenes, the characteristics of hypothetical full-length genes and proteins are provided.
doi:10.1371/journal.pone.0049082.t003

ORFs 13 and 14 encode partitioning proteins ParB and ParA respectively, involved in plasmid segregation during cell division. A putative toxin/antitoxin system (PemK-like and RelB-like proteins) contributing to plasmid stability is encoded by ORFs 16 and 17. pOENI-1 does not encode the full set of proteins required for plasmid conjugation, but only two (ORFs 1 and 3). ORFs 7 and 8 code for a TetR transcriptional regulator and a putative permease of the TauE family. Permeases of this family are known to act as sulfite transporters [47], [48]. It is likely that ORF 7 and ORF 8 encode proteins that are functionally related since pairs of similar genes were detected in *Lactobacillus mali* (Table 3) and in the *O. oeni* phages fOg30, fOgPSU1 and fOg44 [49]. ORFs 10 and 11 encode a LysR transcriptional regulator and a NADH:flavin oxydoreductase of the old yellow enzyme (OYE) family, group 4. The biological role of OYEs is still poorly understood, but they can contribute to the oxidative or general stress response [50], [51], [52]. The rest of pOENI-1 includes a resolvase (ORF 9), a transposase (ORF 12) and five hypothetical proteins of 69 to 128 amino acids in length.

Investigations in public databases revealed that several ORFs of pOENI-1 involved in plasmid maintenance or replication are conserved in other LAB plasmids. The most similar are plasmids pH10 of *Lactobacillus helveticus* H10 (ORFs 1 to 6, ORF 9, ORF 15 and ORFs 18 to 20) and p1 of *Lactobacillus casei* ATCC 334 (ORFs 12 to 20) (Fig. 1B). The TauE and OYE-encoding genes of pOENI-1 were not detected in these plasmids or in other plasmids previously described. Blast analyses revealed that two of the three *O. oeni* genomes available in databases contain sequences related to pOENI-1. The almost complete plasmid sequence is disseminated in four contigs in the draft genome of *O. oeni* AWRIB429, an industrial starter named C10. Only a 104-bp fragment of ORF 15 (*repA*) was not detected in this draft genome. In addition, O. oeni ATCC BAA 1163 – a strain that is known for its poor fermentation capacities in wine [53] – contains chromosomal genes sharing above 98% identity to pOENI-1 ORFs 9 to 12 (Figure 1C). The four genes of *O. oeni* ATCC BAA 1163 are contiguous and form a cluster along with six additional genes unrelated to pOENI-1. This cluster is inserted into the gene OEOE_0812 and bordered by an 8-bp repeated sequence.

Distribution of pOENI-1, pOENI-1v2 and plasmid genes amongst *O. oeni* strains

Forty-four *O. oeni* strains from diverse origins were analyzed to examine the frequency of pOENI-1 in the species: 11 industrial starters from seven companies, 15 laboratory strains collected between 1983 and 2009, and 18 strains isolated from red and white wines during this study. The phylogenetic relationships of the strains were determined by REA-PFGE and MLST analyses. The dendrogram depicted in Figure 2 shows that all strains belong to two major phylogenetic lineages as suggested in previous studies [36], [54]. The presence of pOENI-1 was investigated by a PCR-based strategy targeting the plasmid ORFs 15 (*repA*), 8 (*tauE*) and 11 (*oye*). The three ORFs were detected only in four strains, including three industrial starters (C9, C10, C6) and a new isolate (S11) (Figure 2). A second series of three PCRs targeting large regions of pOENI-1 confirmed that these strains contain complete plasmids (see primers list in Table 2). However, the region extending from ORFs 11 to 13 was 2376-bp long in pOENI-1 of strain C9, whereas it extended over 5942-bp in the other strains (Figure 2). The sequence of this larger fragment was determined by analyzing the genome of *O. oeni* S11 by the 454 technology. Genome sequence analysis revealed that *O. oeni* S11 holds a 21,926-bp plasmid that was named pOENI-1v2. It carries the same 20 ORFs as pOENI-1 and six additional ORFs located between ORFs 12 and 13, which accounts for the larger PCR products described above. These ORFs encode a recombinase, transposases and hypothetical proteins (Figure 1D). All other parts of pOENI-1 and pOENI1-v2 are very similar (>99% sequence identity), except that two mutations disrupting ORFs 4 and 20 in pOENI-1 are not detected in pOENI-1v2 in which these ORFs encode full-length proteins. It is possible that strains C10 and C6 have acquired pOENI-1v2 by vertical inheritance as they occupy close positions in the dendrogram depicted in Figure 2. In contrast, strains C9 and S11 are distantly related, suggesting that plasmids were also transmitted via horizontal transfer events.

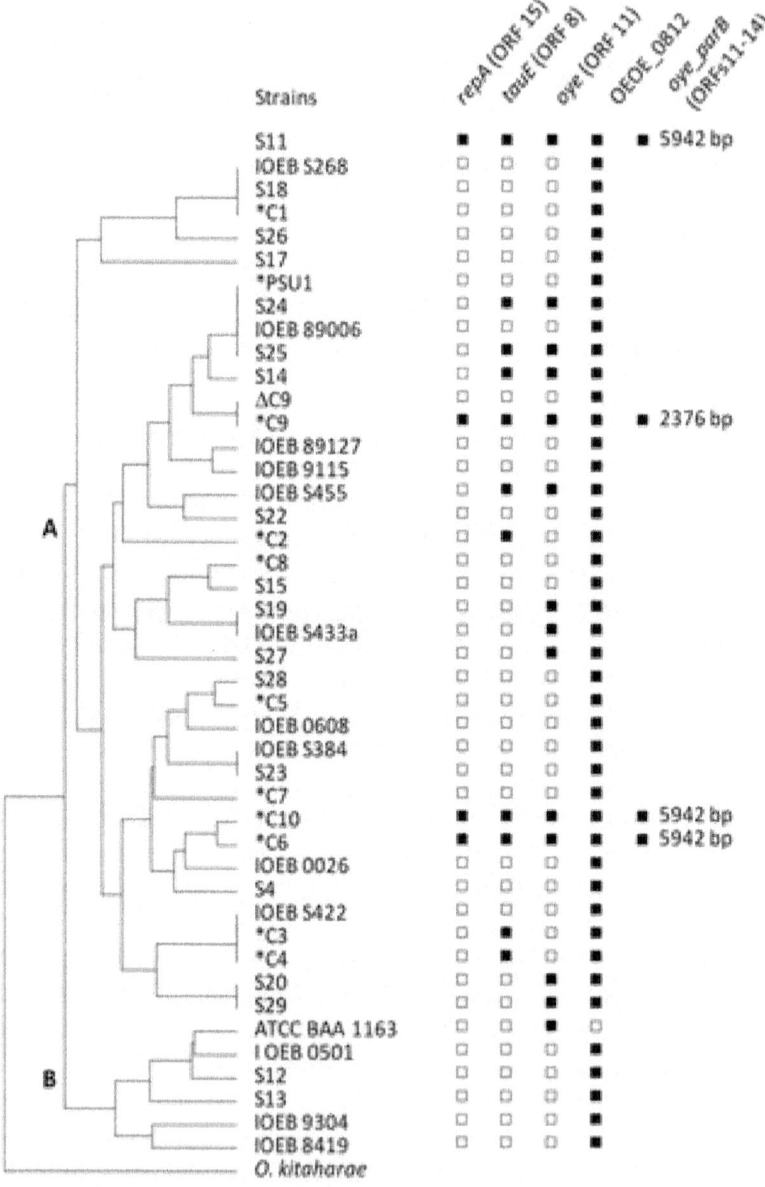

Figure 2: Distribution of pOENI-1 genes in 44 *O. oeni* strains. The dendrogram was constructed from DNA banding patterns obtained by NotI-PFGE analysis of 44 *O. oeni* strains. *Oenococcus kitaharae* was used as outgroup. Strain S11 was positioned on the basis of MLST data since no NotI-PFGE pattern was obtained for this strain. The pres-

ence (filled square) or absence (empty squares) of plasmid genes*repA*, *tauE*, *oye* and of the chromosomal gene OEOE_0812 were determined by PCR. The presence/absence of a region encompassing the *oye* and *parB* genes was also investigated. IOEB: Institute of oenology of Bordeaux, S: SARCO, ATCC: American type culture collection. Indutrial strains are marked with asterisks. Letters A and B in the dendrogram represent two phylogenetic groups of strains [36]. doi:10.1371/journal.pone.0049082.g002

PCR screening did not disclose any other strain containing a full plasmid sequence since *repA*was present only in the four strains above-mentioned. However, *tauE* and *oye* were detected together or separately in 7 and 10 additional strains, respectively (Figure 2). Of the 11 industrial starters analyzed in this work, six contained *tauE* (C9, C10, C6, C4, C3 and C2), while the gene *oye* was found only in the three starters carrying a plasmid. The *tauE* and *oye*genes are randomly distributed among strains, suggesting that they were mostly acquired through horizontal gene transfer events.

The number of plasmids per cell was determined by qPCR analysis of a plasmid gene (*oye*) and a chromosomal gene (*rpoB*). qPCR tests and standard curves were developed for both genes and used to determine the ratios *oye*/*rpoB* in *O. oeni* strains carrying pOENI-1, pOENI-1v2 or the *oye* gene alone (Table 4). pOENI-1 and pOENI-1v2 were present at 3.3 to 4.7 copies per cell, respectively. Bacteria carrying *oye* but no plasmid have approximately one copy of*oye* gene per cell, which is consistent with a chromosomal localization. A complete gene OEOE_0812 was detected in these strains, indicating that their copy of *oye* was not inserted in the same position as in strain ATCC BAA 1163 (Fig. 1C).

Table 4: Plasmid/*oye* copy number. doi:10.1371/journal.pone.0049082.t004

Strain	Plasmid type	*oye* (copies.µl^{-1})	*rpoB* (copies.µl^{-1})	ratio *oye*/*rpoB*
C9	pOENI-1	$7.7 \times 10^3 \pm 2.8 \times 10^3$	$2.1 \times 10^3 \pm 0.7 \times 10^3$	3.7 ± 0.6
C10	pOENI-1v2	$6.6 \times 10^4 \pm 4.9 \times 10^4$	$1.4 \times 10^4 \pm 1.0 \times 10^4$	4.7 ± 0.2
C6	pOENI-1v2	$2.3 \times 10^4 \pm 0.5 \times 10^4$	$6.7 \times 10^3 \pm 2.4 \times 10^3$	3.7 ± 1.2
S11	pOENI-1v2	$6.8 \times 10^4 \pm 1.8 \times 10^4$	$2.1 \times 10^4 \pm 0.4 \times 10^4$	3.3 ± 0.4
Type S14[a]	no plasmid	-	-	1.1 ± 0.2

Comparison of plasmid-containing and plasmid-free cells during wine fermentations

Detection of plasmids pOENI-1 and pOENI-1v2 in three industrial starters prompted us to examine whether they contribute to the technological properties of their hosts. In order to compare strains sharing the same genetic background, we have generated plasmid-less derivatives of strains C9 and C10 by growing cells in liquid GJ medium during approximately 20 generations prior to plate samples and to test colonies using a plasmid-specific PCR test. Analysis of 68 and 30 colonies of *O. oeni* C9 and C10 allowed for the identification of two and one plasmid-less mutants, respectively. Controls performed by REA-PFGE and PCR assays confirmed that the mutants share the same genetic background as parental strains and have lost the plasmids (Figure 3).

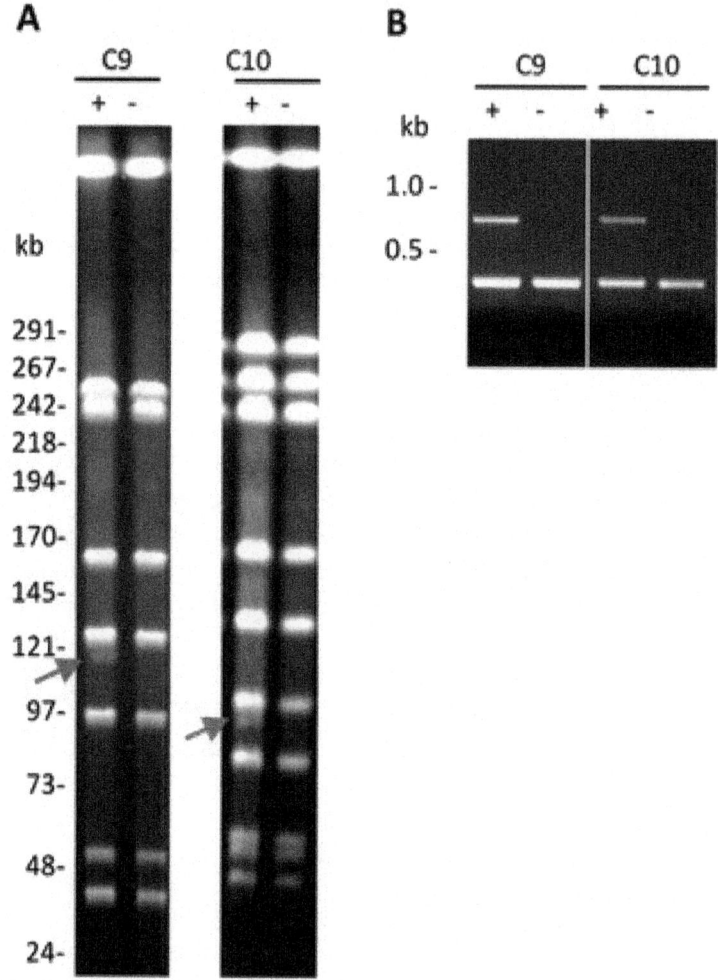

Figure 3: Control of plasmid-free strains. A. Comparison of NotI-PFGE patterns of plasmid containing strains (C9+, C10+) and isogenic plasmid-less derivatives (C9− and C10−). Red arrows indicate bands corresponding to plasmids in strains C9+ and C10+. B. Absence of plasmids in strains C9− and C10− was confirmed by multiplex PCR targeting a plasmid gene (ORF 20, 821-bp PCR product) and a chromosomal gene (*mleA*, 430-bp PCR product). doi:10.1371/journal.pone.0049082.g003

To determine if plasmids confer an advantage during MLF, strains with (C9+, C10+) and without (C9−, C10−) plasmids were produced under industrial conditions, freeze-dried and tested in micro-vinification assays. They were inoculated to $10^7.\text{ml}^{-1}$ in a red wine and consumption of L-malate and bacterial populations were monitored until completion of MLF. Bacterial populations evolved similarly whichever the strain. They declined rapidly after inoculation in wine, started to grow after 5 to 10 days and showed similar growth curves during all the rest of the experiments. Bacteria started to consume significantly L-malate after a lag phase of about 20 days and completed MLF in 37 to 43 days following inoculation (Figure 4). Only for strains C9+ and C9− a slight difference was noticed, strain C9+ being able to achieve MLF six days before strain C9−. This difference was also observed and more pronounced (>10-days difference) when this assay was performed in other wines (data not shown). In contrast, strains C10+ and C10− showed similar kinetics of MLF in all trials.

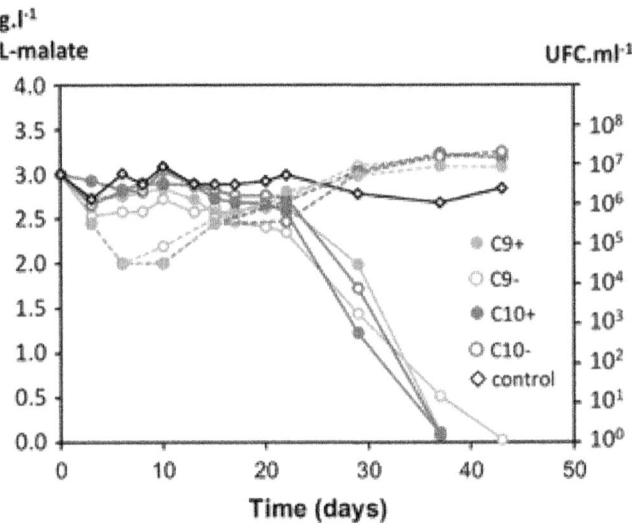

Figure 4: Comparison of MLF kinetics of isogenic strains with/without plasmids. Kinetics of L-malate conversion (solid lines) and monitoring of cell population (dotted lines) were monitored following inoculation of bacteria to 10^7 cells ml^{-1} in a red wine containing 3 g l^{-1} L-malate. A control was performed without added bacteria. Values are means of two biological replicates. doi:10.1371/journal.pone.0049082.g004

A second series of tests was performed to determine if plasmids confer a growth advantage during the phases that precede MLF. Bacteria were inoculated in a sterile grape must to 10^3cells.ml^{-1} at the same time as yeasts to perform alcoholic fermentation (Figure 5). The growth of strains with or without plasmids was similar during AF and the ensuing MLF. Additional trials consisting in mixtures of C9+/C9– or C10+/C10– cells inoculated as above showed the same growth curves. The kinetics of MLF were also very similar in all cases, except that strain C9+ achieved MLF two days prior to C9–. To determine if the plasmids were stable during cell growth in wine, samples were collected at the inoculation time, at the end of AF and at the end of MLF and they were plated to isolate colonies that were tested in PCR assays specific for the plasmids (Table 5). At inoculation, C9+ and C10+ contained only 90% of plasmid-containing cells, which denotes the instability of the plasmids during precultures in laboratory. During AF and MLF (approx. 20 generations), plasmids were stable since they were detected in 90 to 100% of the cells. In samples inoculated with equal amounts of plasmid-carrying and plasmid-free cells, the C9+/C9– ratio increased from 48/52 at inoculation time to 57/43 at the end of MLF. However an opposite tendency was noticed for the mixture C10+/C10– (Table 5). We concluded that plasmids did not confer a clear advantage to their host cells in laboratory trials, although they were stably maintained during growth in wine.

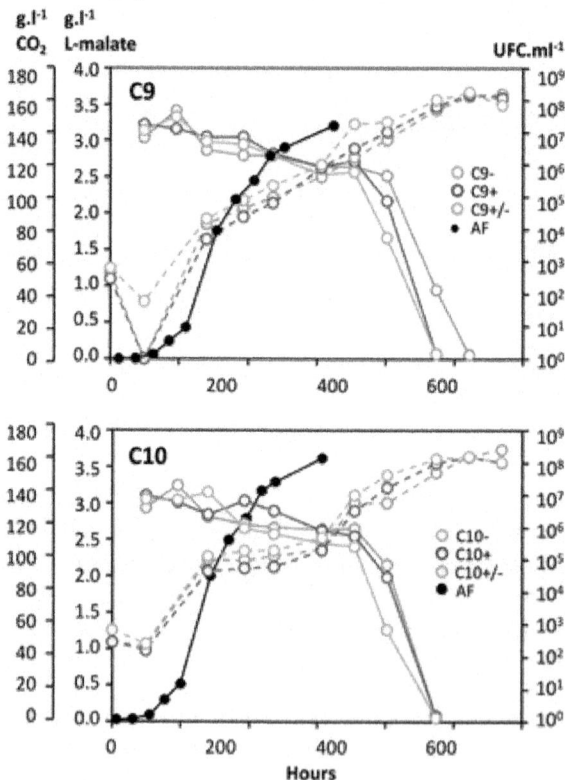

Figure 5: Comparison of growth in wine of isogenic strains with/without plasmids.

Kinetics of alcoholic fermentation (CO_2 released, dark line), MLF (colored solid lines) and bacterial populations (colored dotted lines) were monitored in a sterile grape must inoculated with industrial wine yeasts and 10^3.ml^{-1} bacteria carrying pOENI-1 or pOENI-1v2 (red lines), bacteria without plasmids (blue lines) or a mixture of both (green lines). Kinetics of AF (dark symbols) is the mean of the three experiments. doi:10.1371/journal.pone.0049082.g005

Table 5: Percentage of plasmid-carrying/plasmid-free cells at different times of wine-making. doi:10.1371/journal.pone.0049082.t005

Experiment[a]	Inoculation	End of AF	End of MLF
C9–	100	100	100
C9+	90	70	90
C9+/C9–	42/58	46/54	57/43
C10–	100	100	100
C10+	90	90	100
C10+/C10–	50/50	49/51	45/55

Detection of plasmids and plasmid genes in wines

To determine if the plasmids have a technological significance during real winemaking, we have investigated their presence in bacteria of 95 samples collected in 86 wineries at different phases of wine fermentations (must, AF, MLF). Microbial DNAs were purified from each sample and used as template in quantitative PCR assays to determine the *tauE* and *oye* copy numbers. The chromosomal gene *rpoB* was also quantified in order to assess the total *O. oeni* population. As anticipated, the *O. oeni* population ranged from 10 to 10^5 cells.ml^{-1} in samples collected in must and AF, while it reached up to 10^9 cells.ml^{-1} during MLF (Figure 6). The genes *tauE* and *oye* were detected in all samples. They were present at high copy numbers in the vast majority of samples collected during MLF: above 10^6 copies.ml^{-1} in 55.8% (*tauE*) and 78.9% (*oye*) of samples. The average ratios of *tauE/rpoB* and *oye/rpoB* were calculated in samples collected before and during MLF (Figure 6C). The ratios were below 0.6 in must/AF samples and above 0.9 during MLF. This suggests that bacteria carrying *tauE* and *oye* were underrepresented before MLF but proliferated during AF and became predominant in MLF.

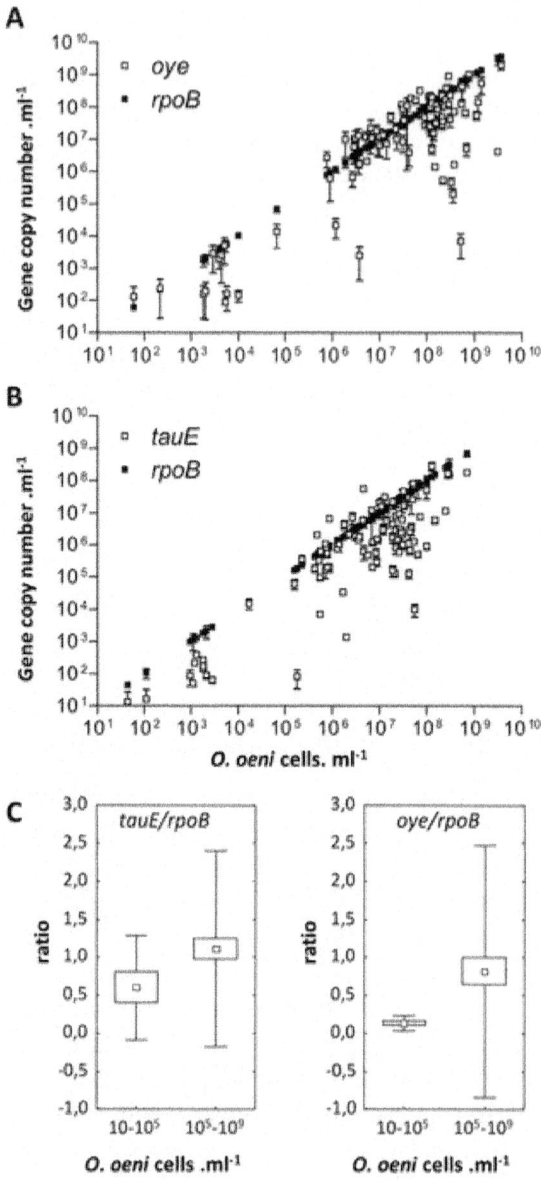

Figure 6: Frequency of *tauE* and *oye* genes during wine fermentations. A. B. The *oye* and *tauE* gene were quantified by qPCR analysis of 95 samples of must or wine collected at different stages of winemaking. Data obtained from *rpoB*quantifications were plotted on the x-axis to appraise the *O. oeni* population and on the y-axis to make easier the comparison between the *O. oeni* population (*rpoB*, filled squares) and the *tauE* or *oye* copy number (empty squares). Data are means of two independent determinations. C. The average ratios of *tauE/rpoB* or *oye/rpoB* were calculated from

samples collected in must or AF (10–10^5 cells.ml^{-1}) and during MLF (10^5–10^9 cells. ml^{-1}). The boxes and lines represent the means (small squares), standard errors (large squares) and standard deviations (lines). doi:10.1371/journal.pone.0049082.g006

Thirty samples of wines were further tested by PCR to determine if the genes *tauE* and *oye* were located on plasmids resembling pOENI-1 or pOENI-1v2. PCR assays were performed using primers bordering a plasmid region that extends from ORF 11 to ORF 13 and extends over 2376-bp in pOENI-1 and 5942-bp in pOENI-1v2. PCR products were obtained for 13 samples (Figure 7). Three samples contained a mixture of products of two different sizes, which denote that cells carrying different plasmids were present together in some samples. Only two and one samples had PCR products of molecular sizes expected for pOENI-1 and pOENI-1-v2, respectively. The other PCR products had intermediate sizes of around 4.0 kb, suggesting that additional forms of pOENI-1 can be encountered. The 17 samples that did not generate PCR products were further analyzed in PCR assays targeting specifically *repA, oye* or *tauE*. They all contain a copy of *repA, oye* and *tauE*, but we could not assess whether the genes were located on plasmids unrelated to pOENI-1 or on the chromosome (not shown).

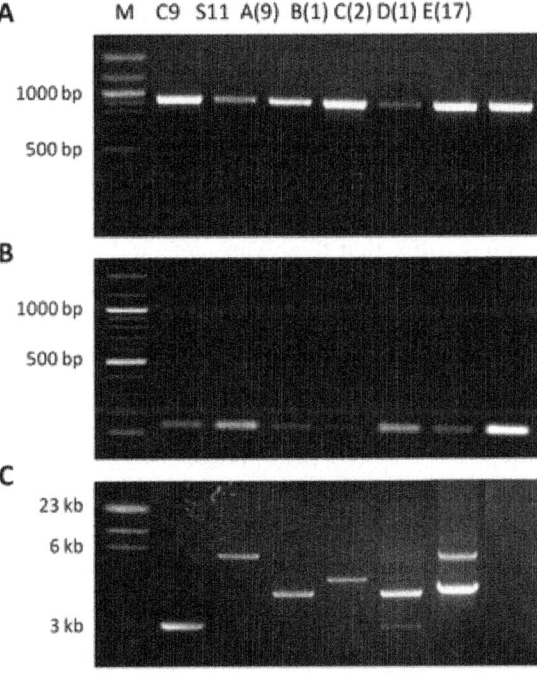

Figure 7: PCR detection of pOENI-1 and related plasmids in wines. PCR assays were performed using DNA templates from *O. oeni* C9 (pOENI-1), *O. oe-*

*ni*S11 (pOENI-1v2) and 30 samples of wine collected during MLF (A–E). The number of samples sharing the same PCR product is indicated in parentheses. The primers allowed detection of pOENI-1 *repA* (panel A), *tauE* (panel B) and a region extending from ORF11 (*oye*) to ORF 13 (*parB*) (panel C). M: DNA size markers.m doi:10.1371/journal.pone.0049082.g007

DISCUSSION

Plasmids of the "pOENI-1 family"

pOENI-1 and pOENI-1v2 are the first large plasmids described in *O. oeni*. The presence of large plasmids in this species was known from previous works [31]–[35], but no sequence was available. However, during the preparation of this manuscript, Borneman and coworkers have reported the sequences of 11 *O. oeni* genomes and found plasmids in four of them (discussed below) [55]. The plasmids pOENI-1 and pOENI-1v2 carry a majority of ORFs involved in maintenance and replication but also a few ones encoding proteins that can benefit to their host cells, such as TauE and Oye. They share a limited sequence similarity with plasmids found in other LAB. The most similar is plasmid p1 of *L. casei* [56] that shares similarities over a 6-kb region comprising the origin of replication and proteins ParA/ParB (partitioning), RepA (replication), RelB/PemK (toxin/antitoxin system of maintenance) and TraA (transfer). pOENI-1 and pOENI-1v2 most likely use a theta-mode of replication since they encode a RepA protein that is conserved in theta-type plasmids described in other LAB [44]. Their origin of replication also is typical of such plasmids [45], [46]. It is noteworthy that they were detected at a low copy number (3 to 5 copies per cell), which is consistent with plasmids using this mode of replication. There is no doubt that pOENI-1 and pOENI-1v2 derive from each other since they share extensive sequence identity (>99% nucleotide sequence identity over the whole pOENI-1 sequence). Their main difference is a 3.5-kb insert that is present between ORFs 12 and 13 in pOENI1-v2 and absent in pOENI-1. This insert encodes recombinase, transposases and hypothetical proteins without apparent functional role. The plasmids also differ at several nucleotide positions. pOENI-1v2 contains full-length ORFs 4 and 20 coding for a glycerate dehydrogenase and a DNA nickase, respectively, whereas these ORFs are interrupted by early stop codons in pOENI-1. Therefore, we suppose that pOENI-1v2 has preceded pOENI-1 during evolution. However, it is difficult to assess whether pOENI-1v2 is the direct progenitor because analysis of wine samples suggest that there are at least two additional plasmids in this "pOENI-1 family". They differ from the former in the insert region which has intermediate sizes between pOENI-1 and pOENI-1v2 (see Figure 7). It seems that one of these intermediate forms is more frequent in wines than pOENI-1 or pOENI-1v2. It was detected in 12 samples of wines collected during MLF, from a total of 13 samples containing the plasmids.

pOENI-1 and pOENI-1v2 are not conjugative because they do not carry the full set of proteins required for conjugation. However, pOENI-1v2 encodes a nickase (ORF 20, disrupted in pOENI-1) that is typical of mobilizable LAB plasmids [44]. This suggests that at least pOENI-1v2 can propagate by mobilization. In fact, the distribution of plasmids in 44 *O. oeni* strains analyzed in this work supports well the hypothesis that they were horizontally exchanged. Of the four strains that carry a copy of pOENI-1 (C9) or pOENI-1v2 (C10, C6, S11), only two are closely related in dendrograms and phylogenetic trees constructed from PFGE or MLST analyses (C10, C6). The third strain carrying pOENI-1v2 (S11) and strain C9 with pOENI-1 are positioned on distant branches. This distribution most likely results from a dissemination of plasmids via horizontal transfer events.

Potential role of plasmids

Previous studies have demonstrated the importance of plasmids in conferring valuable properties to industrial LAB strains [24], [43]. Plasmids of the pOENI-1 family encode the proteins TauE and OYE, which could be useful for wine bacteria. TauE belongs to a family of membrane transporters involved in the import/export of sulfites or sulfur-containing compounds. Enzymes of this family were characterized as exporters in *Cupriavidus necator*and *Neptuniibacter caesariensis*, in which they contribute to the metabolism of taurine (2-aminoethanesulfonate) [48], [57]. Sulfites can be added at different phases of winemaking for their antioxidant and antimicrobial properties. They are also naturally produced by yeasts during alcoholic fermentation. High concentrations of sulfites in wine may prevent the development of bacteria and avoid MLF to occur [58]. The protein OYE could be also advantageous for wine bacteria. It has the functional domains conserved in NADH:flavin oxidoreductases of the large "old yellow enzymes" family. These enzymes are involved in diverse biological functions including stress response in a number of living cells [52], [59]–[64]. They were not characterized in LAB so far, but identified in *Bacillus subtilis* in which they are expressed in response to oxidative stress and acidification of the cytosol [50], [65]. However, our comparison of strains carrying or not the plasmids has not revealed clear phenotypic differences during MLF or during growth in wine. Strain C9 carrying pOENI-1 has repeatedly completed MLF a few days before its plasmid-less derivative, but this was not confirmed by comparing strains C10 carrying or not pOENI-1v2. Therefore it is yet unclear whether the plasmids confer a significant advantage during growth in wine and what could be this advantage. Further analyses of plasmid genes expression should help to solve this issue. It is noteworthy that besides TauE and OYE, the plasmids encode several hypothetical proteins which could be important for bacteria.

Predominance of plasmids in starter strains and indigenous strains performing MLF

Despite the absence of clear evidence for a role of the plasmids, their distribution among *O. oeni* strains and bacteria present in wine suggests that they could contribute to the fitness of bacteria performing MLF. Of the 44 strains of our study collection, which included 11 industrial starters and 33 non-starter strains, the plasmids were present in only four strains: three starters (C9, C10, C6) and a new isolate (S11) that performs well MLF (data not shown). This represents a frequency of 27% in starters (3/11) and 3% in other strains (1/33). In their recent study, Borneman and coworkers have also detected plasmids in the genomes of four of the 11 strains that they have analyzed: two starters (AWRIB419, AWRIB422) and two strains (AWRIB565, AWRIB576) sharing close genetic relationships with the starter AWRIB429 [55]. By examining these new genomes, we have found that they contain pOENI-1 (AWRIB419), pOENI-1v2 (AWRIB565, AWRIB576) and a divergent pOENI-1-like plasmid containing a different RepA protein and apparently lacking the partitioning proteins (AWRIB422). In addition, although this was not mentioned by the authors, the starter AWRIB429 also contains a plasmid. This starter was investigated in our study under the name "C10" and our results prove that it contains the plasmid pOENI-1v2 (Figures 2 and 3). This plasmid sequence is split into four contigs of the published sequence of AWRIB429 (ACSE00000000). These new findings confirm the high frequency of plasmids of the pOENI-1 family in starter strains. They are detected to date in eight strains, among which there are five starters (C6, C9, C10=AWRIB429, AWRIB419, AWRIB422), two strains closely related to one of these starters (AWRIB565, AWRIB576) and one strain performing well MLF (S11). It is noteworthy that we have also detected a large fragment of the plasmid which includes the ORF encoding TauE in the genome sequence of another starter (AWRIB548, named C4 in our study). In this strain, a fragment of the plasmid has possibly been integrated in the chromosome (sequence acc. number ALAH00000000). The same situation was detected in the genome sequence of starter C3 (unpublished data) that is genetically closely related to C4 (Figure 2).

Our results showed also that the plasmids or the plasmid-encoded genes are frequent in indigenous bacteria performing MLF. The genes *tauE* and *oye* were detected in all of the 95 samples of wine analyzed in this work. They were particularly abundant in samples collected during MLF, reaching the same level as the total cell population. It is very unlikely that the plasmid-encoded genes confer the capacity to survive in wine and to perform MLF. Wine is a complex and harsh environment for most microorganisms [1]. The ability of bacteria to survive in wine and to conduct efficiently MLF involves many genes, many of which have already been described [11]–[18]. However, the predominance of pOENI-1-like plasmids and plasmid-encoded genes in starter strains and indigenous bacteria performing spontaneous MLF indicates that they contribute positively to the fitness of these bacteria during winemaking.

ACKNOWLEDGMENTS

The authors are grateful to the SARCO laboratory for its technical contribution, LAFFORT Company for providing wine samples and Olivier Claisse for technical assistance.

AUTHOR CONTRIBUTIONS

Conceived and designed the experiments: MF EB PL. Performed the experiments: MF EB. Analyzed the data: MF EB AL VM PL. Contributed reagents/materials/analysis tools: AL VM PL. Wrote the paper: MF PL.

REFERENCES

1.   Fleet GH, Lafon-Lafourcade S, Ribereau-Gayon P (1984) Evolution of yeasts and lactic acid bacteria during fermentation and storage of Bordeaux wines. Appl Environ Microbiol 48: 1034–1038.

2.   Lafon-Lafourcade S, Carre E, Ribereau-Gayon P (1983) Occurrence of lactic acid bacteria during the different stages of vinification and conservation of wines. Appl Environ Microbiol 46: 874–880.

3.   Lonvaud-Funel A (1999) Lactic acid bacteria in the quality improvement and depreciation of wine. Antonie Van Leeuwenhoek 76: 317–331. doi: 10.1007/978-94-017-2027-4_16

4.   Davis CR, Wibowo D, Eschenbruch R, Lee TH, Fleet GH (1985) Practical implications of malolactic fermentation: A Review. Am J Enol Vitic 36: 290–301.

5.   Davis CR, Wibowo DJ, Lee TH, Fleet GH (1986) Growth and metabolism of lactic acid bacteria during and after malolactic fermentation of wines at different pH. Appl Environ Microbiol 51: 539–545.

6.   Henick-Kling T, Sandine WE, Heatherbell DA (1989) Evaluation of

malolactic bacteria isolated from Oregon wines. Appl Environ Microbiol 55: 2010–2016.

7.   Garcia-Ruiz A, Moreno-Arribas MV, Martin-Alvarez PJ, Bartolome B (2011) Comparative study of the inhibitory effects of wine polyphenols on the growth of enological lactic acid bacteria. Int J Food Microbiol 145: 426–431. doi: 10.1016/j.ijfoodmicro.2011.01.016

8.   Ugliano M, Moio L (2005) Changes in the concentration of yeast-derived volatile compounds of red wine during malolactic fermentation with four commercial starter cultures of *Oenococcus oeni*. J Agric Food Chem 53: 10134–10139. doi: 10.1021/jf0514672

9.   Gagné S, Lucas PM, Perello MC, Claisse O, Lonvaud-Funel A, et al. (2010) Variety and variability of glycosidase activities in an *Oenococcus oeni* strain collection tested with synthetic and natural substrates. J Appl Microbiol 110: 218–228. doi: 10.1111/j.1365-2672.2010.04878.x

10. Torriani S, Felis GE, Fracchetti F (2010) Selection criteria and tools for malolactic starters development: an update. Ann Microbiol 61: 33–39. doi: 10.1007/s13213-010-0072-x

11. Jobin MP, Garmyn D, Divies C, Guzzo J (1999) Expression of the *Oenococcus oeni trxA* gene is induced by hydrogen peroxide and heat shock. Microbiology 145: 1245–1251. doi: 10.1099/13500872-145-5-1245

12. Grandvalet C, Assad-Garcia JS, Chu-Ky S, Tollot M, Guzzo J, et al. (2008) Changes in membrane lipid composition in ethanol- and acid-adapted *Oenococcus oeni* cells: characterization of the *cfa* gene by heterologous complementation. Microbiology 154: 2611–2619. doi: 10.1099/mic.0.2007/016238-0

13. Coucheney F, Gal L, Beney L, Lherminier J, Gervais P, et al. (2005) A small HSP, Lo18, interacts with the cell membrane and modulates lipid physical state under heat shock conditions in a lactic acid bacterium. Biochim Biophys Acta 1720: 92–98. doi: 10.1016/j.bbamem.2005.11.017

14. Weidmann S, Rieu A, Rega M, Coucheney F, Guzzo J (2010) Distinct amino acids of the *Oenococcus oeni* small heat shock protein Lo18 are essential for damaged protein protection and membrane stabilization. FEMS Microbiol Lett 309: 8–15. doi: 10.1111/j.1574-6968.2010.01999.x

15. Bourdineaud JP, Nehme B, Tesse S, Lonvaud-Funel A (2003) The *ftsH* gene of the wine bacterium *Oenococcus oeni* is involved in protection against environmental stress. Appl Environ Microbiol 69: 2512–2520. doi: 10.1128/aem.69.5.2512-2520.2003

16. .Bourdineaud JP, Nehme B, Tesse S, Lonvaud-Funel A (2004) A bacterial

gene homologous to ABC transporters protect *Oenococcus oeni* from ethanol and other stress factors in wine. Int J Food Microbiol 92: 1–14. doi: 10.1016/s0168-1605(03)00162-4

17. Beltramo C, Grandvalet C, Pierre F, Guzzo J (2004) Evidence for multiple levels of regulation of *Oenococcus oeni clpP-clpL* locus expression in response to stress. J Bacteriol 186: 2200–2205. doi: 10.1128/jb.186.7.2200-2205.2003

18. Grandvalet C, Coucheney F, Beltramo C, Guzzo J (2005) *CtsR* is the master regulator of stress response gene expression in *Oenococcus oeni*. J Bacteriol 187: 5614–5623. doi: 10.1128/jb.187.16.5614-5623.2005

19. Mills DA, Rawsthorne H, Parker C, Tamir D, Makarova K (2005) Genomic analysis of*Oenococcus oeni* PSU-1 and its relevance to winemaking. FEMS Microbiol Rev 29: 465–475. doi: 10.1016/j.fmrre.2005.04.011

20. Athane A, Bilhere E, Bon E, Morel G, Lucas P, et al. (2008) Characterization of an acquired dps-containing gene island in the lactic acid bacterium *Oenococcus oeni*. J Appl Microbiol 105: 1866–1875. doi: 10.1111/j.1365-2672.2008.03967.x

21. Bon E, Delaherche A, Bilhere E, De Daruvar A, Lonvaud-Funel A, et al. (2009)*Oenococcus oeni* genome plasticity associated with fitness. Appl Environ Microbiol 75: 2079–2090. doi: 10.1128/aem.02194-08

22. Renouf V, Favier M (2010) Genetic and physiological characterisation of *Oenococcus oeni* strains to perform malolactic fermentation in wines. S Afr J Enol Vitic 31: 75–81.

23. Wang TT, Lee BH (1997) Plasmids in Lactobacillus. Crit Rev Biotechnol 17: 227–272. doi: 10.3109/07388559709146615

24. Mills S, McAuliffe OE, Coffey A, Fitzgerald GF, Ross RP (2006) Plasmids of lactococci – genetic accessories or genetic necessities? FEMS Microbiol Rev 30: 243–273. doi: 10.1111/j.1574-6976.2005.00011.x

25. Siezen RJ, Renckens B, van Swam I, Peters S, van Kranenburg R, et al. (2005) Complete sequences of four plasmids of *Lactococcus lactis subsp. cremoris* SK11 reveal extensive adaptation to the dairy environment. Appl Environ Microbiol 71: 8371–8382. doi: 10.1128/aem.71.12.8371-8382.2005

26. Fremaux C, Aigle M, Lonvaud-Funel A (1993) Sequence analysis of *Leuconostoc oenos* DNA: organization of pLo13, a cryptic plasmid. Plasmid 30: 212–223. doi: 10.1006/plas.1993.1053

27. Zuniga M, Pardo I, Ferrer S (1996) Nucleotide sequence of plasmid p4028, a cryptic plasmid from *Leuconostoc oenos*. Plasmid 36: 67–74.

doi: 10.1006/plas.1996.0034

28. Brito L, Vieira G, Santos MA, Paveia H (1996) Nucleotide sequence analysis of pOg32, a cryptic plasmid from *Leuconostoc oenos*. Plasmid 36: 49–54. doi: 10.1006/plas.1996.0031

29. Alegre MT, Rodriguez MC, Mesas JM (1999) Nucleotide sequence analysis of pRS1, a cryptic plasmid from *Oenococcus oeni*. Plasmid 41: 128–134. doi: 10.1006/plas.1998.1382

30. Mesas JM, Rodriguez MC, Alegre MT (2001) Nucleotide sequence analysis of pRS2 and pRS3, two small cryptic plasmids from *Oenococcus oeni*. Plasmid 46: 149–151. doi: 10.1006/plas.2001.1537

31. Lucas PM, Claisse O, Lonvaud-Funel A (2008) High frequency of histamine-producing bacteria in the enological environment and instability of the histidine decarboxylase production phenotype. Appl Environ Microbiol 74: 811–817. doi: 10.1128/aem.01496-07

32. Brito L, Paveia H (1999) Presence and analysis of large plasmids in *Oenococcus oeni*. Plasmid 41: 260–267. doi: 10.1006/plas.1999.1397

33. Prevost H, Cavin JF, Lamoureux M, Divies C (1995) Plasmid and chromosome characterization of *Leuconostoc oenos* strains. Am J Enol Vitic 46: 43–48.

34. Sgorbati B, Palenzona D, Sozzi T (1985) Plasmidograms in some heterolactic bacteria from alcoholic beverages and their structural relatedness. Microbiol Alim Nutr 3: 21–34.

35. Sgorbati B, Palenzona D, Ercoli L (1987) Characterization of the pesticides-resistance plasmid pBL34 from *Leuconostoc oenos*. Microbiol Alim Nutr 5, 295–301.

36. **36.**Bilhere E, Lucas PM, Claisse O, Lonvaud-Funel A (2009) Multilocus sequence typing of *Oenococcus oeni*: detection of two subpopulations shaped by intergenic recombination. Appl Environ Microbiol 75: 1291–1300. doi: 10.1128/aem.02563-08

37. Tamura K, Dudley J, Nei M, Kumar S (2007) MEGA4: Molecular Evolutionary Genetics Analysis (MEGA) software version 4.0. Mol Biol Evol 24: 1596–1599. doi: 10.1093/molbev/msm092

38. Lucas PM, Wolken WA, Claisse O, Lolkema JS, Lonvaud-Funel A (2005) Histamine-producing pathway encoded on an unstable plasmid in *Lactobacillus hilgardii* 0006. Appl Environ Microbiol 71: 1417–1424. doi: 10.1128/aem.71.3.1417-1424.2005

39. Besemer J, Borodovsky M (2005) GeneMark: web software for gene finding in prokaryotes, eukaryotes and viruses. Nucleic Acids Research

33: W451–W454. doi: 10.1093/nar/gki487

40. Delcher AL, Harmon D, Kasif S, White O, Salzberg SL (1999) Improved microbial gene identification with GLIMMER. Nucleic Acids Research 27: 4636–4641. doi: 10.1093/nar/27.23.4636

41. Hunter S, Jones P, Mitchell A, Apweiler R, Attwood TK, et al. (2012) InterPro in 2011: new developments in the family and domain prediction database. Nucleic Acids Research 40: D306–D312. doi: 10.1093/nar/gkr948

42. Zhang W, Yu D, Sun Z, Chen X, Bao Q, et al. (2008) Complete nucleotide sequence of plasmid plca36 isolated from *Lactobacillus casei* Zhang. Plasmid 60: 131–135. doi: 10.1016/j.plasmid.2008.06.003

43. Ito Y, Kawai Y, Arakawa K, Honme Y, Sasaki T, et al. (2009) Conjugative plasmid from*Lactobacillus gasseri* LA39 that carries genes for production of and immunity to the circular bacteriocin gassericin A. Appl Environ Microbiol. 75: 6340–6351. doi: 10.1128/aem.00195-09

44. Fang F, Flynn S, Li Y, Claesson MJ, van Pijkeren JP, et al. (2008) Characterization of endogenous plasmids from *Lactobacillus salivarius* UCC118. Appl Environ Microbiol 74: 3216–3228. doi: 10.1128/aem.02631-07

45. Chattoraj DK (2000) Control of plasmid DNA replication by iterons: no longer paradoxical. Mol Microbiol 37: 467–476. doi: 10.1046/j.1365-2958.2000.01986.x

46. Rajewska M, Wegrzyn K, Konieczny I (2012) AT-rich region and repeated sequences – the essential elements of replication origins of bacterial replicons. FEMS Microbiol Rev 36: 408–434. doi: 10.1111/j.1574-6976.2011.00300.x

47. Ruckert C, Koch D, Rey D, Albersmeier A, Mormann S, et al. (2005) Functional genomics and expression analysis of the *Corynebacterium glutamicum* fpr2-cysIXHDNYZ gene cluster involved in assimilatory sulphate reduction. BMC Genomics 6: 121.

48. Weinitschke S, Denger K, Cook AM, Smits THM (2007) The DUF81 protein TauE in*Cupriavidus necator* H16, a sulfite exporter in the metabolism of C2 sulfonates. Microbiology 153: 3055–3060. doi: 10.1099/mic.0.2007/009845-0

49. Parreira R, Sao-Jose C, Isidro A, Domingues S, Vieira G, et al. (1999) Gene organization in a central DNA fragment of *Oenococcus oeni* bacteriophage fOg44 encoding lytic, integrative and non-essential functions. Gene 226: 83–93. doi: 10.1016/s0378-1119(98)00554-x

50. Fitzpatrick TB, Amrhein N, Macheroux P (2003) Characterization of YqjM, an Old Yellow Enzyme homolog from *Bacillus subtilis* involved in the oxidative stress response. J Biol Chem 278: 19891–19897. doi: 10.1074/jbc.m211778200

51. van den Hemel D, Brige A, Savvides SN, Van Beeumen J (2006) Ligand-induced conformational changes in the capping subdomain of a bacterial old yellow enzyme homologue and conserved sequence fingerprints provide new insights into substrate binding. J Biol Chem 281: 28152–28161. doi: 10.1074/jbc.m603946200

52. Brige A, Van den Hemel D, Carpentier W, De Smet L, Van Beeumen JJ (2006) Comparative characterization and expression analysis of the four Old Yellow Enzyme homologues from *Shewanella oneidensis* indicate differences in physiological function. Biochem J 394: 335–344. doi: 10.1042/bj20050979

53. Bon E, Delaherche A, Bilhere E, De Daruvar A, Lonvaud-Funel A, et al. (2009)*Oenococcus oeni* genome plasticity associated with fitness. Appl Environ Microbiol 75: 2079–2090. doi: 10.1128/aem.02194-08

54. Bridier J, Claisse O, Coton M, Coton E, Lonvaud-Funel A (2010) Evidence of distinct populations and specific subpopulations within the species *Oenococcus oeni*. Appl Environ Microbiol 76: 7754–7764. doi: 10.1128/aem.01544-10

55. Borneman A, McCarthy J, Chambers P, Bartowsky E (2012) Comparative analysis of the *Oenococcus oeni* pan genome reveals genetic diversity in industrially-relevant pathways. BMC Genomics 13: 373. doi: 10.1186/1471-2164-13-373

56. Makarova K, Slesarev A, Wolf Y, Sorokin A, Mirkin B, et al. (2006) Comparative genomics of the lactic acid bacteria. Proc Natl Acad Sci U S A 103: 15611–15616. doi: 10.1073/pnas.0607117103

57. Krejčík Z, Denger K, Weinitschke S, Hollemeyer K, Pačes V, et al. (2008) Sulfoacetate released during the assimilation of taurine-nitrogen by *Neptuniibacter caesariensis*: purification of sulfoacetaldehyde dehydrogenase. Arch of Microbiol 190: 159–168. doi: 10.1007/s00203-008-0386-2

58. Carrete R, Vidal MT, Bordons A, Constanti M (2002) Inhibitory effect of sulfur dioxide and other stress compounds in wine on the ATPase activity of *Oenococcus oeni*. FEMS Microbiol Lett 211: 155–159. doi: 10.1111/j.1574-6968.2002.tb11218.x

59. Blehert DS, Fox BG, Chambliss GH (1999) Cloning and sequence analysis of two*Pseudomonas* flavoprotein xenobiotic reductases. J Bacteriol 181: 6254–6263.

60. Odat O, Matta S, Khalil H, Kampranis SC, Pfau R, et al. (2007) Old yellow enzymes, highly homologous FMN oxidoreductases with modulating roles in oxidative stress and programmed cell death in yeast. Journal Biol Chem 282: 36010–36023. doi: 10.1074/jbc.m704058200

61. Reekmans R, Smet KD, Chen C, Hummelen PV, Contreras R (2005) Old yellow enzyme interferes with Bax-induced NADPH loss and lipid peroxidation in yeast. FEMS Yeast Research 5: 711–725. doi: 10.1016/j. femsyr.2004.12.010

62. Trotter EW, Collinson EJ, Dawes IW, Grant CM (2006) Old yellow enzymes protect against acrolein toxicity in the yeast *Saccharomyces cerevisiae*. Appl Environ Microbiol 72: 4885–4892. doi: 10.1128/ aem.00526-06

63. Williams RE, Rathbone DA, Scrutton NS, Bruce NC (2004) Biotransformation of explosives by the old yellow enzyme family of flavoproteins. Appl Environ Microbiol 70: 3566–3574. doi: 10.1128/ aem.70.6.3566-3574.2004

64. Yin B, Yang X, Wei G, Ma Y, Wei D (2008) Expression of two old yellow enzyme homologues from *Gluconobacter oxydans* and identification of their citral hydrogenation abilities. Mol Biotechnol 38: 241–245. doi: 10.1007/s12033-007-9022-7

65. Kitko RD, Cleeton RL, Armentrout EI, Lee GE, Noguchi K, et al. (2009) Cytoplasmic acidification and the benzoate transcriptome in *Bacillus subtilis*. PLoS One 4: e8255. doi: 10.1371/journal.pone.0008255

Chapter 6

COPPER TOLERANCE AND BIOSORPTION OF SACCHAROMYCES CEREVISIAE DURING ALCOHOLIC FERMENTATION

Xiang-yu Sun[1], Yu Zhao[2], Ling-ling Liu[1] , Bo Jia[1] , Fang Zhao[1] , Wei-dong Huang[1] , Jicheng Zhan[1]

[1] College of Food Science and Nutritional Engineering, China Agricultural University, Beijing, 100083, P.R. China,

[2] Faculty of Science, University of Copenhagen, København S, Denmark

ABSTRACT

At high levels, copper in grape mash can inhibit yeast activity and cause stuck fermentations. Wine yeast has limited tolerance of copper and can reduce copper levels in wine during fermentation. This study aimed to understand copper tolerance of wine yeast and establish the mechanism by which yeast decreases copper in the must during fermentation. Three strains of *Saccharomyces cerevisiae* (lab selected strain BH8 and industrial strains AWRI R2 and Freddo) and a simple model fermentation system containing 0 to 1.50 mM Cu^{2+} were used. ICP-AES determined Cu ion concentration in the must decreasing differently by strains and initial copper levels during fermentation. Fermentation performance was heavily inhibited under copper stress, paralleled a decrease in viable cell numbers. Strain BH8 showed higher copper-tolerance than strain AWRI R2 and higher adsorption than Freddo. Yeast cell surface depression and intracellular structure deformation after copper treatment were observed by scanning electron microscopy and transmission electron microscopy; electronic differential system detected higher surface Cu and no intracellular Cu on 1.50 mM copper treated yeast cells. It is most probably that surface adsorption dominated the biosorption process of Cu^{2+} for strain BH8, with saturation being accomplished in 24 h. This study demonstrated that *Saccharomyces cerevisiae* strain BH8 has good tolerance and adsorption of Cu, and reduces Cu^{2+} concentrations during fermentation in simple model system mainly through surface adsorption. The results indicate that the strain selected from China's stress-tolerant wine grape is copper tolerant and can reduce copper in must when fermenting in a copper rich simple model system, and provided information for studies on mechanisms of heavy metal stress.

INTRODUCTION

Copper (Cu) is unavoidable in winemaking: long-term use of copper fungicide [1–2] may increase the copper level in soil [3–4] and grape berry; winemaking equipment [5] and copper sulphate or copper citrate addition for eliminating H_2S [6–7] may also increase the copper content in must. In a narrow range of low concentration, copper is an essential trace element in almost all organisms and plays an important positive role for organisms [8–10]. However, it would have inhibitory effect on cell when out of the useful range, even toxicity. A high level of Cu^{2+} such as 0.1 mM [11] in must inhibits yeast growth and activity; and high level Cu^{2+} is generally believed to cause sluggish fermentation (32 mg/l and 64 mg/l) [12] and a reduction in alcohol production (10.24 mg/l and 80.64 mg/l) [13]. At the same time, with the increase of copper content in wines, particularly existence with other heavy metals such as iron, manganese, zinc, nickel, lead, scandium etc, will cause harm to the health of consumers [14]. Maximum residue levels (MSL) of copper in European regulation is 20 mg/kg in grape must and 1 mg/L in wine [7]; China's national regulation of wine, GB15037-2006, claims the same Cu MSL in wine but no limits in grape must.

In order to remove the excessive copper ions in wines, the current method is to add the adsorbent such as glue and then remove it by filtering. OIV allowed to add potassium ferrocyanide, bentonite, gum Arabic, polyvinylimidazole polyvinylpyrrolidone copolymers, chitin, chitosan et al. in wines to reduce the copper content, but these additives will affect wine sensory quality in different degrees, even detrimental to drinkers' health [15–17]. And*Saccharomyces cerevisiae* has the capability of adsorption of copper ions [15–17]. Utilization*Saccharomyces cerevisiae* to complete the alcoholic fermentation and remove the redundant copper ions at the same time can not only ensure the safety of the quality of the wine, but also highly retain the original color and flavor of wine. Also it conforms to the requirements of the organic wine production, so it is a kind of environmental protection and effective method. Both non-living and living wine yeast were proved to have copper-uptake capacity of 0.04 to 0.2 mM Cu/g cell in solutions containing 6.4 to 256 mg/L copper [18–19]. But its mechanism is still under discussion. A major theory is biosorption with two phases: passive extracellular combination and active intracellular transportation [20–21]; in aqueous solution, the former phase was found to happen in the first 10 min and the latter reacted slowly and did not occur when the ratio of Cu^{2+} to biomass was below 100 nmol/mg [22], indicating two mutually independent phases. Ion exchange at plasma membrane and vacuolar accumulation were believed to be potentially important mechanisms for heavy metal tolerance [23]. But no significant difference was found between normal yeast and yeast with defective vacuoles in Cu^{2+} adsorption [24].

Since in a short period of time, Bordeaux mixture pesticides are still difficult to be replaced [25], and some vineyard soils in Germany [2] and Czech [3], and grapes and wines in Italy [7–8] were found exceeded Cu MSL of local regulation. In future, we will probably need to face the wine fermentation under high copper concentration and the extra problem to reduce the copper concentration of wines. With different yeast strains, the copper resistances and the copper adsorption capacity are different [26–27]. In this situation, if we could find some copper resistance strains and figure out the adsorption mechanism of *Saccharomyces cerevisiae*, it would be very helpful to face this problem.

In this research, we chose two industrial *Saccharomyces cerevisiae* strains AWRI R2 and Freddo which were commonly used by Chinese winemakers for its good fermentation performances, and one laboratory *Saccharomyces cerevisiae* strain BH8 which showed a good property to resist under many different stresses [25, 28–29], by analyzing copper concentration in must during fermentation and copper's effect on their growth, fermentation performance, ultrastructure changes, and elemental analysis to study their copper tolerance and interaction with copper. Cu was added into model synthetic medium (MSM) in reference to fermentation tests in YNB medium and white grape must [12], reaching initial concentrations of 0.50, 1.00 and 1.50 mM, which would not be permitted in wine; the levels were only for research to hopefully demonstrate more clearly the absorption mechanism. The results could be used for selecting wine yeast strains with high copper-adsorption and resistance to high level copper in grape must during alcoholic fermentation, and could give a better understanding on the adsorption mechanism of *Saccharomyces cerevisiae* to copper.

MATERIALS AND METHODS

Yeast strains

Three *Saccharomyces cerevisiae* strains were used; one laboratory strain, BH8 (B), separated (from BeiHong grape must) and stored at the laboratory (China Agricultural University, Beijing), identified as *S. cerevisiae* by Institute of Microbiology, Chinese Academy of Sciences [29]; two industrial strains, AWRI R2 (A; Maurivin Co., Australia) and Freddo (F; Erbslöh Co., Germany), commonly used by Chinese winemakers for their good fermentation performances.

Yeasts maintained on slants were pre-cultured aerobically to 6×10^7 cfu/mL in shaking flasks containing 60 mL YPD medium (1% yeast extract, 2% peptone, and 2% glucose) at 28°C, 120 r/m [29].

Medium

Model synthetic medium (MSM) simulating components of standard grape juice [30] was applied in studying fermentation characteristics of wine yeast, containing the following components (g/L): glucose (100), fructose (100), tartaric acid (3), citric acid (0.3), l-malic acid (0.3), MgSO4 (0.2), KH_2PO_4 (2). Nitrogen was adjusted to 190 mg total N/L with $(NH_4)_2SO_4$ (0.3 g/L) and asparagine (0.6 g/L). Mineral salts (mg/L): $MnSO_4$ H_2O (4), $ZnSO_4$ $7H_2O$ (4), KI (1), $CoCl_2$ $6H_2O$ (0.4), $(NH_4)_6Mo_7O_{24}$ $4H_2O$ (1), H_3BO_3 (1). Vitamins (mg/L): eso-inositol (300), biotin (0.04), thiamin (1), pyridoxine (1), nicotinic acid (1), pantothenic acid (1), p-amino benzoic acid (1). Fatty acids (mg/L): palmitic acid (1), palmitoleic acid (0.2), stearic acid (3), oleic acid (0.5), linoleic acid (0.5), linoleic acid (0.2).

Fermentation experiments

$CuSO_4$ $5H_2O$ was added into MSM in a graded Cu^{2+} series of 0 (control), 0.50 mM (32 mg/L), 1.00 mM (64 mg/L) and 1.50 mM (96 mg/L) [12]. 4 mL yeast precultures were inoculated in 500 mL flasks containing 400 mL MSM to obtain a density of 10^6 cells /mL [29]. Flasks were sealed with glass capillary stoppers filled with concentrated H_2SO_4 to prevent weight loss caused by water evaporation Cultures were constantly shaken at 28°C, 120 r/m in thermostatic shaker (SKY-2102C, Shsukun Co. Ltd., Shanghai) [29]. Mass loss caused by CO_2 evolution was monitored by weighing the fermentation flasks every 24 h [26]. Fermentation was considered to have stopped when mass loss was less than 0.02 g for 3 days. Samples of fermentation must were taken before and every 24 h after inoculation. Fermentation experiments were separated into two groups: one group for weighing, another group for sampling, and each group was carried out in triplicate.

Determination of cell growth and viability

Cell growth was followed by measuring OD_{600} of the fermenting MSM [29] with a UV1800 spectrophotometer (Shimadzu, Japan). MSM free of Cu^{2+}

was used as blank control. Viable cell level of strain B was determined by cell counting using the following procedure: 1 μL of five times diluted MSM sample was embedded on a cytometer and dyed with 1μL of 0.1% methylene blue, a dye commonly used in distinguishing viable and dead cells as it only stains dead cells. Total and viable cell were counted using optical microscope (COIC XSZ-3G) with 40× object lens. Survival rate was calculated following the equation viability % = V/T (V: viable cell amount; T: total cell amount).

Analysis of fermentation performance

The remaining reducing sugars and ethanol content in samples taken during alcoholic fermentation were determined by HPLC using Waters 2414 RI Detector and BIO-RAD Aminex HPX-87H resin-based column (300*7.8mm) [31], which was eluted with 5 mM H_2SO_4 at 65°C, 0.6 mL/ min. Statistical differences for cell growth and fermentation performance of the strains were analyzed using single variable general liner model with PASW Statistics 18.

Analysis of Cu biosorption

Cu adsorption of wine yeast strains were determined by measuring remaining copper in their fermenting MSM. A series of sterile MSM with 0.50, 1.00 and 1.50 mM Cu^{2+} were used as blank control. Samples taken during fermentation were filtrated by 0.45-μm cellulose acetate membrane filters; 4 mL of filtrate was dried at 105°C in 50 mL conical flask in dust-free drying oven, then digested with 5 mL HNO_3-$HClO_4$ (4:1, GR) adding in the flask which was then covered with watch glass, and heated on hot plate at 80°C for 2 h, then 120°C for 2 h, and 190°C until no white fog visible in the flask and the remaining liquid being clear and colorless; digested samples were washed by 18.2 MΩ ultrapure water and filtrated to 25 mL. Glassware were soaked overnight in 20% HNO_3 and washed with ultrapure water before used. Cu concentrations of pre-treated samples were determined by ICP-AES (Perkin Elmer Optima 2000DV) at 327.393 nm. Removal ratio η and adsorption efficiency A (mg/g) of copper ion on yeast were calculated according to equations: $\eta = (C_0\text{-}C_1)/C_0$, and $A = (C_0\text{-}C_1) \times V/M$, where C_0 and C_1 are initial and final Cu concentrations of MSM ferment, respectively, and V represents volume of sample, M means dry weight of yeast separated by centrifuge from the sample

Structural analysis of yeast cell

Yeast cells for SEM-EDS and TEM-EDS were harvested by centrifugation of 10 mL of MSM sample at 4000 rpm (4°C, 10 min).

Sem-eds

Scanning electron microscope (SEM) was used to analyze the extracellular structure; and energy dispersive spectrometer (EDS) was used for surface elemental composition analysis. Harvested yeast cells were washed in deionized water three times by centrifugation and re-suspension. The cells were fixed with 2.5% glutaraldehyde-PBS overnight, washed in 30% PBS buffer (0.1 M, pH7.2) for three times (20 min each time), then post-fixed with 1% osmic acid (1 h), and washed three times with PBS. They were then dehydrated using ethanol with increasing concentrations (v/v) (30%, 50%, 70%, 80%, 90%, and 100%, each for three times, 20 min each time, followed by isoamyl acetate exchanging (three times, 20 min each time), critical point drying and gold crystal spraying. Pretreated cell samples were examined with SEM (Hitachi S-3400N) and SEM-EDS (Jeol JSM-6510A) [32].

Tem-Eds

Intracellular structure was assessed by transmission electron microscopy (TEM) and EDS was combined with TEM for the elemental composition analysis of the cells. Harvested cells were fixed with 2% $KMnO_4$ (4°C, overnight), washed in deionized water (six times, 15 min each time), dehydrated in a graded ethanol series (50%, 70%, 80%, and 90% for once, 100% for three times, 10 min once), then exchanged in ethanol-propanone (1:1, 8 min) and in anhydrous propanone (5 min), followed with a propanone-Epon812 mixture macerations (3:1, 1 h; 1:3, overnight; 1:1, 1 h; pure resin, 24 h), then embedded in Epon812 (37°C, 12 h). Pellets were polymerized by heating in an oven (60°C, 36 h), cut into 70 nm slices on an ultramicrotome (Leica EM UC6), double-dyed in uranyl acetate and lead citrate, and then examined with TEM (Hitachi H-7650) and TEM-EDS (FEI Tecnai F20) [33].

RESULTS

Cell growth

Growth of all three yeast strains weakened as copper concentration increased (Fig 1A, 1B and 1C). Their growth in the control MSM was fastest, reaching log-phase and the end of growth after approximately 6 h and 48 h respectively, with maximum OD_{600} reaching about 2.25. For the 0.50 mM Cu^{2+} medium, the growth curve of strain F mimicked control but was delayed by about 12 hours; strains A and B were slower to start and the log-growth phase tapered off sooner than their controls although OD_{600} ultimately reached same levels as control. As copper concentration increased, all three strains were increasingly sluggish in growth with longer lag-phases and extended log-phases with lower growth rates, reaching stable-phase after 72 h with biomass (OD_{600} ratio) of approximately 80% of control. Copper was clearly inhibiting yeast growth. These results demonstrate that strain F had the highest growth and was the earliest one reached stable-phase, indicating a greater copper tolerance. Of the three strains strain B is more copper-tolerant than strain A and less than strain F.

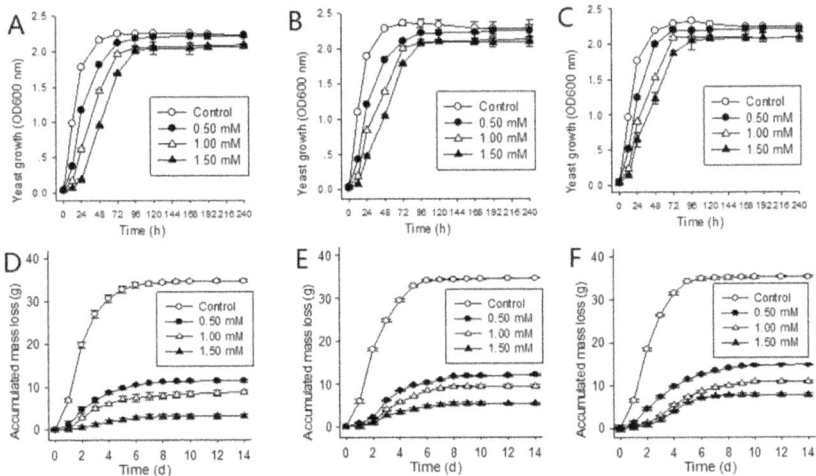

Figure1: Growth curves (A, strain AWRI R2 (A); B, strain BH8 (B); C, strain Freddo (F)) and accumulated fermentation system mass loss(D, strain A; E, strain B; F, strain F) of *S. cerevisiae* strains during fermentation in MSM with 0 (control) (A*, B*, F*), 0.50 (A*, B**, F***), 1.00 (A*, B**, F***) and 1.50 mM (A*, B**, F***) Cu^{2+}. (*, **, and *** represents different statistical significance level, CI = 0.95, n = 3). doi:10.1371/journal.pone.0128611.g001

Fermentation Performance

For all three strains the control ferments progressed most rapidly (Fig 1D, 1E 1F and Fig 2) in the first four days and remained stable after nine days. Total CO_2 evolution was approximately 35g; residual reducing sugars were approximately 4g/L and ethanol concentration approximately 11% (v/v), with no statistical difference (CI = 0.95), indicating good fermentation performance by all strains. However in copper containing MSM, fermentations of all three strains were significantly affected, being sluggish or even becoming stuck, depending on Cu^{2+}concentration. In the 0.50 mM Cu^{2+} MSM, after 12 days total CO_2 evolution of the three strains were less than 50% of their controls with statistically significant differences. Fermentation in 1.50 mM Cu^{2+} MSM lengthened to 14 days with total CO_2 evolution being less than 23% of controls (9.08% (A), 15.59% (B) and 22.25% (F)). Corresponding to the growth curve (Fig 1A, 1B and 1C), the curves for the reducing sugars (Fig 2A, 2B and 2C) and ethanol (Fig 2D, 2E and 2F) became stable earlier and changed less with increasing copper concentration.

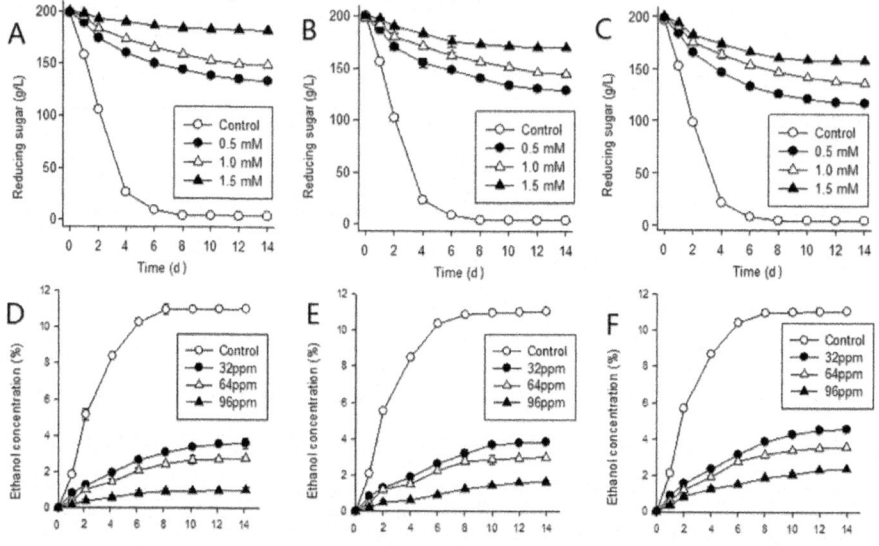

Figure 2: Fermentation must reducing sugar (A, strain A; B, strain B; C, strain F) and fermentation ethanol concentration (D, strain A; E, strain B; F, strain F) for *S. cerevisiae* strains in MSM with 0 (control) (A*, B*, F*), 0.50 (A*, B**, F***), 1.00 (A*, B**, F***) and 1.50 mM (A*, B**, F***) Cu^{2+}. (*, **, and *** represent different statistical significance level, CI = 0.95, n = 3). doi:10.1371/journal.pone.0128611.g002

For the corresponding concentrations of Cu^{2+}, growth activity and fermentation efficiency (**Figs1 and 2**) of strain F was the highest, followed by strain B with strain A being the lowest (such as under 0.5 mM, at 24h, the OD value of strain A was 1.171, strain B 1.208, strain F 1.245; the alcohol production of strain A was 0.81, strain B 0.82, strain F 0.88)

Copper biosorption

It can be seen from Fig 3 that Cu ion concentration of MSM ferments for all three strains went down with time during fermentation, with first four days decreasing rapidly and later period slowly (Fig 3), corresponding to the fermentation performance (Fig 1D, 1E 1F and Fig 2), indicating relations between yeast activity and copper biosorption. Contrarily, Cu ion concentrations in control with no yeast did not reduce significantly (Fig 3). Higher initial copper concentration correlated with lower removal ratio and higher adsorption efficiency; this can been seen from a significant drop of Cu removal ratio (Fig 4A) between groups of 0.50 mM and 1.00 mM initial Cu^{2+} and a leap upward of Cu biosorption by unit yeast (Fig 4B) between groups of 1.00 mM and 1.50 mM initial Cu^{2+} for all three yeast strains. The increase of adsorption efficiency as initial Cu concentration rises could be explained by yeast biomass decrease. Compared with strain A and F, strain B showed a medium removal ratio and adsorption efficiency under all three initial Cu levels (0.50, 1.00 and 1.50 mM; Fig 4). Among three yeast strains, strain A showed the strongest removal ratio (67.37%, Fig 4A) in 0.50 mM Cu and highest adsorption efficiency (15.82 mg/g, Fig 4B) in 1.50 mM Cu.

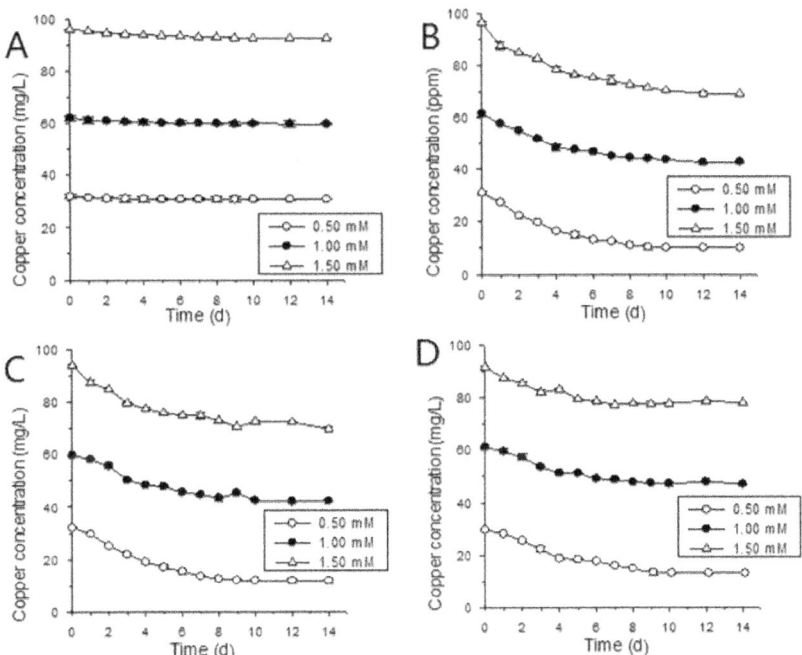

Figure 3: Copper ion concentration (A, Control; B, strain A; C, strain B; D, strain F) of MSM during fermentation for *S. cerevisiae* strains in MSM with 0 (control) (A*, B*, F*), 0.50 (A*, B**, F***), 1.00 (A*, B**, F***) and 1.50 mM (A*, B**, F***) Cu^{2+}. (*, **, and *** represent different statistical significance level, CI = 0.95, n = 3). doi:10.1371/journal.pone.0128611.g003

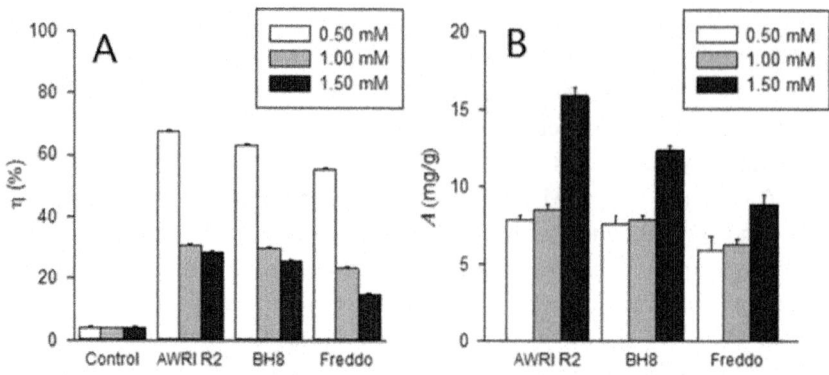

Figure 4: Removal ratio η (A) and adsorption efficiency *A* (B) of Cu^{2+} on *S. cerevisiae*strains AWRI R2 (A), BH8 (B) and Freddo (F) at the end of alcoholic fermentation in MSM with 0.50, 1.00 and 1.50 mM Cu^{2+}. doi:10.1371/journal.pone.0128611.g004

Impacts of Cu on survival rate

To directly reflect copper's lethal effect and the copper tolerance of strain B, living and dead yeasts were counted separately to calculate survival rates at different Cu^{2+} concentrations. The survival rate of strain BH8 (**Table 1**) increased with time in 24 h at each Cu^{2+} level, and decreased as Cu^{2+} concentration increased, indicating a lower reproduction capacity under the stress of high levels of Cu^{2+}.

Table 1: Viability of *S. cerevisiae* BH8 at 12 h and 24 h in MSM with 0, 0.5, 1.0 and 1.5 mM Cu^{2+}. doi:10.1371/journal.pone.0128611.t001

Cu^{2+} level (mM)	viability %	
	12 h	**24 h**
0.00	96.48±0.47[a]	99.97±0.86 [a]
0.50	36.08±0.21 [a]	66.48±0.39 [a]
1.00	16.21±0.06 [a]	23.47±0.45 [a]
1.50	8.58±0.15 [a]	11.13±0.09 [a]

a. Mean value and SD for three independent fermentations.

Impacts of Cu on surface morphology and element

SEM images (Fig 5) show that toxic effects of Cu^{2+} on strain B lead to increasing changes in micromorphology with increasing time and Cu^{2+} concentration. EDS results (Fig 6) indicated cell surface of strain B mainly consists of carbon (C; over 60%) and oxygen (O; over 20%), with the mass fraction (Mass %) and atom% (at %) of Na decreasing with Cu increasing with time after Cu treatment. Gold (peak at 2.00 to 3.00 keV) parameters were not calculated with EDS since gold was sprayed on cell surface during the preparation for SEM testing. As the SEM images showed, yeast cells for the controls (Fig 5A and 5B) were orbicular-ovate, 4 to 6 μm long and 2 to 4 μm wide with smooth surfaces and no intercellular adhesions; besides that, there were also a little fold occurred on some individual cells, which could be a natural consequence of the sample preparation (centrifugal, deionized water washing, and ethanol dehydration) [34]; what's more, buds and bud scars were no more than three for per cell. EDS didn't detected Cu peak (Fig 6). In 0.50 mM Cu^{2+} treatment level, most cells remained regular oval at 24 h (Fig 5C), but cell deformation and pitted surface became obvious and occurred on more cells; after 48 h, there were more bud scars for per cell (Fig 5D). Meanwhile, potassium (K) was undetectable with EDS while Cu was detected with atom% less than 0.05% (Fig 6). At 1.00 mM Cu^{2+}, the cells were slightly deformed and pitted after 24 h (Fig 5E) and significantly stretched with deep pits on most cells after 48 h (Fig 5F). For 1.00 mM Cu^{2+} treatment, the EDS results was resembles to those of 0.50 mM with no K peak; copper was slightly higher in mass% but still low as atom% of 0.05% (Fig 6). In 1.50 mM Cu^{2+} treatment, cells were mostly rough and significantly pitted on the surface with some being stretched with adhesion by 24 h (Fig 5G), and almost all yeast cells were deformed having significantly rough and uneven surfaces by 48 h (Fig 5H); nitrogen (N) was detected in increasing amounts with fermentation time and Cu was higher in the EDS results, K peak was also not detected (Fig 6). Base on these results, we deduced that the disappearance of K peak and decreasing of Na peak with increasing of Cu peak with time after Cu treatment might have certain relations.

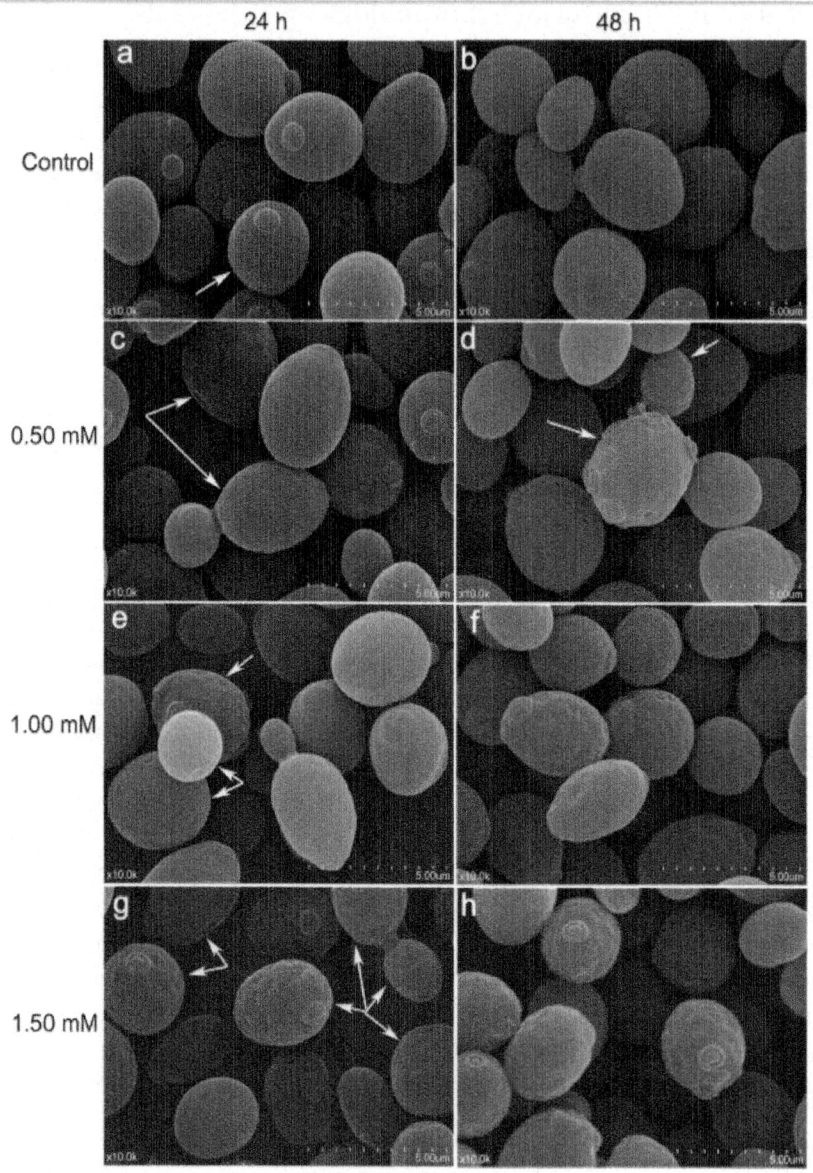

Figure 5: Images (×10000) of yeast surface of *S. cerevisiae* strain BH8 cultivated in MSM with 0 (control), 0.50, 1.00 and 1.50 mM Cu^{2+} for 24 h and 48 h. Arrows indicate pits on individual cell surfaces. doi:10.1371/journal.pone.0128611.g005

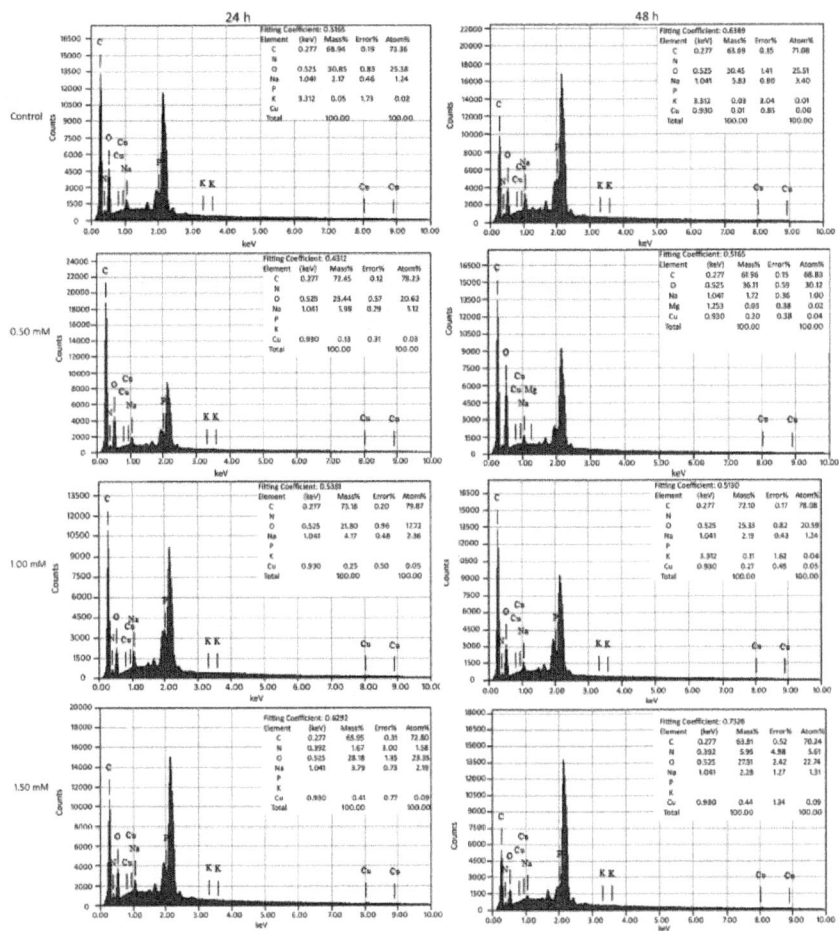

Figure 6: Elemental composition of yeast surface of *S. cerevisiae* strain BH8 cultivated in MSM with 0 (control), 0.50, 1.00 and 1.50 mM Cu^{2+} for 24 h and 48 h. doi:10.1371/journal.pone.0128611.g006

Impacts of Cu on intracellular morphology and element

Fig 7 presents intracellular images (×20000) of strain B examined with TEM. The yeast from the control (Fig 7A) was a normal oval shape with complete cell walls and plasma membranes. The cell wall thickness was even, the organelles dispersed in plasma, and the vacuoles were small and of a similar size. In contrast the yeast taken after 48 h from the MSM ferment containing 1.50 mM Cu^{2+} (Fig 7B), had rough cell walls and plasma membranes which is in agreement with its SEM image (Fig 5H). Plasmolysis occurred with uneven plasma distribution and organelles could not be distinguished. This cytoplasm contraction could be related to Cu^{2+} induced lipid peroxide activity in the plasma membrane.

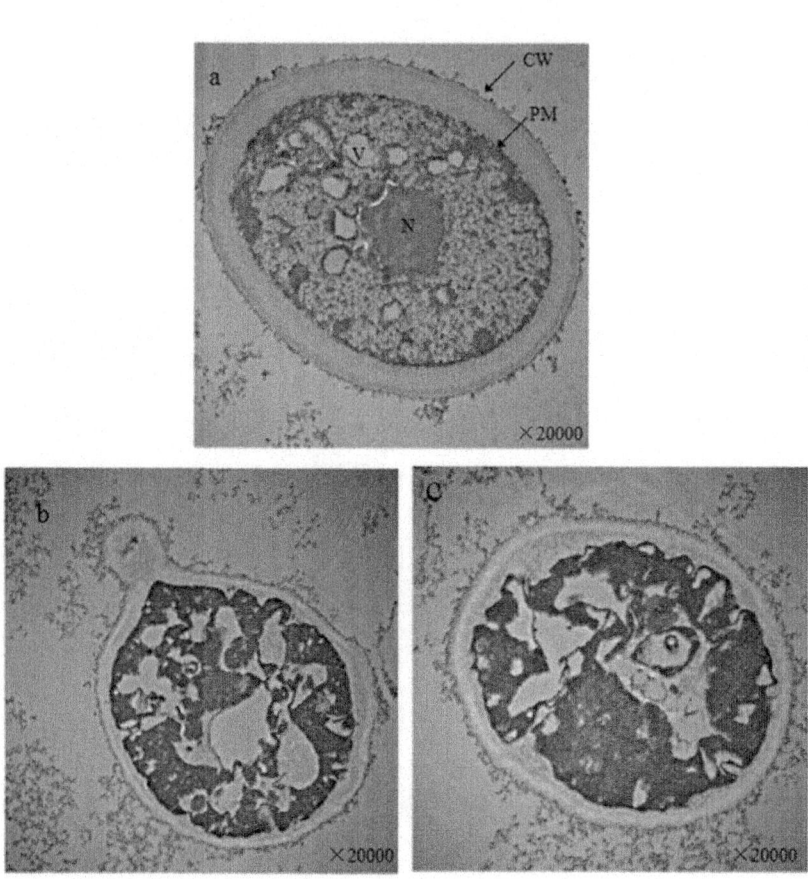

Figure 7: Intracellular images (×20000) of *S. cerevisiae* strain BH8 before (a) and after (b, c) culturing in MSM with 1.50 mM Cu^{2+} for 48 h; CW: cell wall; N: cell nuclear; PM: plasma membrane; V: vacuole. doi:10.1371/journal.pone.0128611.g007

Fig 8 shows the intracellular element ratios determined by TEM-EDS using the same *S.cerevisiae* BH8 sample of Fig 7. Lead (Pb) was introduced into yeast cell from the sample preparation. Yeast mainly consists of carbon and oxygen, which is in agreement with composition of yeast surface (Fig 6). Compared to the control, yeast in 1.50 mM Cu^{2+} for 48 h had no significant changes in C and O although there was some N present. No copper peak was detected, and the atomic ratio (Atomic %) of Ni decreased by 0.54 (Fig 8), indicating Ni^+ leakage after cell membrane deformation. That no Cu was detected inside the yeast cells suggests that strain B does not accumulate Cu²⁺ in its cell and living cells of strain B reduce Cu²⁺ mainly by surface adsorption.

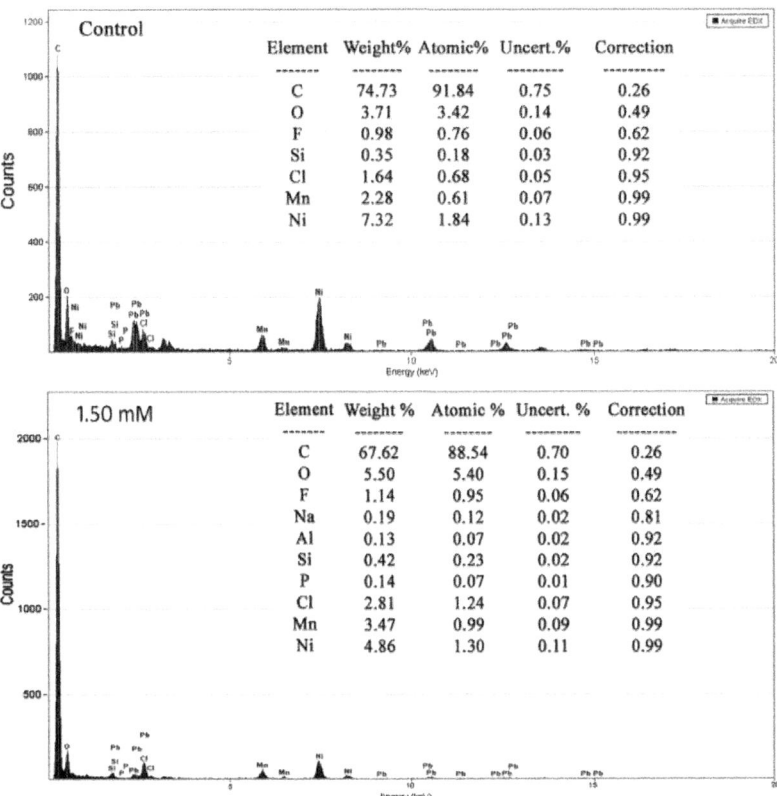

Figure 8: Intracellular elemental composition of *S. cerevisiae* BH8 before (a) and after (b) cultivated in MSM with 1.50 mM Cu²⁺ for 48h. doi:10.1371/journal.pone.0128611. g008

The mechanism of surface Cu²⁺ adsorption could be further studied with atomic force microscopy and confocal laser scanning microscopy. However the sample preparation method needs to be studied since the reported methods for other microorganisms have failed on*Saccharomyces cerevisiae*.

DISCUSSION

As there are many restrictions on natural grape juice, such as the supply of seasonal restrictions, the difference of grape juice composition caused by viticulture region and grape varieties, and the effect of solid composition of natural grape juice on separation of yeast cells, for a long time the studies on

Saccharomyces cerevisiae with different researchers were hard to consistent. And MSM has the advantages of easy using, good reproducibility et al [30], in this study, we chose MSM rather than natural grape juice as fermentation medium.

At a low concentration range, copper is a necessary metal elements for biological growth and metabolism and cofactors for intracellular enzymes metabolism [27]. But once grossing over the beneficial range, it will have inhibitory effect on cells, even toxicity [35]. In wine making, high copper content also affects the wine fermentation process and wine quality. In this experiment, in the MSM medium without copper, all three strains showed a good growth activity and fermentation performance, and the growth curve, CO_2 release quantity, reducing sugar utilization and alcohol production were similar between these three strains (**Figs 1 and2**). Once copper was added into the MSM medium, the growth of *Saccharomyces cerevisiae*was delayed even stagnated, and the effect was positive correlated with the copper concentration. This was in agreement with the result of Shanmuganathan et al [36]. A possible reason for this could be that the yeast cell accumulates large amounts of reactive oxygen species (ROS) at the high concentrations, leading to protein denaturation, membrane order alteration and damage to intracellular enzyme and consequent reduced metabolism, and ultimately cell death [36]. Therefore, anaerobic fermentation was not possible with the reducing sugars not being able to be used for energy.

For the copper biosorption of *Saccharomyces cerevisiae*, in this experiment, even if they didn't add yeast, the copper concentration would reduce slightly with the extension of time (**Fig 3A**). The possible reason was that copper and a small amount of sulfur ions in the solution formed precipitation [25]. After fermentation, the removal rate was 14.86~67.37% and the adsorption efficiency was 5.88~15.82 mg/g. This was in consistent with Volesky et al study [37] and Donmez et al study [38], in their study, the adsorption efficiency was 2~40 mg/g. With different yeast strains, the adsorption quantity of *Saccharomyces cerevisiae* was different. For these three strains, the highest removal rate and the highest adsorption efficiency were all strain A, while strain F was the lowest. In Brandolini et al study [26], the cell growth and fermentation performance of copper resistant yeast strain behaved better than normal yeast strain, and could absorb more copper ions too. By contrast, in this experiment, strain B behaved better on cell growth and fermentation performance under copper stress, but for the copper removal ratio, strain A was better. Liu [39] also reported similar results. This difference might be related with the tested strain features. Also, with different initial copper concentrations, the adsorption quantity of *Saccharomyces cerevisiae* was different. In this experiment, though with higher initial copper concentrations, the toxicity of copper on yeast was higher and leaded to yeast biomass decreased and the removal rate decreased, but the adsorption efficiency increased, which was in agreement with the previous research results [40].

In order to could give a better understanding on the adsorption mechanism of *Saccharomyces cerevisiae* to copper, SEM-OES and TEM-OES were used to observe the ultrastructure change and elemental transformation. In SEM-OES observation, yeast cell surfaces became uneven after copper adsorption together with potassium peak disappeared while copper and nitrogen appeared, but the contents of these elements were very low. It might be under coverage of the high gold peak value, or indicating a small capacity of surface copper adsorption by strain B. With the increasing of the initial copper concentration, the cell surface was more and more roughness, the copper content was getting higher and the K:Cu ratio continued decreased (**Fig 8**), indicated that the adsorption of copper might be associated with the release of potassium from the cell surface. When the copper concentration reached 1.50 mM, nitrogen peak appeared on cell surface, and with the extension of time, the nitrogen peak enhanced. This might because that the copper began to complex with Nitrogen groups of MSM. This was in agreement to Brady et al study [40]. They found 70% of K^+ was rapidly released during Cu^{2+} biosorption in waste water. Hence, it might indicate that ion exchange was involved in the biosorption of Cu^{2+} during fermentation. With the extension of time, though the cell surface was more and more roughness, the copper content was basically remain unchanged, which means the copper adsorption quantity on the yeast surface had reached saturation at the point of 24 h or before. In previous study in waste water, the adsorption of *Saccharomyces cerevisiae* on copper was divided into two stages, the first phase happened quickly on cell surface without energy consumption and the second phase was a long and slow intracellular accumulation and transformation process involving metabolism [38, 40]. The SEM-OES results were fit with the first phase.

Then TEM-EDS was used to observed the intracellular structure and elemental transformation. Under copper stress, the thickness of *Saccharomyces cerevisiae* cell wall was not uniform, the cytoplasm shrank and uneven distribution, organelles couldn't be recognized. The reason of cytoplasm shrank might be related with the lipid over oxidation of cell plasma membrane. When the copper concentration reached 1.50 mM, nitrogen peak appeared on cell surface, the intracellular Potassium content reduced, further illustrated ion exchange was involved in the biosorption of Cu^{2+} during fermentation. And there was no copper detected in intracellular, indicated that *Saccharomyces cerevisiae* could not transport copper into internal. In combination with the results of SEM-OES, the main adsorption mechanism of *Saccharomyces cerevisiae* to copper during alcoholic fermentation was cell surface adsorption. As to whether intracellular accumulation exists, it still needs further studies to confirm.

In conclusion, copper stress could delay even stagnate the growth of *Saccharomyces cerevisiae*, reduce the reducing sugar uptake and ethanol production, and the degree was related to the initial copper concentration and strains. The copper tolerance and copper adsorption ability of strains showed a negative correlation. After *Saccharomyces cerevisiae*adsorbed copper, the yeast surface and intracellular all changed irregularly. Ion exchange was involved in the biosorption of Cu^{2+} during fermentation, and the main adsorption mechanism of*Saccharomyces cerevisiae* to copper during alcoholic fermentation was cell surface adsorption, reaching saturation in 24 h.

SUPPORTING INFORMATION

fermentation	yeast growth (OD 600 nm)			
time (h)	0 mM group	0.5 mM group	1 mM group	1.5 mM group
0	0.028±0.0075	0.033±0.002603	0.04±0.001528	0.044±0.0065
12	0.986±0.0045	0.369±0.013115	0.165±0.014193	0.061±0.016
24	1.777±0.012	1.172±0.008819	0.616±0.018339	0.18±0.019
48	2.161±0.0295	1.812±0.001528	1.438±0.022234	0.958±0.003
72	2.252±0.002	2.12±0.012991	1.964±0.023497	1.693±0.0105
96	2.25±0.005	2.184±0.015431	2.055±0.024835	2.016±0.046
120	2.26±0.012	2.205±0.016093	2.061±0.024265	2.045±0.085
168	2.254±0.007	2.216±0.012914	2.074±0.01837	2.04±0.083
192	2.244±0.004	2.212±0.033287	2.077±0.019519	2.059±0.016093
240	2.23±0.0015	2.218±0.013642	2.091±0.018889	2.07±0.012914

S1 Table. Data for Fig 1A: growth curves of strain A. (DOC)

fermentation	yeast growth (OD 600 nm)			
time (h)	0 mM group	0.5 mM group	1 mM group	1.5 mM group
0	0.023±0.01837	0.033±0.007	0.041±0.012914	0.047±0.023497
12	1.103±0.019519	0.429±0.004	0.189±0.033287	0.076±0.024835
24	1.888±0.018889	1.208±0.0015	0.839±0.013642	0.466±0.024265
48	2.281±0.025887	1.837±0.0065	1.391±0.01392	1.046±0.01837
72	2.357±0.03985	2.102±0.016	1.998±0.033451	1.7859±0.019519
96	2.354±0.059221	2.218±0.019	2.086±0.042226	2.071±0.018889
120	2.337±0.066484	2.2188±0.003	2.101±0.036191	2.089±0.025887
168	2.287±0.089414	2.226±0.0105	2.107±0.033118	2.089±0.03985
192	2.28±0.103233	2.249±0.046	2.122±0.023714	2.086±0.059221
240	2.285±0.124156	2.251±0.085	2.126±0.032002	2.093±0.066484

S2 Table. Data for Fig 1B: growth curves of strain B. (DOC)

fermentation time (h)	yeast growth (OD 600 nm)			
	0 mM group	0.5 mM group	1 mM group	1.5 mM group
0	0.034±0.002	0.037±0.012991	0.041±0.023497	0.045±0.046
12	0.968±0.005	0.513±0.015431	0.235±0.024835	0.138±0.023714
24	1.766±0.012	1.245±0.016093	0.904±0.024265	0.649±0.103233
48	2.192±0.007	1.998±0.012914	1.526±0.01837	1.236±0.085
72	2.29±0.004	2.201±0.033287	2.091±0.019519	1.869±0.032002
96	2.331±0.0015	2.191±0.013642	2.098±0.018889	2.049±0.124156
120	2.281±0.0065	2.203±0.01392	2.098±0.025887	2.073±0.019
168	2.247±0.016	2.213±0.033451	2.094±0.03985	2.083±0.042226
192	2.246±0.019	2.215±0.042226	2.088±0.059221	2.087±0.059221
240	2.242±0.003	2.215±0.036191	2.094±0.066484	2.097±0.018889

S3 Table. Data for Fig 1C: growth curves of strain F. (DOC)

fermentation time (d)	accumulated mass loss (g)			
	0 mM group	0.5 mM group	1 mM group	1.5 mM group
0	0	0	0	0
1	6.92±0.37	1.18±0.86	0.25±0.335	0.06±0.455
2	19.84±0.75	4.71±0.845	2.78±0.325	0.59±0.415
3	27±0.925	6.99±0.58	5.05±0.315	1.15±0.435
4	30.62±0.83	8.68±0.38	6.05±0.285	1.92±0.445
5	32.75±0.76	9.77±0.19	7.23±0.7	2.32±0.48
6	33.79±0.645	10.51±0.05	7.49±0.965	2.8±0.5
7	34.06±0.58	10.93±0.065	7.87±0.91	3.11±0.51
8	34.35±0.355	11.25±0.17	7.97±0.685	3.14±0.52
9	34.56±0.29	11.4±0.235	8.23±0.71	3.15±0.56
10	34.6±0.265	11.43±0.265	8.45±0.66	3.16±0.565
12	34.69±0.31	11.45±0.335	8.67±0.62	3.16±0.56
14	34.71±0.375	11.49±0.33	8.85±0.485	3.15±0.49

S4 Table. Data for Fig 1D: accumulated fermentation system mass loss of strain A. (DOC)

fermentation time (d)	accumulated mass loss (g)			
	0 mM group	0.5 mM group	1 mM group	1.5 mM group
0	0	0	0	0
1	5.93±0.41	0.73±0.135	0.27±0.845	0.14±0.335
2	17.97±0.36	2.07±0.145	1.07±0.58	0.73±0.325
3	24.67±0.315	6.09±0.16	3.99±0.38	2.53±0.315
4	29.4±0.29	8.35±0.17	5.67±0.19	3.31±0.305
5	32.71±0.25	9.66±0.175	7.14±0.05	4.03±0.305
6	34.02±0.23	10.21±0.185	8.39±0.065	4.67±0.31
7	34.16±0.235	10.89±0.23	8.89±0.17	5.03±0.375
8	34.24±0.18	11.68±0.225	9.26±0.235	5.26±0.385
9	34.37±0.17	11.79±0.22	9.32±0.265	5.35±0.385
10	34.43±0.135	11.83±0.225	9.34±0.335	5.37±0.253
12	34.51±0.125	11.89±0.225	9.35±0.33	5.38±0.236
14	34.51±0.115	12.03±0.225	9.35±0.345	5.38±0.221

S5 Table. Data for Fig 1E: accumulated fermentation system mass loss of strain B. (DOC)

fermentation time (d)	accumulated mass loss (g)			
	0 mM group	0.5 mM group	1 mM group	1.5 mM group
0	0	0	0	0
1	6.54±0.325	1.28±0.36	0.52±0.145	0.35±0.58
2	18.5±0.315	4.49±0.315	1.26±0.16	0.82±0.38
3	26.41±0.305	7.36±0.29	2.94±0.17	1.86±0.19
4	31.51±0.305	9.79±0.25	5.47±0.175	4.02±0.05
5	34.29±0.31	11.46±0.23	7.42±0.185	6.04±0.065
6	34.93±0.375	12.6±0.235	8.72±0.23	7.2±0.17
7	35.09±0.385	13.44±0.18	9.56±0.225	7.55±0.235
8	35.2±0.385	14.09±0.17	10.07±0.22	7.79±0.265
9	35.26±0.253	14.52±0.135	10.48±0.225	7.82±0.335
10	35.28±0.236	14.73±0.125	10.88±0.225	7.82±0.33
12	35.29±0.221	14.82±0.115	10.93±0.225	7.84±0.345
14	35.28±0.221	14.87±0.115	10.93±0.225	7.85±0.345

S6 Table. Data for Fig 1F: accumulated fermentation system mass loss of strain F. (DOC)

fermentation time (d)	reducing sugar (g/L)			
	0 mM group	0.5 mM group	1 mM group	1.5 mM group
0	198.5456±1.58689	199.0458±0.89586	199.892±0.05895	200.238±0.225
1	158.26±0.9825	189.256±1.86282	194.856±0.25625	198.256±2.0458
2	105.58±3.5895	174.332±1.25625	183.565±1.5863	193.256±0.2354
4	26.58±0.58312	160.25±3.5891	173.568±0.2635	190.256±4.5689
6	9.586±0.00689	150.256±0.11256	165.658±0.88931	186.583±0.3964
8	4.158±0.1795	144.258±0.002892	159.256±0.1245	184.265±0.712
10	4.025±0.86586	138.256±0.17256	153.256±2.0589	183.256±3.2589
12	3.992±0.11578	135.256±0.8921	150.256±0.0823	182.985±0.15
14	3.956±0.22568	133.485±0.05831	149.398±0.678	182.264±0.823

S7 Table. Data for Fig 2A: fermentation must reducing sugar of strain A. (DOC)

fermentation time (d)	reducing sugar (g/L)			
	0 mM group	0.5 mM group	1 mM group	1.5 mM group
0	200.256±0.6826	198.568±0.1258	198.124±0.05895	199.586±0.1258
1	156.258±0.2568	186.586±0.3985	192.358±0.25625	197.582±0.2231
2	102.156±2.5698	170.256±2.589	180.256±2.856	190.256±0.3125
4	23.561±0.58312	155.235±4.25	171.258±0.45	183.258±0.264
6	8.953±0.069	148.256±0.69	162.358±3.256	175.682±5.328
8	4.123±0.3589	140.235±0.891	156.325±0.125	173.256±0.126
10	4.025±0.1587	133.258±0.263	151.325±0.6945	171.268±0.856
12	3.958±0.369	130.258±0.286	146.283±0.15	170.258±0.961
14	3.948±0.125	128.56±0.145	144.56±0.8411	169.85±0.126

S8 Table. Data for Fig 2B: fermentation must reducing sugar of strain B. (DOC)

fermentation time (d)	reducing sugar (g/L)			
	0 mM group	0.5 mM group	1 mM group	1.5 mM group
0	196.258±0.25948	198.364±0.5495	199.258±1.209	199.856±0.4591
1	152.354±0.648	183.256±0.98491	189.568±0.94156	193.586±0.1651
2	98.256±0.594	165.259±0.95165	175.356±0.149	182.258±1.549
4	21.586±0.8942	146.325±2.3159	163.258±4.159	173.586±2.849
6	7.562±0.12654	132.568±0.19216	153.586±0.941	165.893±0.10651
8	3.925±1.4156	125.345±0.1561	146.586±0.125	160.258±0.4159
10	3.92±0.48941	120.586±0.1981	141.258±0.6945	158.126±0.8941
12	3.918±0.6151	117.256±0.159	137.568±0.9849	157.68±0.156
14	3.92±0.849	116.358±0.849	135.368±0.126	157.35±0.1657

S9 Table. Data for Fig 2C: fermentation must reducing sugar of strain F. (DOC)

fermentation	ethanol concentration (%)			
time (d)	0 mM group	0.5 mM group	1 mM group	1.5 mM group
0	0	0	0	0
1	1.852±0.0085	0.81±0.0859	0.43±0.00549	0.18±0.0059
2	5.158±0.3159	1.26±0.0958	1.03±0.001987	0.39±0.06954
4	8.369±0.08213	1.96±0.1261	1.46±0.129	0.58±0.0984
6	10.258±0.09156	2.66±0.03165	2.09±0.126	0.82±0.159
8	10.9526±0.284	3.08±0.0051	2.46±0.09156	0.93±0.1589
10	10.96852±0.061495	3.38±0.09126	2.68±0.21594	0.95±0.09459
12	10.9826±0.054165	3.55±0.0984	2.73±0.05645	0.98±0.0591
14	11.01±0.0561	3.6±0.2645	2.78±0.08456	1.02±0.168489

S10 Table. Data for Fig 2D: fermentation ethanol concentration of strain A (DOC)

fermentation	ethanol concentration (%)			
time (d)	0 mM group	0.5 mM group	1 mM group	1.5 mM group
0	0	0	0	0
1	2.085±0.0561	0.82±0.0859	0.46±0.00694	0.21±0.001459
2	5.536±0.0626	1.31±0.01465	1.21±0.0981	0.49±0.098489
4	8.485±0.0816	1.91±0.1268	1.53±0.0815	0.62±0.0612
6	10.368±0.1026	2.64±0.0984	2.26±0.006841	0.93±0.008419
8	10.869±0.00846	3.21±0.2159	2.78±0.0781	1.26±0.001859
10	10.968±0.0681	3.69±0.0984	2.89±0.2345	1.46±0.0945
12	11.01±0.0951	3.8±0.00894	2.98±0.00816	1.63±0.084
14	11.05±0.0984	3.85±0.0366	3.021±0.062	1.67±0.121

S11 Table. Data for Fig 2E: fermentation ethanol concentration of strain B. (DOC)

fermentation	ethanol concentration (%)			
time (d)	0 mM group	0.5 mM group	1 mM group	1.5 mM group
0	0	0	0	0
1	2.12±0.0989	0.88±0.0649	0.6±0.0789	0.36±0.01264
2	5.69±0.08416	1.56±0.0165	1.22±0.0489	0.83±0.0984
4	8.69±0.0216	2.35±0.0894	1.93±0.10126	1.28±0.126
6	10.43±0.126	3.16±0.0481	2.79±0.00984	1.52±0.0489
8	10.96±0.0949	3.86±0.0651	3.16±0.0651	1.88±0.0984
10	10.99±0.01296	4.26±0.129	3.39±0.005948	2.06±0.13549
12	11.03±0.0894	4.51±0.0489	3.52±0.00159	2.31±0.0894
14	11.06±0.00894	4.56±0.008948	3.58±0.0654	2.38±0.008459

S12 Table. Data for Fig 2F: fermentation ethanol concentration of strain F. (DOC)

fermentation	copper concentration (mg/L)		
time (d)	0.5 mM group	1 mM group	1.5 mM group
0	31.9563±0.964	62.2158±0.8145	96.2875±0.0258
1	31.3145±0.1785	61.4895±1.1524	95.586±0.1235
2	31.0258±0.8561	61.0785±0.125	94.8562±0.2478
3	30.9325±1.258	60.8523±0.4885	94.3258±0.0878
4	30.7961±0.925	60.5842±1.0568	94.0831±0.05895
5	30.7258±0.5861	60.3148±0.2175	93.8215±0.03547
6	30.7068±0.5892	60.2895±0.6589	93.5847±0.02447
7	30.6952±0.3984	60.1987±0.0985	93.2475±0.0821
8	30.6825±1.1251	60.0657±0.1425	93.1058±0.00258
9	30.6842±0.8569	59.9275±0.9852	92.8452±0.00478
10	30.6838±0.2478	59.9856±0.5698	92.8017±0.089
12	30.6819±0.2588	59.9098±1.2285	92.7482±0.14
14	30.6828±0.0036	59.8762±0.2189	92.7852±0.0047

S13 Table. Data for Fig 3A: copper ion concentration of MSM during fermentation for control group. (DOC)

fermentation	copper concentration (mg/L)		
time (d)	0.5 mM group	1 mM group	1.5 mM group
0	30.95625±0.85	61.48125±0.8958	96.25±0.289
1	27.2563±0.01	57.625±0.5586	87.1875±1.589
2	22.1813±0.489	54.91875±0.5895	84.625±0.25
3	19.7313±0.5892	51.8125±0.4885	82.5±0.068
4	16.375±0.1589	48.21875±1.0568	78.25±0.8958
5	14.9125±1.058	47.45±0.0058	76.125±0.27859
6	13.0154±0.5892	46.575±0.6589	75.125±0.058
7	12.3875±0.3984	44.875±0.0789	74.0625±1.85982
8	10.925±0.0258	44.36875±0.5895	72.4375±0.5865
9	10.2568±0.8569	43.96875±0.3658	71.375±0.03568
10	10.0125±0.2478	43.55±0.1184	70.1875±0.0425
12	9.9897±0.2588	42.49375±0.2225	69.125±0.5836
14	9.9874±0.0036	42.80625±0.2189	69.05±0.0785

S14 Table. Data for Fig 3B: copper ion concentration of MSM during fermentation for strain A. (DOC)

fermentation	copper concentration (mg/L)		
time (d)	0.5 mM group	1 mM group	1.5 mM group
0	32.125±0.058954	59.625±0.00589	93.875±0.5895
1	29.75±0.12589	58.075±0.5896	87.375±0.98658
2	25.1433±0.26895	55.6±0.9865	84.875±0.2589
3	21.9313±0.589075	50.19375±0.25894	79.6875±0.2589
4	19.0625±0.6985	48.25625±0.1256	77.3125±0.4789
5	17.0225±0.8956	47.7±0.23589	75.875±0.2578
6	15.4485±1.0258	45.49375±0.3	74.9375±0.6928
7	13.5821±0.22368	44.6375±0.598	75±1.2568
8	12.5685±0.2895	43.4125±0.9688	72.9375±0.2589
9	12.0895±0.004895	45.29375±0.4885	70.75±0.1478
10	11.9861±0.0589	42.44375±0.1568	72.5625±0.2816
12	11.8612±0.8964	42.08125±0.2568	72.375±0.1367
14	11.8489±0.7892	42.2±0.15978	69.75±0.8592

S15 Table. Data for Fig 3C: copper ion concentration of MSM during fermentation for strain B. (DOC)

fermentation	copper concentration (mg/L)		
time (d)	0.5 mM group	1 mM group	1.5 mM group
0	29.73±0.2589	61.19±0.5897	91.63±0.3698
1	28.25±0.1898	59.49±0.9858	87.5±0.2658
2	25.66±0.5895	57.21±1.2569	85.44±3.549
3	22.25±0.8978	53.51±0.25	82±0.8698
4	18.87±0.257	51.24±0.0369	83.19±0.2584
5	18.4±0.6589	51.13±0.0098	79.44±0.00369
6	17.71±0.00859	49.08±0.247	78.56±0.36548
7	16.05±0.0458	48.8±0.3697	77.31±0.2156
8	14.95±0.15798	47.69±0.15	77.88±0.259
9	13.63±0.0879	47.39±0.0872	77.55±1.0326
10	13.43±0.4589	47.16±0.6985	77.81±0.2594
12	13.33±0.2258	47.89±0.1235	78.69±0.4895
14	13.28±0.267	47.03±0.3694	78.01±0.2354

S16 Table. Data for Fig 3D: copper ion concentration of MSM during fermentation for strain F. (DOC)

	removal ratio η (%)		
	0.5 mM group	1 mM group	1.5 mM group
Control	3.9851±0.162	3.7604±0.324	3.6373±0.585
AWRI R2	67.371±0.697	30.38±0.462	28.26±0.235
BH8	63.1163±0.398	29.56±0.495	25.69±0.126
Freddo	55.33±0.495	23.14±0.346	14.86±0.427

S17 Table. Data for Fig 4A: removal ratio ηof Cu^{2+} on S. cerevisiae strains AWRI R2 (A), BH8 (B) and Freddo (F) at the end of alcoholic fermentation in MSM with 0.50, 1.00 and 1.50 mM Cu^{2+} (DOC)

	adsorption efficiency A (mg/g)		
	0.5 mM group	1 mM group	1.5 mM group
AWRI R2	7.83±0.323	8.46±0.363	15.82±0.604
BH8	7.58±0.49	7.84±0.258	12.31±0.307
Freddo	5.88±0.89	6.24±0.349	8.86±0.591

S18 Table. Data for Fig 4B: adsorption efficiency A of Cu^{2+} on S. cerevisiae strains AWRI R2 (A), BH8 (B) and Freddo (F) at the end of alcoholic fermentation in MSM with 0.50, 1.00 and 1.50 mM Cu^{2+} (DOC)

ACKNOWLEDGMENTS

This research was supported by the grants from the National Natural Science Foundation of China (31471835), and the National "Twelfth Five-Year" Plan for Science & Technology Support (2012BAD31B07). The authors would like to thank Mr. Malcolm J. Reeves (Institute of Food, Nutrition and Human Health, Massey University, Palmerston North 4442, New Zealand) for his assistance in improving the English of this paper.

AUTHOR CONTRIBUTIONS

Conceived and designed the experiments: WH JZ. Performed the experiments: XS LL YZ BJ FZ. Analyzed the data: XS LL YZ BJ FZ. Contributed reagents/materials/analysis tools: XS LL YZ BJ. Wrote the paper: XS LL YZ.

REFERENCES

1.   Brun LA, Maillet J, Richarte J, Herrmannb P, Remyb JC. (1998) Relationships between extractable copper, soil properties and copper

uptake by wild plants in vineyard soils. Enviro Pollut 102: 151–161. doi: 10.1016/s0269-7491(98)00120-1

2. Probst B, Schüler C, Joergensen R. (2008) Vineyard soils under organic and conventional management—microbial biomass and activity indices and their relation to soil chemical properties. Biol Fertil Soils 44: 443–450. doi: 10.1007/s00374-007-0225-7

3. Ash C, Vacek O, Jakšík O, Tejnecky V. (2012) Elevated soil copper content in a bohemian vineyard as a result of fungicide application. Soil Water Res 7: 151–158. doi: 10.4161/psb.18936. pmid:22415043

4. Provenzano M, Bilali HE, Simeone V. (2010) Copper contents in grapes and wines from a Mediterranean organic vineyard. Food Chem 122: 1338–1343. doi: 10.1016/j.foodchem.2010.03.103

5. Volpe MG, Cara FL, Volpe F, Mattia AD, Serino V, Petitto F, et al. (2009) Heavy metal uptake in the enological food chain. Food Chem 117: 553–560. doi: 10.1016/j.foodchem.2009.04.033

6. Tamasi G, Pagni D, Carapelli C, Justice NB, Cini R. (2010) Investigation on possible relationships between the content of sulfate and selected metals in Chianti wines. J Food Compost Anal 23: 333–339. doi: 10.1016/j.jfca.2009.12.011

7. .García-Esparza M, Capri E. (2006) Copper content of grape and wine from Italian farms. Food Addit Contam 233: 274–280. doi: 10.1080/02652030500429117

8. Nogueirol R, Alleoni L, Nachtigall G, Meloc G. (2010) Sequential extraction and availability of copper in Cu fungicide-amended vineyard soil from Southern Brazil. J Hazard Mater 181: 931–937. doi: 10.1016/j.jhazmat.2010.05.102. pmid:20579811

9. Kim B, Nevitt T, Thiele D. (2008) Mechanisms for copper acquisition, distribution and regulation. Nat Chem Biol 4: 176–185. doi: 10.1038/nchembio.72. pmid:18277979

10. Claus H. (2010) Copper-containing oxidases: occurrence in soil microorganisms, properties and applications. Soil Heavy Met 19: 281–313. doi: 10.1007/978-3-642-02436-8_13

11. Ohsumi Y, Kitamoto K, Anraku Y. (1988) Changes induced in the permeability barrier of the yeast plasma membrane by cupric ion. J Bacteriol 170: 2676–2682. pmid:3286617

12. Azenha M, Vasconcelos M, Moradas-Ferreira P. (2000) The influence of Cu concentration on ethanolic fermentation by Saccharomyces cerevisiae. J Biosci Bioeng 90: 163–167. pmid:16232836 doi: 10.1263/jbb.90.163

13. Mrvcic J, Stanzer D, Stehlik-Tomas V, Skevin D, Grb S. (2007) Optimization of bioprocess for production of copper-enriched biomass of industrially important microorganism Saccharomyces cerevisiae. J Biosci Bioeng 103: 331–337. pmid:17502274 doi: 10.1263/jbb.103.331

14. Naughton D, Petróczi A. (2008) Heavy metal ions in wines: meta-analysis of target hazard quotients reveal health risks. Chem Cent J 3: 1–7. doi: 10.1186/1752-153x-2-22

15. Benítez P, Castro R, Barroso C. (2002) Removal of iron, copper and manganese from white wines through ion exchange techniques: effects on their organoleptic characteristics and susceptibility to browning. Anal Chim Acta 458: 197–202. doi: 10.1016/s0003-2670(01)01499-4

16. Mira H, Leite P, Catarino S, Ricardo S, Curvelo G. (2007) Metal reduction in wine using PVI-PVP copolymer and its effects on chemical and sensory characters. Vitis 46: 138–147.

17. Schubert M, Glomb M. (2010) Analysis and chemistry of migrants from wine fining polymers. J Agr Food Chem 58: 8300–8304. doi: 10.1021/jf101127t. pmid:20568775

18. Veglio F, Beolchini F. (1997) Removal of metals by biosorption: review. Hydrometallurgy 44: 301–316. doi: 10.1016/s0304-386x(96)00059-x

19. Naja G, Murphy V, Volesky B. Biosorption, Metals. (2010) In Encyclopedia of Industrial Biotechnology: Bioprocess, Bioseparation, and Cell Technology. M.C. Flickinger (ed.): 1–29p.

20. Norris P, Kelly D. (1977) Accumulation of cadmium and cobalt by Saccharomyces cerevisiae. J Gen Microbiol 99: 317–324. doi: 10.1099/00221287-99-2-317

21. Eide D. (1998) The molecular biology of medallion transport in *Saccharomyces cerevisiae*. Annu Rev Nutr 18: 441–469. pmid:9706232 doi: 10.1146/annurev.nutr.18.1.441

22. Brady D, Stoll A, Starke L, Duncan J. (1994) Chemical and enzymatic extraction of heavy metal binding polymers from isolated cell walls of *Saccharomyces cerevisiae*. Biotechnol Bioeng 44: 297–302. pmid:18618746 doi: 10.1002/bit.260440307

23. Hall J. (2002) Cellular mechanisms for heavy metal detoxification and tolerance. J Exp Bot 366: 1–11. pmid:11741035 doi: 10.1093/jexbot/53.366.1

24. .Ramsay L, Gadd G. (1997) Mutants of *Saccharomyces cerevisiae* defective in vacuolar function confirm a role for the vacuole in toxic metal ion detoxification. FEMS Microbiol Lett 152: 293–298. pmid:9231423 doi: 10.1111/j.1574-6968.1997.tb10442.x

25. Li H, Guo A, Wang H. (2008) Mechanisms of oxidative browning of wine. Food Chem 108: 1–13. doi: 10.1016/j.foodchem.2007.10.065

26. Brandolini V, Tedeschi P, Capece A, Maietti A, Mazzotta D, Salzano G, et al. (2002)*Saccharomyces cerevisiae* wine strains differing in copper resistance exhibit different capability to reduce copper content in wine. World J Microbiol Biotechnol 18: 499–503.

27. Ferreira J, Toit M, Toit W. (2006) The effects of copper and high sugar concentrations on growth, fermentation efficiency and volatile acidity production of different commercial wine yeast strains. Aust J Grape Wine Res 12: 50–56. doi: 10.1111/j.1755-0238.2006.tb00043.x

28. Li H, Du G, Li H, Wang H, Yan G, Zhan J, et al. (2010) Physiological response of different wine yeasts to hyperosmotic stress. Am J Enol Vitic 61: 529–535. doi: 10.5344/ajev.2010.09136

29. Du G, Zhan JC, Li JY, You YL, Zhao Y, Huang WD. (2012) Effect of fermentation temperature and culture medium on glycerol and ethanol during wine fermentation. Am J Enol Vitic 63: 132–138. doi: 10.5344/ajev.2011.11067

30. Marullo P, Bely M, Masneuf-Pomarede I, Aigle M, Dubourdieu D. (2004) Inheritable nature of enological quantitative traits is demonstrated by meiotic segregation of industrial wine yeast strains. FEMS Yeast Res 7: 711–719. pmid:15093774 doi: 10.1016/j.femsyr.2004.01.006

31. Ciani M, Beco L, Cornitini F. (2006) Fermentation behaviour and metabolic interactions of multistarter wine yeast fermentations. Int J Food Microbiol 108: 239–245. pmid:16487611 doi: 10.1016/j.ijfoodmicro.2005.11.012

32. Moreno I, González-Weller V, Gutierrez M, Marino A, Cameán A, González A, et al. (2008) Determination of Al, Ba, Ca, Cu, Fe, K, Mg, Mn, Na, Sr and Zn in red wine samples by inductively coupled plasma optical emission spectroscopy: evaluation of preliminary sample treatments. Microchem J 88: 56–61. doi: 10.1016/j.microc.2007.09.005

33. Zu YG, Zhao XH, Hu M. (2006) Biosorption effects of copper ions on Candida utilis under negative pressure cavitation. J Environ Sci 18: 1254–1259. pmid:17294974 doi: 10.1016/s1001-0742(06)60071-5

34. Chen C, Wang JL. (2007) Characteristics of Zn2+ biosorption by Saccharomyces cerevisiae. Biomed Environ Sci 20: 478–482. pmid:18348406

35. Robinson N, Winge D. (2010) Copper Metallochaperones. Annu Rev Biochem 79: 537–562. doi: 10.1146/annurev-biochem-030409-143539. pmid:20205585

36. Shanmuganathan A, Avery S, Willetts S, Houghton J. (2004) Copper-induced oxidative stress in Saccharomyces cerevisiae targets enzymes of the glycolytic pathway. FEBS Lett 556: 253–259. pmid:14706859 doi: 10.1016/s0014-5793(03)01428-5

37. Volesky B, Holan Z. (1995) Biosorption of heavy metals. Biotechnol Progr 11: 235–250. pmid:7619394 doi: 10.1021/bp00033a001

38. properties of some yeasts. Process Biochem 35: 135–142. pmid:10102182 doi: 10.1016/s0032-9592(99)00044-8

39. Liu L. (2012) Copper biosorption bechanism of *Saccharomyces cerevisiae* during alcoholic fermentation. Beijing: China Agricultural University.

40. Brady D, Duncan JR. (1994) Bioaccumulation of metal cations by Saccharomyces cerevisiae. Appl Microbiol Biotechnol 41: 149–154. doi: 10.1007/bf00166098

Chapter 7

THE ROLE OF LACTIC ACID BACTERIA IN MILK FERMENTATION

Yantyati Widyastuti[1], Rohmatussolihat[1], Andi Febrisiantosa[2]

[1]Research Center for Biotechnology, Indonesian Institute of Sciences, Cibinong, Indonesia;

[2]Technical Implementation Unit for Development of Chemical Engineering Processes, Indonesian Institute of Sciences, Yogyakarta, Indonesia.

ABSTRACT

Species of lactic acid bacteria (LAB) represent as potential microorganisms and have been widely applied in food fermentation worldwide. Milk fermentation process has been relied on the activity of LAB, where transformation of milk to good quality of fermented milk products made possible. The presence of LAB in milk fermentation can be either as spontaneous or inoculated starter cultures. Both of them are promising cultures to be explored in fermented milk manufacture. LAB have a role in milk fermentation to produce acid which is important as preservative agents and generating flavour of the products. They also produce exopolysaccharides which are essential as texture formation. Considering the existing reports on several health-promoting properties as well as their generally recognized as safe (GRAS) status of LAB, they can be widely used in the developing of new fermented milk products.

INTRODUCTION

Species of lactic acid bacteria (LAB) belong to numerous genus under the family of Lactobacillaceae. They represent as potential microorganisms and have been widely applied in food fermentation worldwide due to their well-known status as generally recognized as safe (GRAS) microorganisms. They are also recognised for their fermentative ability and thus enhancing food safety, improving organoleptic attributes, enriching nutrients and increasing health benefits [1-4]. Fermentation is generally considered as a safe and acceptable preservation technology of food and fermentation using LAB can be categorized into two groups based on the raw material used, non-dairy and dairy fermentation. Milk from different mammalian animals can be used in dairy fermentation to produce several products. Milk of cow followed by milk of goat and sheep are the most widely used raw materials to produce particular economic value fermented milk products worldwide. Due to the characteristics of milk that is highly perishable, the main purpose of milk fermentation using LAB is to prolong its shelf-life as well as to preserve the nutritious component of milk. It is also recognised that fermentation of milk using LAB will undoubtedly produce good quality of products with highly appreciated organoleptic attributes. Recently, there is a growing interest to develop a variety of fermented milk products for other beneficial purposes, particularly for health purposes and preventing of toxins produced by foodborne pathogens and spoilage bacteria that enter human body [1, 2, 5, 6]. The beneficial effects of fermented milk products are produced by a variety of bioactive compounds of LAB [7].

Lactic acid bacteria represent as the most extensively studied microorganisms for milk fermentation [8-10]. The presence of LAB in milk fermentation can be either as spontaneous or inoculated starter cultures. Milk itself is known as one of the natural habitats of LAB [11, 12]. Although under spontaneous fermentations the growth of LAB cannot be predicted or controlled, but this procedure has been practised and carried out traditionally for years. A procedure called as backslopping is often used. There are some examples of fermented milk by LAB produced under this procedure such as those of artisanal cheese klila [13], kumis [14], iben [15] and kurut [16]. In general, the technology of milk fermentation is relatively simple and cost-effective. On the other hand, standardized fermented milk products are produced and manufactured in large-scale production under controlled conditions and become an important industrial application of LAB as starter cultures. There are some important features of LAB starters in fermented milk products. A single potential starter culture will dominate and reduce the diversity of microorganisms in fermented milk products compare to that of products under natural fermentation. In this review we focus on the potential role of LAB in milk fermentation based on their properties that support the development of fermented milk products.

PROPERTIES OF LACTIC ACID BACTERIA

Milk fermentation process has been relied on the activity of LAB, which play a crucial role in converting milk as raw material to fermented milk products. In milk fermentation industry, various industrial strains of LAB are used as starter cultures. Starter cultures of LAB were obtained from a sequence activities and passed a process of isolation, selection and confirmation. Several behaviour as the characteristics of each individual selected strains of LAB has been established and used in the production of fermented milk products industrially. The most important properties of LAB are their ability to acidify milk [17] and to generate flavour and texture, by converting milk protein due to their proteolytic activities [7, 18]. The mild acid taste and pleasant fresh, are characteristics of fermented milk products such as yoghurt and cheese.

Preservative Property of Lactic Acid Bacteria

Milk and fermented milk products are favorable substrates for the growth of microorganisms that may bring to spoilage condition. The most well-known characteristics of LAB related to preservative property is their ability to produce acid, which in turn exhibit antimicrobial activity. Acidification of the milk protects the milk against spoilage microorganisms and proliferation of pathogens. LAB also release antimicrobial metabolites so called bacteriocins [19]. Both acids and bacteriocins are great potential to be used in food preservation, which are considered as safe natural preservatives.

Acid Production

Organic acids is the end product of carbohydrate metabolism produced by BAL. Homofermentative species of LAB convert sugars in milk mostly into lactic acid, whereas the heterofermentative species convert lactose into lactic acid, acetic acid, ethanol and CO_2. Production of lactic acid by LAB is strain dependend. The newly identified Lactobacillus paracasei subsp. paracasei CHB2121 was reported to produce high concentrations of L (+)-lactic acid efficiently. It produced 192 g/L lactic acid from medium containing 200 g/L of glucose, with 3.99 g/(L.h) productivity, and 0.96 g/g yield. In addition, the optical purity of the produced lactic acid was estimated to be 96.6% L(+)-lactic acid. This strain may be suitable for use in the industrial production of lactic acid [20]. A study using 10 strains of Lactobacillus showed that organic acid production was considerably influenced by media used. Three different media including skimmed milk, de Mann Rogosa Sharpe (MRS) broth and Jerusalem artichoke were used. The highest acidity was obtained in MRS broth and the weakest acidification was found in skimmed milk. Lactobacillus casei Shirota produced the highest and Lactobacillus rhamnosus VT1 the lowest amount of substances being estimated as titratable acidity. The study pointed at the dissimilarity of organic acid production of Lactobacillus strains [21]. Two Lactococcus lactis strains were studied for their ability to repress the growth of bacterial pathogens. The study showed that L. lactis subsp. lactis biovar. diacetylactis strongly inhibits the pathogenic E. coli and Salmonella enteritidis strains tested. The main inhibitory effect seemed to be associated with fast acid production which resulted in rapid pH reduction. Given these good attributes of L. lactis subsp. lactis biovar. diacetylactis, it can be recommended for use as a starter culture, preferably as a freeze-dried culture, to prepare cultured milk similar to naturally fermented milk [22].

In a yoghurt preservation period experiment and mould-proof accelerated testing at 4°C, addition of 2% (v/v) Lactobacillus casei AST18 in yogurt completely inhibited the growth of Penicillium sp., which was used as indicator fungi. L. casei AST18 produced lactic acid and cyclo-(Leu-Pro) as antifungal compounds. The addition of L. casei AST18 improved the quantity of Lactobacillus, but the number of Streptococcus lactis in 2% AST18-added yoghurt decreased by 1.0 Log (cfu/mL) compared with that in the blank group. Direct use of antifungal strains as protective cultures presents important application value to the food industry [23]. LAB from raw milk of cow, goat and ewe are detected to have antifungal activity against 4 spoilage fungi, Penicillium expansum, Mucor plumbeus, Kluyveromyces lactis and Pichia anomala. Raw milk of cow and goat can be considered as reservoir of antifungal LAB. The most active colonies with antifungal activity belonged to Lactobacillus spp. It is suggested that their apparent specialization may be linked to organic acids and/or ethanol produced, and the tested fungi are sensitive to those molecules. Acetic acid produced by heterofermentative Lactobacilli certainly played a role in antifungal activity [11].

Bacteriocins Production

Bacteriocins are substances of protein structure, either proteins or polypeptides, that possessing antimicrobial activities and produced during the primary phase of bacterial growth. Generally bacteriocins only active against closely related bacterial species. Most of LAB bacteriocins are small (<10 kDa) cationic, heat-stable, amphiphilic and membrane permeabilizing peptides. Bacteriocins of LAB can be divided into 3 classes, 1) lantibiotics, small (<5 kDa) heat stable of peptide substances that contain the characteristic polycyclic thioether amino acids lanthionine or methyllanthionine, as well as the unsaturated amino acids dehydroalanine and 2-aminoisobutyric acid; 2) non-lantibiotics, small (<10 kDa) relatively heat stable, non-lanthionine containing membrane active peptides; and 3) bacteriocins, heat labile proteins which are in general of large molecular weight (>30 kDa) [19]. Bacteriocins can be used as partially purified or purified concentrates and supplemented to food products, although he application of bacteriocins require specific approval as food preservatives.

Among the bacteriocins produced by LAB, nisin produced by Lactococcus lactis spp., is the only bacteriocin that has been officially employed in the food industry and its use has been approved worldwide [19,24, 25]. Bacteriocins produced by wild L. lactis strains isolated from traditional starter free-cheese made from raw milk were reported as nisin A, nisin Z and lactoccocin 972 [26]. L. lactis W8 produced nisin concomitantly when used to produce dahi, a traditional Indian fermented milk. Dahi prepared using L. lactis W8 showed inhibitory against L. monocytogenes ATCC 19111, Salmonella typhimurium ATCC 23565, Enterobacter aerogenes ATCC 13048 and Vibrio cholerae, however there was no inhibitory activity when cell-free supernatant of heat-treated dahi was used. L. lactis W8 appeared as a potent starter culture for production of fermented milk products of safe quality. Milk fermented with L. lactis W8 can be used as a rich source of nisin for commercial purposes [27]. Pediocin PA-1 produced by Pediococcus acidilactici is an equally promising biopreservative in foods as nisin. However, its indusrial scale production has not been taken up yet due to lack of a comparable scale of production. To improve pediocin production heterologous systems have been studied which have used a variety of promoters for enhanced expression, secretory proteins for fusion and peptide tags to facilitate purification. Pediocin is also an attractive antimicrobial agent against many pathogenic bacteria and hence has pharmaceutic application [28].

Flavor Formation

There are variety of fermented milk products available in the market from different parts of the world. Variation may due to the different technology applied and strains of LAB used. Cheese is among fermented milk products with high variety and may be classified based on different criteria that include the contribution of LAB strains in the ripening process. Starter cultures of LAB are responsible for the formation of cheese flavor. Several LAB are widely used and their role can be divided into starters, and non-starters, including adjunct, cultures. Main role of starter cultures is to produce acid during manufacture and also contribute to the ripening process. Non-starter cultures do not responsible to the production of acid, but they contribute more during ripening process. Flavor formation and the characteristics flavor of individual cheese varieties are develop during ripening process by both starter and non-starter LAB. All LAB are active sinergistically to produce specific flavor of the cheese products (Table 1). Several steps in cheese flavor development by LAB including metabolism of lactose, lactate and citrate; lypolysis that liberate free fatty acids and proteolysis where degradation of casein followed by amino acid catabolism occured [30]. Effort to improve the quality of cheese by producing cheese with specific flavor has been of increasing interest and researches focusing on the use of potential starters derived new and powerful technology have been widely carried out [18,31-33].

The most common LAB cultures used in yoghurt manufacture is Streptococcus thermophilus and Lactobacillus bulgaricus. They, in association and synegistically, produce volatile metabolites that determine the flavour of yoghurt. The mutual benefit between them occured by releasing the amino acids from the milk as well as organic acids and therefore they produced more lactic acid and aromatic compounds [34]. Flavour of yoghurt are supported by various compounds, in which lactic acid represents as the major contributor, and other aroma compounds (Figure 1).

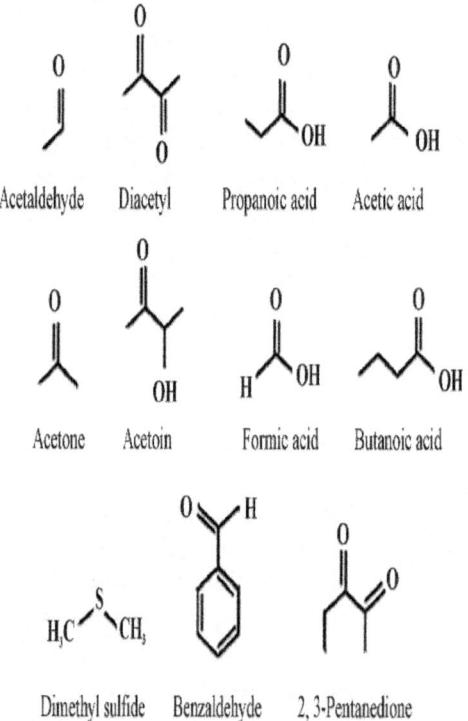

Figure 1: Major aroma compounds in yoghurt [34].

Table 1: Major aroma compounds in cheese derived from amino acids [29]

Amino acids	Aldehydes	Alcohols	Carboxylic acids	Thyol/divers
Leucine	3-Methylbutanal or Isovaleraldehyde	3-Methylbutanol	3-Methylbutanoic acid or isovaleric acid	
Isoleucine	2-Methylbutanal	2-Methylbutanol	2-Methylbutanoic acid	
Valine	2-Methylpropanal or isobutyraldehyde	2-Methylpropanol	2-Methylpropanoic acid or isobutyric acid	
Phenylalanine	Phenylacetaldehyde, benzaldehyde (-2C)	Phenylethanol	Phenylacetic acid	
Tyrosine	OH-Phenylacetaldehyde, OH-benzaldehyde (-2C)	OH-Phenylethanol	OH-Phenylacetic acid	p-cresol, phenol
Tryptophane	Indol-3-acetaldehyde, indol-3-aldehyde	Tryptophol	Indol-3-acetic acid	Skatole, indole
Methionine	3-Methylthiopropanal, or methional	3-Methylthiopropanol	3-Methylthiopropionic acid	Methanethiol

Starter Cultures of Lactic Acid Bacteria for Cheese Production

Starter cultures of LAB can be either mesophilic from the genera of Lactococcus and Leuconostoc or thermophilic from the genera of Streptococcus and Lactobacillus [35]. Among species, Lactococcus lactis [36,37], Streptococcus thermophilus [30,38] and Lactobacillus helveticus [36-42] are intensively studied. L. helveticus is specialized of milk species and belong to the member of dairy niche species [43,44] Several cheese products are based on L. helveticus as starter (Table 2). It is also known that L. helveticus have significant role in production of specific flavor compounds in Italian cheese types [46,47] and debittering of cheese [48,49].

Non Starter Cultures of Lactic Acid Bacteria

Nonstarter LAB play important role in cheese ripening. Nonstarter LAB released enzymes that participate in the basic role of the transformation of curd in cheese. Since the population of nonstarter LAB is uncontrol, it is suggested that selection of strains to be developed to maintain a certain cheese flavour [50]. In traditional cheeses, their flavor intensity also from the non-starter LAB [51].

Texture Development

Youghurt is a special fermented milk product which have a texture of soft and thicker compare to that of its raw material, the milk. The mild acid with pleasant fresh taste supported characteristics of yoghurt as an extraordinary fermented milk product. The texture of yoghurt is supported by production of exopolysaccharides (EPS), as viscosifying agent, produced by LAB. It is also suggested that is contributed by coagulation, as a result of neutralisation of the negative charges on the milk proteins, as another effect of acid produced by LAB. EPS is produced by some of LAB, depending on the strain [52].

Health-promoting Property of Lactic Acid Bacteria

The high demand of fermented milk products is due to the health property generating from consumption of fermented milk products. Bioactive peptides produced from hydrolysis of casein in milk generated by L. helveticus have been reviewed and showed effect of antihypertensive, immunomodulatory activity, anti-cancer and calcium binding ability.

Table 2. The use of L. helveticus as cheese starter [45]

No	Cheese product	Type of starter
1.	Asiago	Natural whey and milk culture
2.	Canestrato Pugliese	Natural whey culture
3.	Emmental	Commercial culture
4.	Grana Padano	Natural whey culture
5.	Gruyère	Commercial culture
6.	Montasio	Natural whey culture
7.	Mozzarella	Natural culture and commercial culture
8.	Parmigiano Reggiano	Natural culture
9.	Pecorino Romano	Natural culture in scotta
10.	Pecorino Sardo	Natural whey and milk culture
11.	Pecorino Siciliano	Natural whey culture
12.	Provolone Italiano	Natural whey culture
13.	Sbrinz	Commercial culture
14.	Taleggio	Commercial culture

L. helveticus is known as one of LAB which has efficient proteolytic system [7]. Fermented milk products are reported to contribute to human health through several mechanisms [2] LAB are in the first rank of listed organisms as species used in probiotic preparation. LAB, in general, showed to possess most of the requirements for strains to be called as probiotics. [53-57]. Fermented milk products with their beneficial effect are presented in Table 3.

CONCLUDING REMARKS

LAB are widely applied to several milk products due to their specific properties.

Table 3: Functional benefit of fermented milk products using lactic acid bacteria.

Product name	Origin	Culture	Functional benefit	Reference
Probiotic yogurt	Ontario, Canada	*L. rhamnosus* CAN-1	Nutrition and immune function for people living with HIV	[58]
Mix ewe's and goat's milk yoghurt	Antakya-Hatay, Turkey	*S. thermophilus* and *L. delbrueckii* subsp. *bulgaricus* (codes: CH-1 and YF-333)	High short chain free fatty acids	[59]
Ayran (yoghurt from goat milk)	Turkey	*L. plantarum, L. brevis L. paracasei* subsp. *paracasei, L. casei* subsp. *pseudoplantarum*	High exopolysacharide	[60]
Gioddu, traditional fermented sheep or goat milk	Sardinian, Italy	*S. thermophilus, L. lactis* subsp. *lactis L. delbrueckii* subsp. *bulgaricus, L. casei* subsp. *casei, L. mesenteroides* subsp. *mesenteroides*	Probiotic	[61]
Tarag	Mongolia	*L. helveticus, L. lactis* subsp. *lactis, L. casei*	Probiotic	[62]
Fermented milk	Japan	*L. casei* strain Shirota	Maintenance treatment for myelopathy/tropical spastic paraparesis (HAM/TSP) patients	[63]
Koumiss from mare's milk	Italy	*L. delbrueckii* subsp. *bulgaricus S. thermophilus*	Antiallergic	[64]
Lben	Marocco	Spontaneously/not identified	Low fat and high calcium traditional product	[65]
Functional fermented milk	Italy	*L. lactis* DIBCA2, *L. plantarum* PU11	Enriched of Angiotensin-I Converting Enzyme (ACE)-inhibitory peptides and G-amino butyric acid (GABA)	[66]
Kumis	West Colombia	*E. faecalis, E. faecium*	ACE Inhibitor	[67]
Ewe milk, traditional yoghurt	Iran	*L. brevis*	cholesterol reduction	[68]
Maasai	Kenya	*L. plantarum, L. fermentum, L. actdophillus, L. paracasei*	Diarrhoea and constipation	[69]
Suusac	Kenya	*L. curvatus, L. plantarum, L. salivarius, L. raffinolactis Leuconostoc mesenteroides* subsp. *mesenteroides.*		[70]

Both starter and non-starter are promising cultures to be explored in fermented milk manufacture. Considering the important status as GRAS microorganisms, LAB can be used widely in the developing of new fermented milk product.

REFERENCES

1. P. S. Panesar, "Fermented Dairy Products: Starter Cultures and Potential Nutritional Benefits," Food and Nutrition Sciences, Vol. 2, No. 1, 2011, pp. 47-51.http://dx.doi.org/10.4236/fns.2011.21006
2. R. Sharma, B. S. Sanodiya, D. Bagrodia, M. Pandey, A. Sharma and P. S. Bisen, "Efficacy and Potential of Lactic Acid Bacteria Modulating Human Health," International Journal of Pharma and Bio Sciences, Vol.

3, No. 4, 2012, pp. 935-948.

3.   J. Steele, J. Broadbent and J. Kok, "Perspective on the Contribution of Lactic Acid Bacteria to Cheese Flavor Development," Current Opinion in Biotechnology, Vol. 24, No. 2, 2013, pp. 135-141. http://dx.doi.org/10.1016/j.copbio.2012.12.001

4.   S.-N. Liu, Y. Han and Z.-J. Zhou, "Lactic Acid Bacteria in Traditional Fermented Chinese Foods," Food Research International, Vol. 44, No. 3, 2011, pp. 643-651.http://dx.doi.org/10.1016/j.foodres.2010.12.034

5.   N. P. Shah, "Functional Cultures and Health Benefits," International Dairy Journal, Vol. 17, No. 11, 2007, pp. 1262-1277. http://dx.doi.org/10.1016/j.idairyj.2007.01.014

6.   A. A. Ali, "Beneficial Role of Lactic Acid Bacteria in Food Preservation and Human Health," Research Journal of Microbiology, Vol. 5, No. 12, 2010, pp. 1213-1221.http://dx.doi.org/10.3923/jm.2010.1213.1221

7.   M. W. Griffiths and A. M. Tellez, "Lactobacillus helveticus: The Proteolytic System," Frontiers in Microbiology, Vol. 4, 2013, pp. 1-9. http://dx.doi.org/10.3389/fmicb.2013.00030

8.   N. F. Olson, "The Impact of Lactic Acid Bacteria on Cheese Flavor," FEMS Microbiology Reviews, Vol. 87, No. 1-2, 1990, pp. 131-148. http://dx.doi.org/10.1111/j.1574-6968.1990.tb04884.x

9.   G. Urbach, "Contribution of Lactic Acid Bacteria to Flavour Compound Formation in Dairy Products," International Dairy Journal, Vol. 5, No. 8, 1995, pp. 877-903.http://dx.doi.org/10.1016/0958-6946(95)00037-2

10.   P. A. Maragkoudakis, C. Miaris, P. Rojez, N. Manalis, F. Magkanari, G. Kalantzopoulos and E. Tsakalidou, "Production of Traditional Greek Yoghurt Using Lactobacillus Strains with Probiotic Potential as Starter Adjuncts," International Dairy Journal, Vol. 16, No. 1, 2006, pp. 52-60. http://dx.doi.org/10.1016/j.idairyj.2004.12.013

11.   E. Delavenne, J. Mounier, F. Déniel, G. Barbier and G. Le Blay, "Biodiversity of Antifungal Lactic Acid Bacteria Isolated from Raw Milk Samples from Cow, Ewe and Goat over One-Year Period," International Journal of Food Microbiology, Vol. 155, No. 3, 2012, pp. 185-190. http://dx.doi.org/10.1016/j.ijfoodmicro.2012.02.003

12.   J. T. M. Wouters, E. H. E. Ayad, J. Hugenholtz and G. Smit, "Microbes

from Raw Milk for Fermented Dairy Products," International Dairy Journal, Vol. 12, No. 2-3, 2002, pp. 91-109. http://dx.doi.org/10.1016/S0958-6946(01)00151-0

13. Z. Mennane, K. Khedid, A. Zinedine, M. Lagzouli, M. Ouhssine and M. Elyachioui, "Microbial Characteristics of Klila and Jben Traditionnal Moroccan Cheese from Raw Cow's Milk," World Journal of Dairy & Food Sciences, Vol. 2, No. 1, 2007, pp. 23-27.

14. C. Chaves-López, A. Serio, M. Martuscelli, A. Paparella, E. Osorio-Cadavid and G. Suzzi, "Microbiological Characteristics of Kumis, a Traditional Fermented Colombian Milk, with Particular Emphasis on Enterococci Population," Food Microbiology, Vol. 28, No. 5, 2011, pp. 1041-1047. http://dx.doi.org/10.1016/j.fm.2011.02.006

15. M. Ouadghiri, M. Vancanneyt, P. Vandamme, S. Naser, D. Gevers, K. Lefebvre, J. Swings and M. Amar, "Identification of Lactic Acid Bacteria in Moroccan Raw Milk and Traditionally Fermented Skimmed Milk 'Lben'," Journal of Applied Microbiology, Vol. 106, No. 2, 2008, pp. 486-495. http://dx.doi.org/10.1111/j.1365-2672.2008.04016.x

16. Z. Sun, W. Liu, W. Gao, M. Yang, J. Zhang, L. Wu, J. Wang, B. Menghe, T. Sun and H. Zhang, "Identification and Characterization of the Dominant Lactic Acid Bacteria from Kurut: The Naturally Fermented Yak Milk in Qinghai, China," Journal of General and Appllied Microbiology, Vol. 56, No. 1, 2010, pp. 1-10.

17. A. Mäyrä-Mäkinen and M. Bigret, "Industrial Use and Production of Lactic Acid Bacteria," In: S. Salminen, A. von Wright and A. Ouwehand, Eds., Lactic Acid Bacteria Microbiological and Functional Aspects, Marcel Dekker, Inc., New York, 2004, pp. 175-198.

18. J. M. Kongo, "Lactic Acid Bacteria as Starter-Cultures for Cheese Processing: Past, Present and Future Developments," Chapter 1, 2013. http://creativecommons.org/licenses/by/3.0

19. M. P. Zacharof and R. W. Lovitt, "Bacteriocins Produced by Lactic Acid Bacteria," APCBEE Procedia, Vol. 2, 2012, pp. 50-56. http://dx.doi.org/10.1016/j.apcbee.2012.06.010

20. S.-K. Moon, Y.-J. Wee and G.-W. Choi, "A Novel Lactic Acid Bacterium for the Production of High Purity LLactic Acid, Lactobacillus paracasei subsp. paracasei CHB2121," Journal of Bioscience and Bioengineering, Vol. 114, No. 2, 2012, pp. 155-159. http://dx.doi.org/10.1016/j.jbiosc.2012.03.016

21. Z. Zalán, J. Hudáček, J. Štětina, J. Chumchalová, A. Halász, "Production

of Organic Acids by Lactobacillus Strains in Three Different Media," European Food Research Technology, Vol. 230, No. 3, 2010, pp. 395-404. http://dx.doi.org/10.1007/s00217-009-1179-9

22. J. Mufandaedza, B. C. Viljoen, S. B. Feresu and T. H. Gadaga, "Antimicrobial Properties of Lactic Acid Bacteria and Yeast-LAB Cultures Isolated from Traditional Fermented Milk against Pathogenic Escherichia coli and Salmonella enteritidis Strains," International Journal of Food Microbiology, Vol. 108, No. 1, 2006, pp. 147-152.http://dx.doi.org/10.1016/j.ijfoodmicro.2005.11.005

23. H. Li, L. Liu, S. Zhang, H. Uluko, W. Cui and J. Lv, "Potential Use of Lactobacillus casei AST18 as a Bioprotective Culture in Yogurt," Food Control, Vol. 34, No. 2, 2013, pp. 675-680. http://dx.doi.org/10.1016/j.foodcont.2013.06.023

24. E. M. Balciunas, F. A. C. Martinez, S. D. Todorov, B. D. G. de Melo Franco, A. Converti and R. P. de Souza Oliveira, "Novel Biotechnological Applications of Bacteriocins: A Review," Food Control, Vol. 32, No. 1, 2013, pp. 134-142.http://dx.doi.org/10.1016/j.foodcont.2012.11.025

25. S. Mitra, P. K. Chakrabartty and S. R. Biswas, "Potential Production and Preservation of Dahi by Lactococcus lactis W8 a Nisin-Producing Strain," LWT-Food Science and Technology, Vol. 43, No. 2, 2010, pp. 337-342.http://dx.doi.org/10.1016/j.lwt.2009.08.013

26. Á. Alegría, S. Delgado, C. Roces, B. López and B. Mayo, "Bacteriocins Produced by Wild Lactococcus lactis Strains Isolated from Traditional Starter-Free Cheeses Made of Raw Milk," International Journal of Food Microbiology, Vol. 143, No. 1, 2010, pp. 61-66.http://dx.doi.org/10.1016/j.ijfoodmicro.2010.07.029

27. S. Mitra, P. K. Chakrabartty and S. R. Biswas, "Potential Production and Preservation of Dahi by Lactococcus lactis W8, a Nisin-Producing Strain," LWT-Food Science and Technology, Vol. 43, No. 2, 2010, pp. 337-342.http://dx.doi.org/10.1016/j.lwt.2009.08.013

28. B. Kumar, P. P. Balgir, B. Kaur and N.Garg, "Cloning and Expression of Bacteriocins of Pediococcus spp.: A Review," Archives of Clinical Microbiology, Vol. 2, No. 3, 2011, p. 4.

29. M. Yvon and L. Rijnen, "Cheese Flavour Formation by Amino Acid Catabolism," International Dairy Journal, Vol. 11, No. 4-7, 2001, pp. 185-201.http://dx.doi.org/10.1016/S0958-6946(01)00049-8

30. P. L. H. McSweeney and M. J. Sousa, "Biochemical Pathways for the Production of Flavour Compounds in Cheese during Ripening: A Review," Lait, Vol. 80, No. 3, 2000, pp. 293-324. http://dx.doi.

org/10.1051/lait:2000127

31. P. Hols, F. Hancy, L. Fontaine, B. Grossiord, D. Prozzi, N. Leblond-Bourget, B. Decaris, A. Bolotin, C. Delorme, S. D. Ehrlich, E. Guédon, V. Monnet, P. Renault and M. Kleerebezem, "New Insights in the Molecular Biology and Physiology of Streptococcus thermophilus Revealed by Comparative Genomics," FEMS Microbiology Review, Vol. 29, No. 3, 2005, pp. 435-463.

32. J. R. Broadbent, H. Cai, R. L. Larsen, J. E. Hughes, D. L. Welker, V. G. De Carvalho, T. A. Tompkins, Y. Ardo, F. Vogensen, A. De Lorentiis, M. Gatti, E. Neviani and J. L. Steele, "Genetic Diversity in Proteolytic Enzymes and Amino Acid Metabolism among Lactobacillus helveticus Strains," Journal of Dairy Science, Vol. 94, No. 9, 2011, pp. 4313-4328. http://dx.doi.org/10.3168/jds.2010-4068

33. J. Steele, J. Broadbent and J. Kok, "Perspective on the Contribution of Lactic Acid Bacteria to Cheese Flavor Development," Current Opinion in Biotechnology, Vol. 24, No. 2, 2013, pp. 135-141. http://dx.doi.org/10.1016/j.copbio.2012.12.001

34. W. Routray and H. N. Mishra, "Scientific and Technical Aspects of Yogurt Aroma and Taste: A Review," Comprehensive Reviews in Food Science and Food Safety, Vol. 10, No. 4, 2011, pp. 208-220. http://dx.doi.org/10.1111/j.1541-4337.2011.00151.x

35. P. F. Fox, P. L. H. McSweeney, T. M. Cogan and T. P. Guinee, "Cheese: Chemistry, Physics and Microbiology," Elsevier, 2004.

36. B. Dias and B. Weimer, "Conversion of Methionine to Thiols by Lactococci, Lactobacilli, and Brevibacteria," Applied and Environmental Microbiology, Vol. 64, No. 9, 1998, pp. 3320-3326.

37. J. A. Hannon, K. N. Kilcawley, M. G. Wilkinson, C. M. Delahunty and T. P. Beresford, "Flavor Precursor Development in Cheddar Cheese Due to Lactococcal Starters and the Presence and Lysis of Lactobacillus helveticus," International Dairy Journal, Vol. 17, No. 4, 2007, pp. 316-327.

38. S. Helinck, D. L. Bars, D. Moreau and M. Yvon, "Ability of Thermophilic Lactic Acid Bacteria to Produce Aroma Compounds from Amino Acids," Applied and Environmental Microbiology, Vol. 70, No. 7, 2004, pp. 3855-3681.http://dx.doi.org/10.1128/AEM.70.7.3855-3861.2004

39. N. Klein, M. B. Maillard, A. Thierry and S. Lortal, "Conversion of Amino Acids into Aroma Compounds by CellFree Extracts of Lactobacillus helveticus," Journal of Applied Microbiology, Vol. 91, No. 3, 2001, pp. 404-411. http://dx.doi.org/10.1046/j.1365-2672.2001.01391.x

40. O. Kenny, R. J. FitzGerald, G. O'Cuinn, T. Beresford and K. Jordan, "Autolysis of Selected Lactobacillus helveticus Adjunct Strains during Cheddar Cheese Ripening," International Dairy Journal, Vol. 16, No. 7, 2006, pp. 797-804.http://dx.doi.org/10.1016/j.idairyj.2005.07.008

41. W. J. Lee, D. S. Banavara, J. E. Hughes, J. K. Christiansen, J. L. Steele, J. R. Broadbent and S. A. Rankin, "Role of Cystathionine ß-Lyase in Catabolism of Amino Acids to Sulfur Volatiles by Genetic Variants of Lactobacillus helveticus CNRZ 32," Applied and Environmental Microbiology, Vol. 73, No. 9, 2007, pp. 3034-3039.http://dx.doi.org/10.1128/AEM.02290-06

42. J. K. Christiansen, J. E. Hughes, D. L. Welker, B. T. Rodríguez, J. L. Steele and J. R. Broadbent, "Phenotypic and Genotypic Analysis of Amino Acid Auxotrophy in Lactobacillus helveticus CNRZ 32," Applied and Environmental Microbiology, Vol. 74, No. 2, 2008, pp. 416-423. http://dx.doi.org/10.1128/AEM.01174-07

43. M. Callanan, P. Kaleta, J. O'Callaghan, O. O'Sullivan, K. Jordan, O. McAuliffe, A. Sangrador-Vegas, L. Slattery, G. F. Fitgerald, T. Beresford and R. P. Ross, "Genome Sequence of Lactobacillus helveticus, an Organism Distinguished by Selective Gene Loss and Insertion Sequence Element Expansion," Journal of Bacteriology, Vol. 190, No. 2, 2008, pp. 727-735. http://dx.doi.org/10.1128/JB.01295-07

44. L. Slaterry, J. O. Callaghan, G. F. Fitgerald, T. Beresford and R. P. Ross, "Invited Review: Lactobacillus helveticusA Thermophilic Dairy Starter Related to Gut Bacteria," Journal of Dairy Science, Vol. 93, No. 10, 2010, pp. 4435- 4454.http://dx.doi.org/10.3168/jds.2010-3327

45. M. Gobbetti, M. De Angelis, R. Di Cagno and C. G. Rizzello, "The Relative Contributions of Starter Cultures and Non-Starter Bacteria to the Flavour of Cheese," In: B. C. Weimer, Ed., Improving the Flavor of Cheese, CRC Press Boca Raton, 2007, pp. 121-156.

46. M. Gatti, C. Lazzi, L. Rossetti, G. Mucchetti and E. Neviani, "Biodiversity in Lactobacillus helveticus Strains Present in Natural Whey Starter Used for Parmigiano Reggiano Cheese," Journal of Applied Microbiology, Vol. 95, No. 3, 2003, pp. 463-470.http://dx.doi.org/10.1046/j.1365-2672.2003.01997.x

47. L. Rossetti, M. E. Fornasari, M. Gatti, C. Lazzi, E. Neviani and G. Giraffa, "Grana Padano Cheese Whey Starters: Microbial Composition and Strain Distribution," International Journal of Food Microbiology, Vol. 127, No. 1-2, 2008, pp. 168-171.http://dx.doi.org/10.1016/j.ijfoodmicro.2008.06.005

48. L. Fernández, T. Bhowmik and J. Steele, "Characterization of the Lactobacillus helveticus CNRZ32 pepC Gene," Applied and Environmental Microbiology, Vol. 60, No. 1, 1994, pp. 333-336.

49. E. Soeryapranata, J. R. Powers and G. Ünlü, "Cloning and Characterization of Debittering Peptidases, PepE, PepO, PepO2, PepO3 and PepN, of Lactobacillus helveticus WSU 19," International Dairy Journal, Vol. 17, No. 9, 2007, pp. 1096-1106.

50. L. Settani and G. Moschetti., "Non-Starter Lactic Acid Bacteria Used to Improve Cheese Quality and Provide Health Benefits," Food Microbiology, Vol. 27, No. 6, 2010, pp. 691-697. http://dx.doi.org/10.1016/j.fm.2010.05.023

51. T. P. Beresford, N. A. Fitzsimons, N. L. Brennan and T. M. Cogan, "Recent Advances in Cheese Microbiology," International Dairy Journal, Vol. 11, No. 4-7, 2001, pp. 259-274.http://dx.doi.org/10.1016/S0958-6946(01)00056-5

52. T. X. Yang, K. Y. Wu, F. Wang, X. L. Liang, Q. S. Liu, G. Li and Q. Y. Li, "Effect of Exopolysaccharides from Lactic Acid Bacteria on the Texture and Microstructure of Buffalo Yoghurt," International Dairy Journal, Vol. 34, No. 2, 2014, pp. 252-256.http://dx.doi.org/10.1016/j.idairyj.2013.08.007

53. J. Fioramonti, V. Theodorou and L. Bueno, "Probiotics: What Are They? What Are Their Effects on Gut Physiology?" Best Practice & Research Clinical Gastroenterology, Vol. 17, No. 5, 2003, pp. 711-724. http://dx.doi.org/10.1016/S1521-6918(03)00075-1

54. I. P. Kaur, K. Chopra and A. Saini, "Probiotics: Potential Pharmaceutical Applications," European Journal of Pharmaceutical Sciences, Vol. 15, No. 1, 2002, pp. 1-9.http://dx.doi.org/10.1016/S0928-0987(01)00209-3

55. T. Vasiljevic and N. P. Shah, "Probiotics—From Metchnikoff to Bioactives," International Dairy Journal, Vol. 18, No. 7, 2008, pp. 714-728.http://dx.doi.org/10.1016/j.idairyj.2008.03.004

56. G. Giraffa, N. Chanishvili and Y. Widyastuti, "Importance of Lactobacilli in Food and Feed Biotechnology," Research in Microbiology, Vol. 161, No. 6, 2010, pp. 480- 487.http://dx.doi.org/10.1016/j.resmic.2010.03.001

57. K. Singh, B. Kallali, A. Kumar and V. Thaker, "Probiotics: A Review," Asian Pacific Journal of Tropical Biomedicine, Vol. 1, No. 2, 2011, pp. S287-S290.

58. J. Hemsworth, S. Hekmat and G. Reid., "The Development of Micronutrient Supplemented Probiotic Yogurt for People Living with HIV: Laboratory Testing and Sensory Evaluation," Innovative Food

Science and Emerging Technologies, Vol. 12, No. 1, 2011, pp. 79-84. http://dx.doi.org/10.1016/j.ifset.2010.11.004

59. Z. Guler and A. C. Gürsoy-Balcı, "Evaluation of Volatile Compounds and Free Fatty Acids in Set Types Yogurts Made of Ewes', Goats' Milk and Their Mixture Using Two Different Commercial Starter Cultures during Refrigerated Storage," Food Chemistry, Vol. 127, No. 3, 2011, pp. 1065-1071. http://dx.doi.org/10.1016/j.foodchem.2011.01.090

60. [61] F. Altay, F. Karbancıoglu-Güler, C. Daskaya-Dikmen and D. Heperkan, "A Review on Traditional Turkish Fermented Non-Alcoholic Beverages: Microbiota, Fermentation Process and Quality Characteristics," International Journal of Food Microbiology, Vol. 167, No. 1, 2013, pp. 44-56. http://dx.doi.org/10.1016/j.ijfoodmicro.2013.06.016

61. [62] S. Ortu, G. E. Felis, M. Marzotto, A. Deriu, P. Molicotti, L. A. Sechi, F. Dellaglio and S. Zanetti, "Identification and Functional Characterization of Lactobacillus Strains Isolated from Milk and Gioddu, a Traditional Sardinian Fermented Milk," International Dairy Journal, Vol. 17, No. 11, 2007, pp. 1312-1320. http://dx.doi.org/10.1016/j.idairyj.2007.02.008

62. [63] W. Liu, Q. Bao, Jirimutu, M. Qing, Siriguleng, X. Chen, T. Sun, M. Li, J. C. Zhang, M. Bilige, T. S. Sun and H. P. Zhang, "Isolation and Identification of Lactic Acid Bacteria from Tarag in Eastern Inner Mongolia of China by 16S rRNA Sequences and DGGE Analysis," Microbiological Research, Vol. 167, No. 2, 2012, pp. 110-115. http://dx.doi.org/10.1016/j.micres.2011.05.001

63. [64] T. Matsuzaki, M. Saito, K. Usuku, H. Nose, S. Izumo, K. Arimura and M. Osame, "A Prospective Uncontrolled Trial of Fermented Milk Drink Containing Viable Lactobacillus casei Strain Shirota in the Treatment of HTLV-1 Associated Myelopathy/Tropical Spastic Paraparesis," Journal of the Neurological Sciences, Vol. 237, No. 1, 2005, pp. 75-81. http://dx.doi.org/10.1016/j.jns.2005.05.011

64. [65] R. D. Cagno, A. Tamborrino, G. Gallo, C. Leone, M. De Angelis, M. Faccia, P. Amiranteb and M. Gobbettia, "Uses of Mares' Milk in Manufacture of Fermented Milks," International Dairy Journal, Vol. 14, No. 9, 2004, pp. 767-775. http://dx.doi.org/10.1016/j.idairyj.2004.02.005

65. [66] M. Ouadghiri, M. Vancanneyt, P. Vandamme, S. Naser, D. Gevers, K. Lefebvre, J. Swings and M. Amar, "Identification of Lactic Acid Bacteria in Moroccan Raw Milk and Traditionally Fermented Skimmed Milk 'Lben'," Journal of Applied Microbiology, Vol. 106, No. 2, 2008, pp. 486-495. http://dx.doi.org/10.1111/j.1365-2672.2008.04016.x

66. [67] F. Nejati, C. G. Rizzello, R. Di Cagno, M. Sheikh-Zeinoddin, A.

Diviccaro, F. Minervini and M. Gobbetti, "Manufacture of a Functional Fermented Milk Enriched of Angiotensin-I Converting Enzyme (ACE)-Inhibitory Peptides and g-Amino Butyric Acid (GABA)," LWT-Food Science and Technology, Vol. 51, No. 1, 2013, pp. 183-189.http://dx.doi.org/10.1016/j.lwt.2012.09.017

67. [68] C. Chaves-López, A. Serio, M. Martuscelli, A. Paparella, E. Osorio-Cadavid and G. Suzzi, "Microbiological Characteristics of Kumis, a Traditional Fermented Colombian Milk, with Particular Emphasis on Enterococci Population," Food Microbiology, Vol. 28, No. 5, 2011, pp. 1041- 1047. http://dx.doi.org/10.1016/j.fm.2011.02.006

68. [69] M. Iranmesh, H. Ezzatpanah and N. Mojgani, "Antibacterial Activity and Cholesterol Assimilation of Lactic Acid Bacteria Isolated from Traditional Iranian Dairy Products," LWT-Food Science and Technology, 2013, pp. 1-5.

69. [70] J. M. Mathara, U. Schillinger, P. M. Kutima, S. K. Mbugua and W. H. Holzapfel, "Isolation, Identification and Characterisation of the Dominant Microorganisms of Kule Naoto: The Maasai Traditional Fermented Milk in Kenya," International Journal of Food Microbiology, Vol. 94, No. 3, 2004, pp. 269-278.http://dx.doi.org/10.1016/j.ijfoodmicro.2004.01.008

70. [71] T. A. Lore, S. K. Mbugua and J. Wangoh, "Enumeration and Identification of Icroflora in Suusac, a Kenyan Traditional Fermented Camel Milk Product," LWT-Food Science and Technology, Vol. 38, No. 2, 2005, pp. 125- 130.http://dx.doi.org/10.1016/j.lwt.2004.05.008

Chapter 8

HYGIENISATION AND NUTRIENT CONSERVATION OF SEWAGE SLUDGE OR CATTLE MANURE BY LACTIC ACID FERMENTATION

Hendrik A. Scheinemann[1,4], Katja Dittmar[2] , Frank S. Stöckel[2] , Hermann Müller[3] , Monika E. Krüger[1]

[1] Institute of Bacteriology and Mycology, University of Leipzig, Faculty of Veterinary medicine, An den Tierkliniken 29, 04103 Leipzig, Germany,

[2] Institute of Parasitology, University of Leipzig, Faculty of Veterinary medicine, An den Tierkliniken 35, 04103 Leipzig, Germany,

[3] Institute of Virology, University of Leipzig, Faculty of Veterinary medicine, An den Tierkliniken 29, 04103 Leipzig, Germany,

[4] Gesellschaft zur Förderung von Medizin-, Bio- und Umwelttechnologien e. V. Erich-Neuß-Weg 5, 06120 Halle (Saale), Germany

ABSTRACT

Manure from animal farms and sewage sludge contain pathogens and opportunistic organisms in various concentrations depending on the health of the herds and human sources. Other than for the presence of pathogens, these waste substances are excellent nutrient sources and constitute a preferred organic fertilizer. However, because of the pathogens, the risks of infection of animals or humans increase with the indiscriminate use of manure, especially liquid manure or sludge, for agriculture. This potential problem can increase with the global connectedness of animal herds fed imported feed grown on fields fertilized with local manures. This paper describes a simple, easy-to-use, low-tech hygienization method which conserves nutrients and does not require large investments in infrastructure. The proposed method uses the microbiotic shift during mesophilic fermentation of cow manure or sewage sludge during which gram-negative bacteria, enterococci and yeasts were inactivated below the detection limit of $3 \log_{10}$ cfu/g while lactobacilli increased up to a thousand fold. Pathogens like *Salmonella*, *Listeria monocytogenes*, *Staphylococcus aureus, E. coli* EHEC O: 157 and vegetative *Clostridium perfringens* were inactivated within 3 days of fermentation. In addition, ECBO-viruses and eggs of *Ascaris suum* were inactivated within 7 and 56 days, respectively. Compared to the mass lost through composting (15–57%), the loss of mass during fermentation (< 2.45%) is very low and provides strong economic and ecological benefits for this process. This method might be an acceptable hygienization method for developed as well as undeveloped countries, and could play a key role in public and animal health while safely closing the nutrient cycle by reducing the necessity of using energy-inefficient inorganic fertilizer for crop production.

INTRODUCTION

Pathogens and various facultative organisms can be found in high concentrations in the faeces of animals and humans. Used as fertilizers, these materials can contaminate soil [1,2] as well as field crops [3–7]. Contaminated crops can, in turn, infect consumers [8] to complete the cycle of infection [9] as shown in Fig. 1. Interruption of these cycles is one objective of our research in order to facilitate better usage of these organic nutrient resources. Techniques commonly used to decrease the infectious potential of manures vary in effectiveness. Bagge et al. describe findings of low concentrations of enterococci in after digestion in biogas plants and a resettlement of enterococci and coliforms during storage of digested material [10]. Pourcher et al. describe a reduction of coliforms and enterococci during composting within months, but they still maintain a medium residual concentration [11]. Besides composting, with its huge mass loss [12] and greenhouse gas production [13] during the self-heating phase, or pasteurization before methanogene or hydrogen digestion, no other methods are widely used in Germany.

Lactic acid fermentation has been used since the Stone and Bronze Age [14,15] to make dough and food, as well as to conserve fodder [16]. The developing *Lactobacillaceae* flora prevents oxidation and the development of pathogens within the fermented goods [17] by producing volatile fatty acids (VFA). The VFA can be bactericidal [18–21], depending on pH [22]. The lactic acid faeces fermentation used in this study is modelled after the suppositional conception of the native Amazonian [23] inhabitants who produced an explicitly good soil, the so-called *terra preta do indio*, thousands of years ago [24]. Moreover, they lived in large settlements [25] that imply the obvious need for an adequate hygienisation and nutrient recycling-system which could work in the humid tropics. Even today, some German farmers remember an old country saying of their grandparents which strongly suggests that earlier generations profited from lactic fermentation in their dunghills:

"Schicht in gut, halt ihn feucht, tritt ihn fest, das ist für den Mist das Allerbest."
It means: "Layer it well, keep it wet, tread it well; that is best for dunghills."
Hygienisation via lactic acid fermentation has not been investigated for faeces
previously.

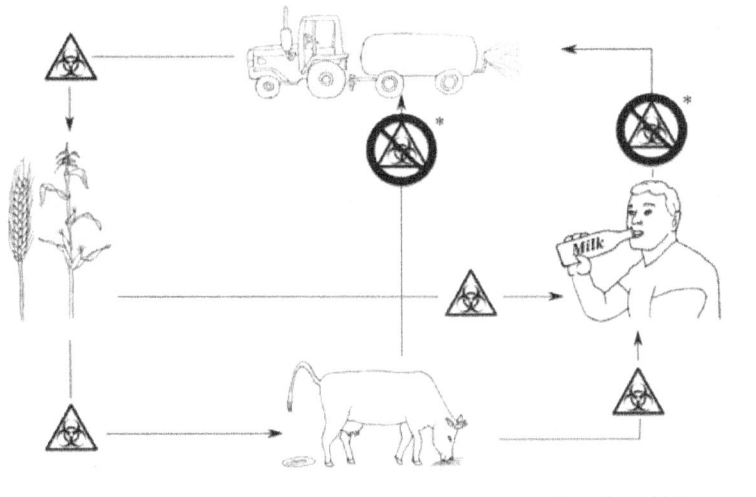

Figure 1: Graphical abstract. doi:10.1371/journal.pone.0118230.g001

Therefore a stand-alone method for the inactivation of pathogens was
tested, independent of the later direct use as a fertilizer or as biogas-plant
substrate.

To test the hygienisation potential of this biological method, bacterial, viral
and parasitic pathogens were fermented in animal manure or sewage sludge
for various times. Furthermore VFA concentrations, pH and mass loss during
fermentation were measured.

MATERIALS AND METHODS

As shown in Fig. 2 sketchily, different faecal materials were fermented in 50
ml tubes (TPP, Switzerland, Trasadingen) and 27 g of each waste material
plus 3 ml of pathogen or control suspension were added to each tube. It was
incubated at 37°C under anaerobic conditions in Anaerocult pots and kits
(Merck Germany, Darmstadt) for 3 to 56 days. After incubation, we analyzed
the bacterial flora, virus titre and rate of embryonation of *Ascaris* eggs.

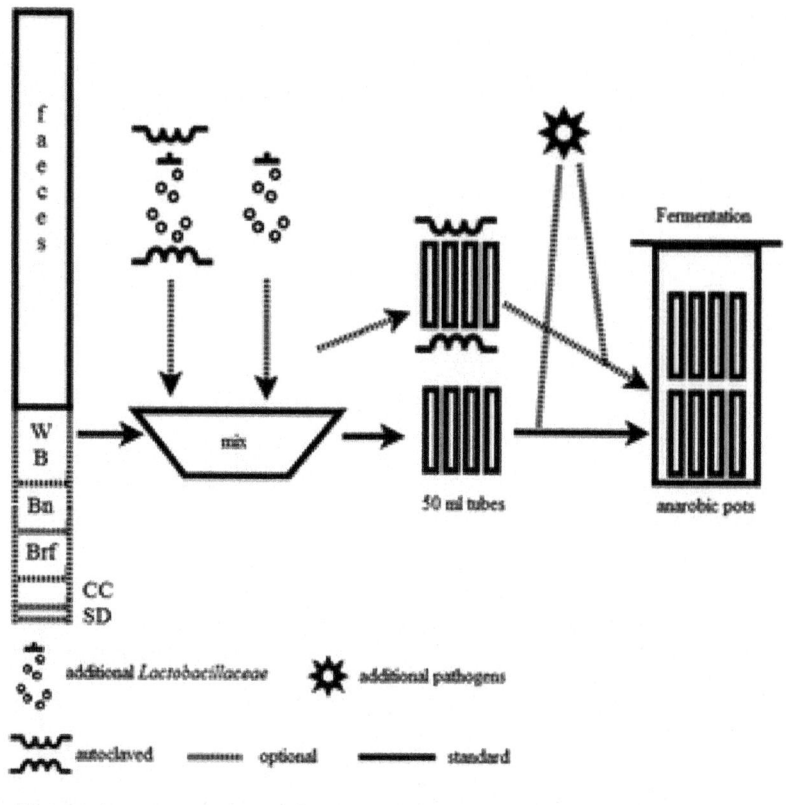

Figure 2: Experimental setup. doi:10.1371/journal.pone.0118230.g002

Up- and Down- scaling

For safety reasons in working with pathogens an experimental size of 27 g was chosen. Moreover satellite experiments without pathogens using 1 litre (37°C) and 60 litre (27°C) containers were performed.

Treatments

Archaeological findings show, that the terra preta made by native Amazonian culture consists basicaly of organic wastes and residue of fire, namely ash, fish, game and human bones, shells, urine and feces. Furthermore, pottery and stone implements covered the ground of the terra preta sites [26] so it seems obvious that they may have transported their organic waste within crocks to those sites. A consequence of storing organic wastes and residues of fire within crocks could have been a self-establishing lactic acid fermentation process.

The design of the faecal material should resemble a typical waste mixture of a veterinary hospital. The treatments consisted mainly of manure from a dairy farm (Versuchsgut Oberholz, Großpösna, Germany) (Table 1 Treatment M). Subsequently, we designed another composition with sewage sludge filter cakes from 10 different sewage plants (AZV Leisnig, Leisnig, Germany) instead of manure to test another application of this method (Table 1 Treatments S).

Table 1: Composition of treatments 1 to 11. doi:10.1371/journal.pone.0118230.t001

Nr.	Treatment	Ingredients	Description
1	M	500 g manure from a dairy farm plus 94 g wheat bran, 38.5 g charcoal, 12 g sawdust, 61 g Bentonite clay, and 61.5 g basalt rock flour	basic manure treatment
2	S	500g different sewage sludge filter cake from 10 different plants 94 g wheat bran, 38.5 g charcoal, 12 g sawdust, 61 g Bentonite lay, and 61.5 g basalt rock flour	basic sludge treatment
3	MEM	384 g of treatment M 125 ml EM = "Effective Microorganism"	Basic manure treatment with additional *Lactobacillaceae* added
4	M(EM)$_a$	384 g of treatment M 125 ml autoclaved EM	control treatment for comparison with MEM to test the influence of additional *Lactobacillaceae*
5	(MEM)$_a$	Treatment MEM autoclaved	control treatment for comparison with MEM to test influence of the whole bacterial flora
6	(M(EM)$_a$)$_a$	Treatment M(EM)$_a$ autoclaved	control treatment to compare with M(EM)$_a$ to test the influence of the whole bacterial flora
7	S$_{Pool}$	Treatment S, but the filter cake was pooled from 5 processing plants	pooled treatment S to reduce the sample number
8	(S$_{Pool}$)$_a$	Treatment S$_{Pool}$ autoclaved	control treatment to compare with S$_{Pool}$, to test the influence of the indigenous fecal flora
9	S$_{Pool}$-WB	Treatment S$_{Pool}$ without (−) wheat bran	sludge treatment S$_{Pool}$ without wheat bran, to test the influence of the bran
10	S$_{Pool}$-WB +SD	Treatment S$_{Pool}$ without (−) wheat bran but with (+) 36 g more sawdust to have comparable dry matter	treatment S$_{Pool}$ with sawdust substituted (v/v) for wheat bran to provide comparable dry matter
11	Fe+WB	500 g feces plus 190g wheat bran	reverse experiment to treatments S$_{Pool}$-WB and S$_{Pool}$-WB+SD

The final dry matter content should be around 20–40%

The "Lehr- und Versuchsgut Oberholtz" is a farm near Leipzig, integrated within the veterinarian studies and belongs to the veterinarian faculty of the University of Leipzig. Samples were taken during routine and/or education examinations, permitted by the administration (http://dekanat.vetmed.uni-leipzig.de/de/lvg).

The sewage sludge was delivered by the waste management company themselves (http://www.azv-leisnig.com/).

Following ingredients to the different faeces were added: wheat bran (LHG, Schmölln, Germany) to provide an easily digestible C-source to promote *Lactobacillaceae* growth and to decompose fodder residue; charcoal (Köhlerei, Tornau, Germany) as a non-rotting soil improver [24,27,28] for later use of the waste product as a fertilizer; sawdust (HVT, Germany, Dittersdorf), commonly used as litter in animal barns and as a low-digestible C-source within the subsequent fertilizer; Bentonite clay to provide nutrients (Beckmann and Brehm, Beckeln, Germany) and basalt rock flour (ABC-Baustoffe, Erfurt, Germany) as a mineral soil supplement [29] to promote the development of clay-humus complexes consistent with the archaeological findings.

The pooled filtrated sewage sludge had a pH of 6.2, a dry matter content of ~ 11.9% containing of 9.4 organic matter and 2.5% inorganic matter. After adding the other components the dry matter content was 39,2% and comparable to the the dry matter of the manure matrix (~ 40%) (Data kindly provided from Prüf- und Entwicklungsinstitut für Abwassertechnik an der RWTH Aachen e.V., Germany)

The initial concentration of lactate producing bacteria was compared with an optional*Lactobacillaceae* source called "Effective Microorganism" (EMa, Multikraft, Pichl/Wels, Austria) (treatments MEM and M(EM)$_a$) and microbial activity of the indigenous faecal flora was investigated by autoclaving the mixtures (treatments (MEM)$_a$ and (M(EM)$_a$)$_a$).

To identify the necessary components for the hygienisation process substitution experiments were carried out. Wheat bran was eliminated (treatment S$_{Pool}$-WB) or substituted with sawdust, (treatment S$_{Pool}$-WB+SD). Finally faeces and wheat bran only (Fe+WB) were investigated. Based on the results of treatment S$_{Pool}$-WB and S$_{Pool}$-WB+SD, enterococci, Gram-negative and yeasts were tested only.

Bacteria, Viruses and Protozoa

Experiments were carried out with the following zoonotic microorganisms [30–34] *Salmonella*Senftenberg (DSMZ No. 10062), *Salmonella* Anatum (Lab No. 12602/2), *Listeria monocytogenes* (Lab No. 12602/5), *Staphylococcus aureus* (Lab No. 12602/4), *E.coli* O:157 (Lab No. 12602/3) and *Clostridium perfringens* (wild strain from manure). The effect of fermentation on spore forming pathogens was tested as well. Because of the close relatedness of *Clostridium sporogenes* (ATCC 3584) to *C. botulinum* A and F [35], we chose the less harmful *C. sporogenes* for these tests. In addition, enteric cytopathogenic bovine orphan virus (ECBO virus) and *Ascaris suum* eggs were evaluated to represent model pathogens with high environmental tenacity [36].

Preparation of resistant mutants

A specially formulated agar for counting pathogens without interference from faecal flora. was developed using streptomycin and rifampicin resistant strains of *Salmonella* Anatum, *Listeria monocytogenes*, *E.coli* O:157 and *Staphylococcus aureus* according to Linde [37,38]. Single antibiotic-resistant mutants were selected by streaking approximately 10^{10} cfu of fresh bacterial mass on nutrient agar 1 (Sifin, Germany, Berlin) supplemented with 400 mg streptomycin (Roth, Germany, Karlsruhe) and 300 mg rifampicin (Infecto Pharm, Germany, Heppenheim) per litre, incubating the plates aerobically at

37°C for 24 h, picking individual colonies from the plate and subculturing them twice in antibiotic supplemented nutrient broth.

Community analyses

The following culture media were used for bacterial counts: Columbia blood agar (Oxoid, Germany, Wesel) for aerobic and anaerobic organisms, Water-blue-Metachrome-yellow Lactose agar (acc. to Gassner, Sifin) for Gram-negative organisms, Citrate-Azide-Tween-Carbonate Agar (CATC, Sifin) for enterococci, Sabouraud-Agar (Oxoid) for yeasts and moulds, Neomycin Polymyxin blood agar [NeoP, neomycin 200 mg/l (Roth, Germany, Karlsruhe), polymyxin B 100 mg/l (Fluka, Switzerland, Basel), blood 10% (cattle), glucose 5g/l and nutrient agar 1 (Sifin)] for Clostridia and MRS Agar (Oxoid) for lactic acid bacteria. Bacterial counts were performed using the plate count method and the most probable number (MPN) [39] for sulphite-reducing, spore-forming bacteria (SRSF). To achieve this, the samples were heated for 10 minutes at 80°C [40,41], diluted by a factor of 10 and cultivated in liquid differential reinforced Clostridia medium (DRCM, Oxoid). Pasteurization at 80°C guaranteed death of enterococci which survived a 60°C treatment. The isolated colonies were analyzed with MALDI-TOF MS (Bruker, USA, Billerica) as described by Shehata [42].

Inactivation of bacterial pathogens

To test inactivation of pathogenic bacteria, 3 ml suspensions (10^7 to 10^9 cfu/ml) of *E.coli* O:157, *S.* Anatum, *Listeria monocytogenes*, *Clostridium (C.) sporogenes* or *C. perfringens* were added to 27 g of treatments MEM and $M(EM)_a$. To test the influence of the native bacterial flora, we autoclaved control samples before adding the pathogen suspension (treatments $(MEM)_a$ and $(M(EM)_a)_a$). We incubated samples under anaerobic conditions (Anaerocult, Merck) at 37°C for 21 days. The concentration of the spiked pathogens was determined at 0, 3, 7, 14 and 21 days incubation by taking 0.5 g of the treatment, diluting it in 4.5 ml PBS and plating on culture media (as described above) and also nutrient agar 1 (Sifin) supplemented with 400 mg streptomycin, 300 mg rifampicin and 1 g glucose per l. The detection limit was set at 100 cfu/g with the plate count method and approximately 10–100 cfu/g in the direct smear and in the *Salmonella* enrichment (acc. to Preuss, Merck).

To simulate waste containing several pathogens, we prepared four pathogen mixtures. Preliminary experiments had previously shown inactivation of non-sporulated pathogens (unpublished data):

- **Mix 1** (*Salmonella* Anatum, *Listeria monocytogenes*, *E.coli* O:157, and *C. perfringens*),
- **Mix 2** (*Salmonella* Senftenberg, *Staphylococcus aureus*, *C. sporogenes*)
- **Mix 3** (*Salmonella* Senftenberg, *Listeria monocytogenes*, *Staphylococcus aureus*, *E.coli* O:157, *C. perfringens*, *C. sporogenes*)
- **Mix 4** (autoclaved cultivation broth).

We designed Mixes 1 to 4 to facilitate rapid differentiation on the plates. Because *Salmonella* serotypes are not differentiable on a plate, they were evaluated separately. In addition to *S.* Anatum, the heat-tolerant *Salmonella* Senftenberg [43] were tested since it is a common serotype in waste treatment processes [44, 45]. From the experience with *S.* Anatum, we did not modify *S.* Senftenberg for antibiotic resistance, because of the well-working selective gassner media. In order to keep the concentration of each pathogen in the treatment group as high as possible, mix 1 and 2 contained only three to four pathogens, whereas control mix 3 contained six. The initial inoculum was from fresh (overnight) cultures so that the final concentration depended on the specific generation time in nutrient broth I (Sifin) or RCM-broth (Sifin) for Clostridia. After gently mixing 3 samples per treatment, we counted them before and after 3 days of fermentation.

Mix 1 and 2 were added to the non-autoclaved treatments while mix 3 was added to the autoclaved treatments. We added Mix 4 to the non-autoclaved treatments as a control sample to investigate VFA development without additional bacteria. The rate of sporulation of *C.perfringens* and *C. sporogenes* were analyzed by counting the inocula before and after heating at 80°C for 10 minutes by the MPN method. Based on the experiences with manure treatments MEM, $M(EM)_a$, $(MEM)_a$, $(M(EM)_a)_a$ and the high natural *C. perfringens* concentration within the sludge, the experimental design of treatments S_{Pool} and $(S_{Pool})_a$ were simplified in order to test the disinfection effect of treatments 7 & 8 S_{Pool} and $(S_{Pool})_a$ with non-spore forming bacteria only. We counted spiked samples before and after 3 days of incubation.

Weight loss during fermentation

We measured mass loss during sludge-fermentation in 50 ml tubes by weighing before and after 21 days of fermentation and subtracting the difference from the initial weight (27 g). This data was compared with compost mass loss reported in the literature.

pH Value

Since bactericidal effectiveness of VFA depends on pH, the pH were measured by placing 5 ml of the sample in 20 ml KCl—solution (1 mol/l), shaking for 60 minutes, and then measuring the pH with a pH-meter (GPH14 Greisinger electronic, Germany, Regenstauf) according to DIN ISO 10390.

Volatile fatty acids (VFA), lactate and alcohol analyses

During the bacterial inactivation process of treatments MEM, M(EM)$_a$, (MEM)$_a$ and (M(EM)$_a)_a$, we measured acetic acid, propionic acid, i- and n-butyric acid, i- and n-valeric acid, n-caproic acid, ethanol, propanol, butanol, 2,3-butanediol, and 1,2-propanediol concentrations, according to VDLUFA [36]. Lactic acid concentration were analyzed according to Haacker [46]. The mean value from 3 samples per treatment before and after 3 days of fermentation were calculated. Based on the information obtained with one sampling for experiments with treatments 3–6, we added several time slots while monitoring treatment S$_{Pool}$. The VFA observations of treatment S$_{Pool}$ were made without spiking the samples with additional pathogens (3 samples each).

Inactivation of ECBO

The non-coated bovine enterovirus (BEV), also known as enteric cytopathogenic bovine orphan virus (ECBO), is commonly used for testing disinfection products due to its high tenacity. Pre-swollen 1 cm^2 x 1 mm lime-wood platelets were incubated for one hour with an ECBO virus (LCR-4) suspension and added it to treatment MEM and M(EM)$_a$. (acc. to DVG, [47]. After mixing the matrices with 3 platelets per tube, treatments MEM and M(EM)$_a$ were incubated at 37°C. The virus titre was measured directly after mixing and also after 3, 7, 14 and 21 days by 4-MPN-rows on foetal calf lung cells. Different starting concentrations of lactic acid cultures were tested (treatments MEM, M(EM)$_a$).

VFA- Bioassay experiments

Bioassay experiments with S. aureus tested the influence of single VFA at ph 5.3 as well as a VFA mixture with the same concentration as the 4 most highly concentrated VFAs found in this study (Table 2) and 0.15% ethanol (Roth) at pH 5.3 (adjusted with NaOH). The VFA mixture consisted of nutrient broth I (Sifin), lactic acid 1% (Roth), acetic acid 0.46% (Roth), n-butyric acid 0.12% (Roth), propionic acid 0.1% (Merck) and i-butyric acid 0.01% (Alfa Aesar, USA, Wardhill, MA) as the single VFA at the same concentration. The test

tubes containing 5 ml broth were inoculated with 10 μl *S. aureus* suspension
($6.15 * 10^5$ cfu/ml) and incubated them at 37°C for 3 days.

Table 2: Concentration (g/kg fresh weight) of VFA and lactic acid, in treatment S_{Pool}
during the fermentation process. doi:10.1371/journal.pone.0118230.t002

Treatment (time)	acetic acid	propionic acid	i- butyric acid	n- butyric acid	i- valeric acid	n- valeric acid	lactic acid
S_{Pool} (start)	0.86	1.32	0.07	0.11	0.09	0.07	4.90
S_{Pool} (20 h)	3.82	1.12	0.07	1.25	0.08	0.07	9.83
S_{Pool} (47 h 15')	4.17	1.08	0.07	1.19	0.08	0.07	12.17
S_{Pool} (3d 16h)	4.60	1.03	0.06	1.17	0.08	0.06	12.47

Inactivation of *Ascaris suum*

The viability of *Ascaris suum* eggs in treatments MEM and M(EM)$_a$ was tested
by adding 3 ml of a tap water suspension containing 27,000 eggs per ml to
27 g of the matrix in tap water and gently mixing. We tested the influence
of different starting concentrations of lactic acid cultures (treatments MEM,
M(EM)$_a$) after 0, 7, 21 and 56 days, respectively, by taking three samples per
time slot and treatment. *Ascaris* eggs were isolated from the manure matrices
by flotation, re-suspending in tap water, transferring to 6 well tissue culture
plates (TPP) and incubating at 25°C for 20 days with aeration every second
day. To determine the rate of embryonation, 300 eggs were analyzed from each
sample (acc. to DVG). Since ascariasis does not occur in cattle in northern
Europe, the matrices were not investigated for eggs before spiking.

RESULTS

Community analyses

The mean values and standard deviations of the bacterial counts of sewage
sludge matrices (treatment S) from 10 different decentralized sewage plants
are shown in Fig. 3. The typical faecal flora before fermentation was changed
dramatically after fermentation. A general overview of a typical community
shift in all of the tested faecal treatments is presented inTable 3. Species
relevance during fermentation of treatment S and M is shown in the. Similar
community shifts also occurred in medium—scale experiments (1 and 60
litre containers) and also at lower incubation temperatures (27°C, data not
shown), but took a few days longer at the lower temperature. Furthermore,
these experiments showed that this method is even easier to handle on a larger
scale. Because anaerobic conditions are self-establishing as long as the matrix

is packed well, the air column within the bowl is kept small (approximately <10–5%) and a fermentation lock is installed. A small air column is needed because the material swells and may plug the lock.

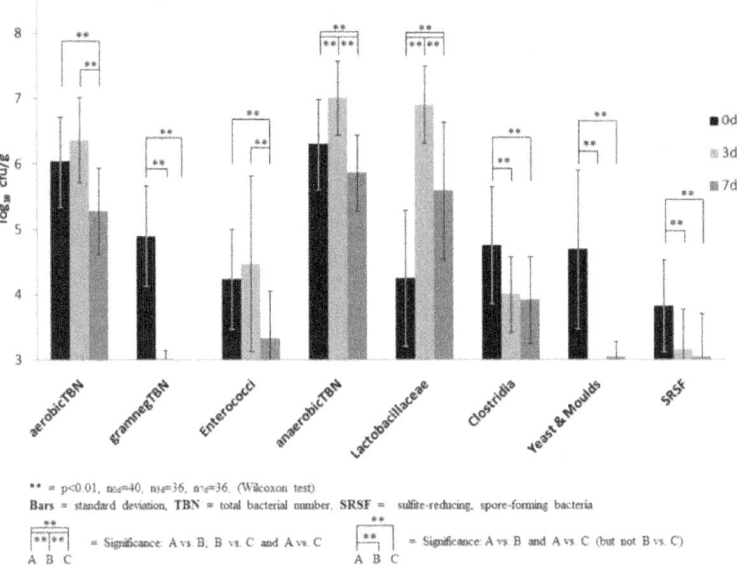

Figure 3: Typical bacterial community shift during fermentation of treatment S. doi:10.1371/journal.pone.0118230.g003

Table 3: General overview of a typical community shift (cfu/g). doi:10.1371/journal.pone.0118230.t003

Group	Start	3 d fermentation	7 d fermentation
Gram-positive, non-spore-forming (Lactobacillaceae, Enterococci excluded)	10^6–10^8	$< 10^2$	$< 10^2$
Bacillaceae	10^3–10^5	10^3–10^5	10^3–10^5
Gram-negative	10^5–10^7	$< 10^2$	$< 10^2$
Enterococci	10^3–10^6	10^3–10^6	$< 10^3$–10^4
Lactobacillaceae	10^3–10^4	10^5–10^7	10^3–10^6
Yeasts / Molds	10^3–10^4	$< 10^2$	$< 10^2$
Clostridiaceae	10^3–10^6	10^3–10^6	10^3–10^6

The smell of the treatments changed from a typical faecal smell to a sour, silage-like odour with fermentation. Gram-negative bacteria, enterococci or yeasts and moulds could not be found in treatments S and Fe+WB, but were present in treatments without the wheat bran (S_{Pool}-WB and S_{Pool}-WB+SD, Table 4).

Table 4: Community shifts within treatments S, S_{Pool}-WB, S_{Pool}-WB+SD, Fe+WB. doi:10.1371/journal.pone.0118230.t004

Group	start	3 d fermentation		7 d fermentation				
Treatment:	S, S_{Pool}-WB, S_{Pool}-WB+SD	S	S_{Pool}-WB	S_{Pool}-WB+SD	S	S_{Pool}-WB	S_{Pool}-WB+SD	Fe+WB
Gram-positive, non-spore-forming (Lactobacillaceae, Enterococci excluded)	10^6	$< 10^2$	10^4	10^4	$< 10^2$	10^4	10^4	nt.
Bacillaceae	10^4	10^4	10^4	10^4	10^4	10^4	10^4	nt.
Gram-negative	10^8	$< 10^2$	10^5	10^6	$< 10^2$	10^4	10^6	$< 10^2$
Enterococci	10^6	10^4	10^7	10^7	$< 10^2$	10^6	10^6	$< 10^2$
Lactobacillaceae	10^6	10^7	10^7	10^7	10^6	10^6	10^4	nt.
Yeasts and Moulds	10^4	$< 10^2$	10^4	10^4	$< 10^2$	$< 10^2$	$< 10^2$	$< 10^2$
Clostridiaceae	10^6	10^5	10^6	10^6	10^6	10^6	10^6	nt.

nt. = not tested

Inactivation of bacterial pathogens

All non-sporulated pathogens where inactivated in the non-autoclaved treatments (treatment MEM, Fig. 4), but sporulated pathogens survived fermentation. *Listeria monocytogenes* and *C.sporogenes* grew in autoclaved treatment 5, while the concentration of the other pathogens decreased but still remained above the minimum detection level. There were only slight differences between treatments M(EM)$_a$ and (M(EM)$_a$)$_a$ compared to 3 and 5. Also the sewage sludge data did not change. The hygienisation ability of non-autoclaved sludge (treatment S) was the same as with non-autoclaved cow manure (treatments M, MEM and M(EM)$_a$).

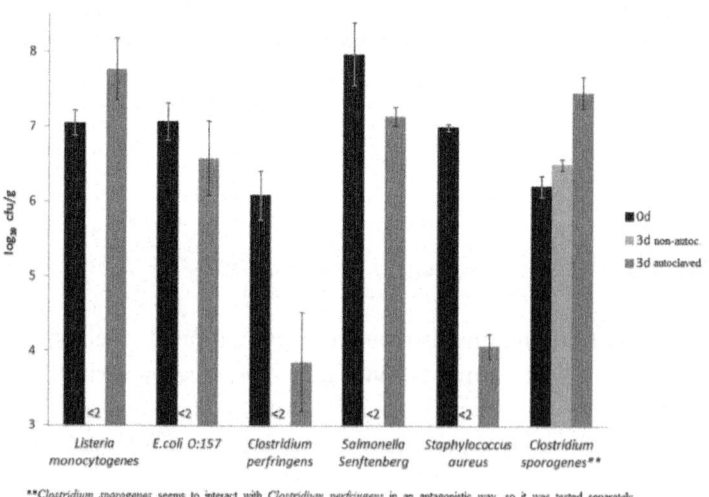

**Clostridium sporogenes* seems to interact with *Clostridium perfringens* in an antagonistic way, so it was tested separately.
Bars = standard deviation.

Figure 4: Development of bacterial pathogens in treatments MEM and (MEM)$_a$. doi:10.1371/journal.pone.0118230.g004

Weight loss

After 21 days of fermentation, the mean weight loss of the sludge treatment (S) was 0.66 g (standard deviation = 0.052 g) which was 2.44% of the total mass of 27 g (p = 0.031 n = 6, Wilcoxon test).

pH Value

The pH decreased in each group independent of treatment during the 3 days fermentation (Table 6) with only the autoclaved controls staying at nearly the same pH. The pH of the non-autoclaved treatments decreased to a much lower level than the autoclaved treatments. Treatment MEM dropped the pH lower than treatment $M(EM)_a$. The drop in pH of treatment S was similar to the unspiked treatment $M(EM)_a$ that dropped to 5.3.

Volatile fatty acids (VFA), lactate and alcohols

The concentration of VFA and alcohols (g/kg) in the different manure treatments spiked with bacterial pathogens (treatments MEM, $M(EM)_a$, $(MEM)_a$ and $(M(EM)_a)_a$) during the hygienisation process are shown in Table 5. N-valeric acid, n-caproic acid, butanol, 2,3-butanediol, 1,2-propanediol and lactate were below the detection limit of 0.05 g/kg and 0.1 g/kg for lactic acid. After 3 days, the highest concentration of n-butyric acid (93 mM) was found in treatment MEM, mix 1 alone with 28 mM acetic acid and a pH of 5.50. The lowest concentration of n-butyric acid (5 mM) was in treatment MEM, mix 4 which also had 47 mM acetic acid and a pH of 4.94.

Table 5: Concentration (g / kg fresh weight) of VFA and alcohols in treatments MEM, $M(EM)_a$, $(MEM)_a$ and $(M(EM)_a)_a$ at the start and after 3 days fermentation. doi:10.1371/journal.pone.0118230.t005

Treatment (time in days)	Bacteria Mix	acetic acid	propionic acid	i- butyric acid	n- butyric acid	ethanol	propanol	i- valeric acid
$(M(EM)_a)_a$ (0d)	Mix 4	1.22	0.29	0.06	0.08	< 0.05	< 0.05	< 0.05
$(MEM)_a$ (0d)	Mix 4	1.35	0.28	0.05	0.09	< 0.05	< 0.05	< 0.05
$(M(EM)_a)_a$ (3d)	Mix 4	1.13	0.26	0.05	0.07	< 0.05	< 0.05	< 0.05
$(MEM)_a$ (3d)	Mix 4	1.33	0.27	0.05	0.09	< 0.05	< 0.05	< 0.05
$(M(EM)_a)_a$ (3d)	Mix 3	2.17	0.24	0.09	0.55	0.80	0.11	0.11
$(MEM)_a$ (3d)	Mix 3	2.06	0.23	0.09	0.47	0.76	0.11	0.11
$M(EM)_a$ (0d)	Mix 4	2.20	0.20	0.06	0.07	1.85	< 0.05	< 0.05
MEM (0d)	Mix 4	2.79	0.26	0.06	0.09	1.99	< 0.05	< 0.05
$M(EM)_a$ (3d)	Mix 4	1.68	0.26	0.05	4.42	1.28	0.04	< 0.05
MEM (3d)	Mix 4	2.85	0.25	0.05	0.46	1.33	0.14	< 0.05
$M(EM)_a$ (3d)	Mix 1	2.63	0.25	0.05	4.30	1.55	0.14	< 0.05
MEM (3d)	Mix 1	1.67	0.27	0.05	8.23	1.45	0.13	< 0.05
$M(EM)_a$ (3d)	Mix 1	1.92	0.28	0.10	4.21	1.29	< 0.05	0.12
MEM Mix 2 (3d)	Mix 1	3.14	0.29	0.10	2.13	1.34	< 0.05	0.12

Shifts in VFA and lactic acid (g/kg) in a pooled sample of treatment S_{Pool} during fermentation are shown in Table 2. While several materials stayed at the same concentration, acetic acid and n-butyric acid increased, and propionic acid decreased. In contrast to the non-measurable lactic acid concentrations (Table 5) Lactic acid increased to 1.25% (wt/wt) in treatment S_{Pool} (Table 2).

Bioassay experiments

Most of the VFAs slightly inhibit growth of *S. aureus* but were not bactericidal at this pH and concentration. Although the most substantial, nearly bacteriostatic, effect of the mixture was caused by acetic acid, a combination of all VFAs plus ethanol tended to be more effective than acetic acid alone. Lactic acid alone at pH 5.3 appeared to support the growth of *S.aureus*.

Inactivation of the ECBO virus

Active ECBO virus was reduced from 5.49 $\log_{10}$ cpe/ ml to below the detection limit of 3 $\log_{10}$cpe/ml within 14 and 7 days fermentation of non-autoclaved manure treatments MEM and M(EM)$_a$, respectively, either with or without EM supplementation in starter cultures.

Inactivation of *Ascaris suum*

There was no reduction in the rate of embryonation of *Ascaris suum* eggs during fermentation of non-autoclaved manure (treatment MEM, 96.8%, SD: 1,90% and treatment M(EM)$_a$, 97.1%, SD: 2,10%) within the first 3 weeks; however, after 56 days of fermentation, embryonation dropped to 0%. A similar drop was observed in tap water after 56 days of anaerobic incubation at 37° C.

DISCUSSION

Community shift

As shown in Fig. 5 and in detail in the apendix massive changes within the bacterial flora were observed. All Gram-negative bacteria and yeasts decreased to levels beneath the detection limit after 3 days of fermentation. Changes in the concentration of Enterococci and Lactobacilli were striking within the first 7 days. This system seems to be specific for certain groups or shifts too fast to be effectively observed at the chosen points of time. Additional experiments showed that Enterococci were not detectable after 21 days (data not shown). The starting concentration of Lactobacilli and Enterococci probably will have a major influence. All non-spore-forming bacteria except acid-tolerant organisms like *Leuconostoc mesenteroides*, were inactivated beneath the detection limit. Spore-forming bacteria such as *B. licheniformis*, *B.cereus* or *Paenibacillus sp.* survived fermentation. It is interesting that a standardized flora was not detected with the methods used. *Pediococcus pentosaceus* was found frequently in treatment S, although not in all samples. Different species may be competing for similar ecological niches. Similarities within Lactobacilli, Enterococci, coliforms and pH shifts could be found within Hrubants work with fermentation of feedlot waste, but the yeast shift data they observed is completely different [48].

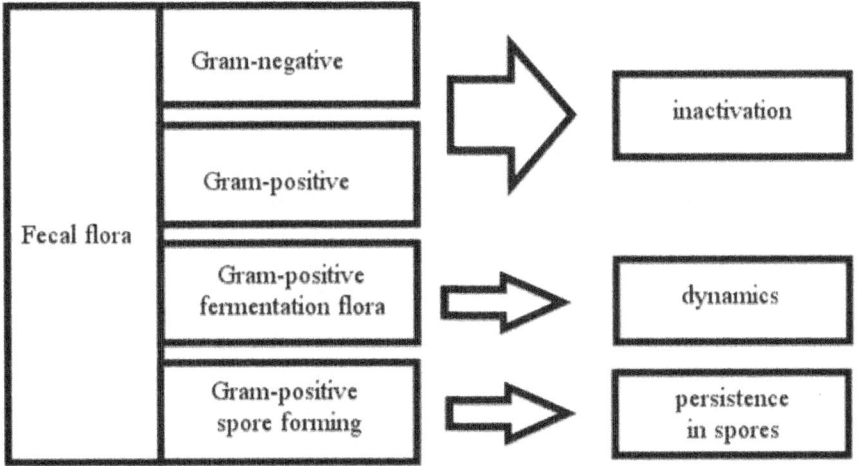

Figure 5: Flow chart of a typical community shift during fermentation. doi:10.1371/journal.pone.0118230.g005

Elimination and substitution experiments showed that the presence of an easily digestible C-source like wheat bran is essential for the inactivation of Gram-negative bacteria and suppression of *Clostridia*. The developing flora without wheat bran was very different from those with wheat bran. Neither Charcoal nor any other additive except wheat bran was necessary for hygienisation, but might improve the soil by absorbing toxins, stabilizing nutrients and forming clay-mould-complexes [24,29,49].

Enterococci should be the preferred microorganism for qualitative tests of successful bacterial hygienisation of fermented faecal matrices because of their obligatory presence in tested faecal matrices and their higher tenacity compared to Enterobacteria. The sour silage-like smell of fermented matrices could be a no-tech indicator of successful fermentation by an experienced user of the method.

Hygienisation of bacterial pathogens

Since all added non-sporulated pathogens died in the non-autoclaved treatments, this hygienisation method seems to be much more effective than composting [11,50,51] because all Gram-negative and Enterococci were inactivated within a few days. So the reaction of the added pathogens was quite the same as of the inherent flora. Gram-positive and Gram-negative bacteria were inactivated whether they were unsporulated or not belonging to the *Lactobacillaceae*. It also is more effective than 60°C/1h pasteurization within biogas plants [10,44], because all Enterococci were inactivated and re-colonisation of gram-negative or Enterococci was never observed during storage. This is probably because the environment is more stable as long as it is not aerated. There is no need for additional technical equipment such as pasteurisation units or compost-turners with this hygeinization process. Every farmer who produces silage will be able to 'ensilage' manure. The effective inactivation of *C.perfringens* may be attributed to the fact that this pathogen did not sporulate (< 10 cfu/ml) in the RCM-media (Oxoid) or other matrices tested. In contrast, the inactivation of *C. perfringens* and *S. aureus* in the autoclaved treatment $(MEM)_a$ may have been caused by interactions with the other organisms of mix 3. Another example of pathogen interactions was that *C.sporogenes* was not even detectable after incubation with *C. perfringens* in treatment $(MEM)_a$. Because of this potential interaction, *C. sporogenes* was tested separately and then grown in treatment $(MEM)_a$. In contrast to the minor sporulation of *C. perfringens*, about 50% of *C.sporogenes* cells sporulated and were not inactivated in non-autoclaved or autoclaved matrices. The growth of *C. sporogenes* demonstrate what may happen in the absence of the inherent antagonistic flora. Moreover a slight growth occurred even within the non autoclaved flora. This might be discussed as a result out of the inaccuracy of the plate count method, but it was sporadically observed in other inherent clostridia grow-cases in matrix S as well. The mean concentration of Clostridia sunk strongly and significantly $p < 0.01$ as shown in Fig. 3, but nevertheless a growth occurred in 6 of 37 cases. It is not clear why the growth occurred but we suspect that the wheat bran with its high amount of proteins ($\sim 15\%$ [52]) may be a suitable resource for Clostridia. Therefore and due to the fact that wheat bran is an edible product we strongly suggest a nother carbon source e.g. green waste ore else.

The slight difference of ~ 0.3 $\log_{10}$ between the start and the 3rd day of fermentation of *C.sporogenes* in the non-autoclaved matrices (treatments MEM & $M(EM)_a$) is not statistically different with the colony count method. The SRSF and Clostridia numbers dropped significantly (~ 0.65 $\log_{10}$ resp. ~ 0.75 $\log_{10}$) within the first 3 days of fermentation in sludge matrices (treatment S, Fig. 3). There might be some ingredients or detergents which facilitate the inactivation of spores in the sludge. Since indigenous sludge-borne *C. perfringens* was not strongly inactivated, it indicates they were sporulated. Sporulated pathogens which survived the 80°C heat treatment before counting, will obviously survive 60°C during composting or pasteurisation also. The method used in this study reduced pathogens more effectively than the fermentation and vermicomposting reported by Factura [53].

It was not determined whether heat sterilization of the control samples (which surely affected the matrix) improved or decreased the growing conditions for the tested pathogens. Gamma sterilization may have been a more gentle treatment for the control samples and should be tested. Nevertheless, the substitution experiments with treatments S_{Pool}, S_{Pool}-WB, S_{Pool}-WB+SD and Fe+WB show the need for the addition of wheat bran to direct development of the original faecal flora.

Although concentrations of the initial infectious pathogens were high when tested individually (data not shown), the concentrations ranged from 3 to 6 times higher because they did not get diluted by the other pathogens in the mix. Independent of the concentration, they were inactivated within 3 days. The main focus of this work was the inactivation of fresh, highly concentrationed pathogens to well below their detection limit. It did not evaluate individual effects occurring between the various microbes.

We recognize that the use of mutants may result in reduced environmental tolerance of the microorganism; however, Linde [54] reported the opposite is possible as well. Linde described mutants dying more slowly because of their slower metabolism, which caused an indirect proliferation in harsh environments. The mutant pathogens we used were tested after 24 hours of incubation in a hypotonic solution which should have improved their stress tolerance by stabilizing their cell walls with oligosaccharides [55]. Even with this treatment, they were inactivated within 3 days along with all of the other added or inhabiting (opportunistic) pathogens that were excreted by cattle or humans.

Loss of weight during fermentation

The 2.44% weight loss in the first 21 days was smaller than the mass loss with composting which may vary from 15% [56] to 57% [12] after 42 and 132 days respectively. In general, composting time periods are longer than 21 days, but the highest emissions of CO_2 [57], NH_3 and heat [13] are generally observed in the first 21 days of composting to indicate that the loss of mass probably is also highest in the first 3 weeks. Although the anaerobic pots (2.4 dm^3) are closed systems, gas production of the matrices was small enough that there was no detectable pressure when opening the pots. Instead, the caps on the pots did not lift by themselves after unlocking the clamps. Medium-scale experiments with gas traps and a gas analysis could give more information about mass flows within the matrices.

pH value

Only the control group treatment MEM, mix 4 reduced the pH below 5 as shown in Table 6. Most non-autoclaved matrices with pathogens dropped to a pH of around 5.5 to 5.2. Those values alone do not explain the hygienisation effect since *Salmonella* and *E. coli* can grow at pH 4.5 [58] and *S. aureus* survives at pH 2.5 [20].

Table 6: pH of treatments MEM, $M(EM)_a$, $(MEM)_a$ and $(M(EM)_a)_a$ at the start and after 3 days fermentation with different pathogen mixes. doi:10.1371/journal. pone.0118230.t006

Treatment	Bacteria Mix	0 d	3 d
$(M(EM)_a)_a$	broth only	6,72 (0,05)	6,70 (0,09)
$(MEM)_a$	broth only	6,70 (0,02)	6,63 (0,04)
$M(EM)_a$	broth only	6,68 (0,04)	5,37 (0,17)
MEM	broth only	6,70 (0,04)	4,94 (0,02)
MEM	Mix 1	n.a.	5,50 (0,03)
$M(EM)_a$	Mix 1	n.a.	5,64 (0,10)
MEM	Mix 2	n.a.	5,10 (0,17)
$M(EM)_a$	Mix 2	n.a.	5,55 (0,04)
$(MEM)_a$	Mix 3	n.a.	6,08 (0,02)
$(M(EM)_a)_a$	Mix 3	n.a.	6,10 (0,03)

(Standard deviation).

Volatile fatty acids

VFA become bactericidal at a certain pH [22]. The chemical conditions we measured were comparable to the reports of Presser [21] and Knarreborg [19] who showed that bactericidal concentrations were not present and, therefore, could not explain a total loss of 10^8 pathogens within 3 days. Knarreborg reported that it takes 100 mM of butyric acid at a pH of 5.5 and 37°C under anaerobic conditions in a gut simulator to reduce coliform bacteria by 6% per hour. This concentration was not reached in our study, but even if this had happened, the residual coliforms would not have fallen below 10^6 cfu after 72h.

The higher concentration of lactic acid during fermentation of treatment S_{Pool} may have been caused by the higher dry matter content since 125 ml of

EM suspension was not added. The measured concentration of 138.5 mM/l and pH of 5.3 are still adequate growing conditions for *E.coli* according to Presser and Knarreborg [19,21].

Goepfert [18] reported the greatest bactericidal effect of acetic acid, and lesser effects of propionic acid and butyric acid to *S. typhimurium* at 0.5%, with exposure periods ranging from several hours to 2 days at 37°C and a pH of 5.0–5.5. Lind [20] described a bacteristatic effect of 0.3% acetic acid or 0.5% lactic acid on the extremely pH tolerant *Staphylococcus aureus* within 2 days. In contrast, it took 0.5% and 1.0%, respectively, for a bactericidal effect within 24h in the presence of proteins within a pH range of 2.5 to 3.5. In summary, we found > 0,5% VFA after 3 days in each measured group except for the group in treatment MEM, mix 4. The pH in this group was the only one lower than pH 5 which increases the bactericidal effectiveness of VFA. In addition, there seems to be a synergistic effect with a mixture of VFA and a low alcohol concentration as observed with *S. aureus*, although this mixture was bacteriostatic rather than bactericidal.

If pH, in combination with the measured fermentic acid concentrations, are as efficient in our matrices (similar to what Goepfert described within aqueos solution), they could be assumed to be the reason for the hygienisation observed. Even if they are not solely responsible for the succesful disinfection, other factors made them effective, e. g. adsorbtion of higher VFA concentrations by charcoal. Various concentrations of VFA are reported to have an effect, but were not measurable in our study after adsorbtion by charcoal or the development of a VFA-digesting bactcrial flora. The bacteriostatic, but not bactericidal, effects of VFAs measured in bioassay experiments indicate that unknown interactions or other chemical parameters could be involved. Perhaps bacterial interactions involving the production of bacteriocins, e.g. pediocin of *Pediococcus pentosaceus*, could be responsible for inactivation of the non-spored bacteria. Many species of known bacteriocin-forming bacteria beside *P. pentosaceus* were found in the matrices during the fermentation process. Non-culturable bacteria also may play a role and should be investigated by T/DGGE or metagenomic analysis and rtPCR for example.

Inactivation of ECBO

This virus tends to be inactivated about 4 days earlier in treatment $M(EM)_a$ than in MEM, which indicates an influence of the matrix. Boegel [59] reported that BEV decreased $0.5 \log_{10}$ per day at 37°C and that it survived better at lower temperatures. Nazir [60] showed that the tenacity of the ECBO virus depends strongly on both temperature and matrix effects. Similarly, Lund [61] reported a $4 \log_{10}$ reduction of BEV at 35°C in physiological saline within 219 hours or within 23 hours in manure containing bleaching clay under anaerobic conditions at pH 8.0. In contrast, Biermann [62] reported very strong temperature stability of the ECBO virus, which lost only 2 $\log_{10}$ infectious units in 26 weeks at 20°C in liquid cow manure. The effect of autoclaving was not tested in this study because of the obvious need of non-autoclaved treatments to inactivate the obligatory bacteria in faeces. More investigations should be performed in the future to discover the matrix and temperature effects on inactivation of the ECBO virus in fermentation processes.

Inactivation of *Ascaris suum*

Ascaris suum eggs remained generally unaffected by incubation the first 21 days. The loss of pathogenicity of *A. suum* eggs between 21 and 56 days, as well as in the tubes spiked with EM (MEM) and control tubes (M(EM)$_a$) and tap water tubes will probably be attributed to the 37°C temperature. Earlier studies with eggs of the related species, *A. lumbricoides*, showed that a temperature of 37°C over a long period is harmful to eggs [63]. Seamster [64] reported the optimum temperature for embryonation of *Ascaris suum* eggs was 31.3°C and that development slowed down as the temperature increased. Only 10% of the eggs developed a motile embryo at 34.4°C, and all ova died in early cleavage stages between 35.6 and 37.8°C. In a mesophilic anaerobic digester maintained at 35°C, the viability of *A. suum* eggs dropped from 95% after one week to 50% after 5 weeks [65]. Even the development at 32°C appeared to be suboptimal since more than 50% of the infective eggs showed a gradual decline in larval motility and healthy protoplasmic appearance [66]. Maya [67] also reported that the viability of *Ascaris* eggs stored in sludge under various conditions decreased as the temperature increased. Storage temperature was found to be the most important factor affecting the viability of *Ascaris* eggs in sludge [68]. Other factors, such as type of sludge digestion, storage in soil versus sludge, pH and egg species had only minor effects. Thus, the influence of fermentation on *A. suum* eggs seems to be minor under the investigated conditions even though the core temperatures of 60–70°C that develop during composting are high enough to kill those eggs within several hours [12,63]. Nevertheless, the survival of *Ascaris sp.* during composting for several months has been reported previously [51], but this may result from different temperature zones in compost piles. There is an obvious need to test the fermentation method under field conditions since daily temperature shifts or heating (e. g. storing the fermented product in black barrels in direct sunlight) could overcome the limitation in hygienisation of parasite eggs during fermentation. The combined positive aspects of heat from composting and the effects of fermentation should be tested also. There were concerns about the effectiveness of temperature on bacterial pathogens but, as shown in these experiments, the C-source and living faecal flora seem to provide the predominant influence on bacterial disinfection.

CONCLUSION

The lactic acid fermentation is a hygienisation method which may be comparable to a pasteurization in the result except for the survival of the lactic acid bacteria. It is a stand-alone possibility for inactivation of pathogens independent of a further use as fertilizer or further biogas production, which is possible as well, as shown by Herrmann [69].

Fermentation was able to hygienise faecal waste from small-scale sludge plants and animal farms much more effectively and faster than composting. Since the lack of self-heating and with only a small amount of gasproduction, this method also conserves nutrients during processing much more effectively than aerobic composting methods [12].

The conserved nutrients could be used to promote a more efficient agriculture. All in all, fermentation seems to be a suitable, fast, easy-to-use, low-tech method to break viral and bacterial infection cycles by producing a hygienized organic fertilizer within 7 days. This study does not support a need for additional microorganisms, such as the tested EM, as long as a living fecal microbiota and a fermentable substrate are present.

The principle of how hygienisation works is not completely clear, although VFA and temperature were shown to have an important influence on bacterial survival. These experiments showed that bacterial hygienisation also worked at 27°C. Further investigation, especially at lower temperatures (21°C, 15°C and 5°C) are necessary to test fermentation benefits under field conditions.

ACKNOWLEDGMENTS

Special thanks to Klaus Linde, Elke Dittmann, Ingo Schneider, Thomas Vahlenkamp, Awad Shehata, Juliane Liepe and very special thanks to Norman Ständer and Don Huber for recommendations, and Tom Harman and Anne Seyffert for proofreading the manuscript. Katrin Erfurt for technical support.

AUTHOR CONTRIBUTIONS

Conceived and designed the experiments: HM KD MEK HAS. Performed the experiments: HAS KD FSS. Analyzed the data: HAS KD FSS. Contributed reagents/materials/analysis tools: HAS KD FSS. Wrote the paper: HAS FSS.

REFERENCES

1. Glathe H, Makawi AAM (1963) Über die Wirkung von Klärschlamm auf Boden und Mikroorganismen. Zeitschrift für Pflanzenernährung, Düngung, Bodenkunde 101: 109–121. doi: 10.1002/jpln.19631010205. pmid:19349661

2. Jiang X, Morgan J, Doyle MP (2002) Fate of Escherichia coli O157:H7 in Manure-Amended Soil. Applied and Environmental Microbiology 68. doi: 10.1128/aem.68.5.2605-2609.2002

3. Al-Ghazali MR, Al-Azawi SK (1990) Listeria monocytogenes contamination of crops on soil treated with sewage sludge cake. Journal of Applied Bacteriology 69. doi: 10.1111/j.1365-2672.1990.tb01557.x

4. the Roots of Wheat (Triticum aestivum L.) by Terminal Restriction Fragment Length Polymorphism and Sequencing of 16S rRNA Clones. Applied and Environmental Microbiology 70: 1787–1794. doi: 10.1128/AEM.70.3.1787-1794.2004. pmid:15006805

5. Minamisawa K, Nishioka K, Miyaki T, Ye B, Miyamoto T, et al. (2004) Anaerobic Nitrogen-Fixing Consortia Consisting of Clostridia Isolated from Gramineous Plants. Applied and Environmental Microbiology 70: 3096–3102. doi: 10.1128/AEM.70.5.3096-3102.2004. pmid:15128572

6. Schikora A, Carreri A, Charpentier E, Hirt H, Ojcius DM (2008) The Dark Side of the Salad: Salmonella typhimurium Overcomes the Innate Immune Response of Arabidopsis thaliana and Shows an Endopathogenic Lifestyle. PLoS ONE 3: e2279. doi: 10.1371/journal.pone.0002279. pmid:18509467

7. Tyler HL, Triplett EW (2008) Plants as a Habitat for Beneficial and/or Human Pathogenic Bacteria. Annual Review of Phytopathology 46: 53–73. doi: 10.1146/annurev.phyto.011708.103102. pmid:18680423

8. Buchholz U, Bernard H, Werber D, Böhmer MM, Remschmidt C, et al. (2011) German Outbreak of Escherichia coli O104:H4 Associated with Sprouts. New England Journal of Medicine 365: 1763–1770. doi: 10.1056/NEJMoa1106482. pmid:22029753

9. Appel B, Böl GF, Greiner M, Lahrssen-Wiederholt M, Hensel A (2012) EHEC outbreak 2011: Investigation of the outbreak along the food chain. Berlin: Bundesinstitut für Risikobewertung.

10. Bagge E, Sahlström L, Albihn A (2005) The effect of hygienic treatment on the microbial flora of biowaste at biogas plants. Water Research 39: 4879–4886. doi: 10.1016/j.watres.2005.03.016. pmid:16297957

11. Pourcher A-M, Morand P, Picard-Bonnaud F, Billaudel S, Monpoeho

S, et al. (2005) Decrease of enteric microorganisms from rural sewage sludge during their composting in straw mixture. Journal of Applied Microbiology 99: 528–539. pmid:16108794 doi: 10.1111/j.1365-2672.2005.02642.x

12. Tiquia SM, Richard TL, Honeyman MS (2002) Carbon, nutrient, and mass loss during composting. Nutrient Cycling in Agroecosystems 62. doi: 10.1023/a:1015137922816

13. Sommer SG (2001) Effect of composting on nutrient loss and nitrogen availability of cattle deep litter. European Journal of Agronomy 14. doi: 10.1016/s1161-0301(00)00087-3

14. Batmanglij N (2000) Milk and its By-products in Ancient Persia and Modern Iran. In: Walker H, editor. Milk—beyond the dairy. Totnes and Devon and Eng: Prospect Books. pp. 64–73.

15. Kulp K, Lorenz KJ (2003) Handbook of dough fermentations. New York (N.Y.) and Basel: M. Dekker.

16. Wilkinson J, Bolsen K, Lin C (2003) History of Silage. In: Buxton DR, Muck RE, Harrison JH, editors. Silage science and technology. Madison and Wis: American Society of Agronomy and Crop Science Society of America and Soil Science Society of America. pp. 1–30.

17. Hammes W., Weiss N, Holzapfel W (1992) The Genera Lactobacillus and Carnobacterium. In: Balows A, Trüpfer HG, Dworkin M, Harder W, Schleifer K-H, editors. The prokaryotes. New York and NY and Berlin and Heidelberg: Springer. pp. 1535–1595.

18. .Goepfert JM, Hicks R (1969) Effect of volatile fatty acids on Salmonella typhimurium. Journal of bacteriology 97: 956–958. pmid:4886302

19. Knarreborg A, Miquel N, Granli T, Jensen BB (2002) Establishment and application of an in vitro methodology to study the effects of organic acids on coliform and lactic acid bacteria in the proximal part of the gastrointestinal tract of piglets. Animal Feed Science and Technology 99: 131–140. doi: 10.1016/s0377-8401(02)00069-x

20. Lind O (1937) Untersuchungen über die bakteriziden Eigenschaften der Milchsäure, Essigsäure und Salzsäure und ihre Verwendung als Konservierungsmittel für Fleisch Giessen: Buch-,Verlags- und Akzidenzdruckerei Eduart Seibert.

21. Presser KA, Ross T, Ratkowsky DA (1998) Modelling the Growth Limits (Growth/No Growth Interface) of Escherichia coli as a Function of Temperature, pH, Lactic Acid Concentration, and Water Activity. Applied and Environmental Microbiology 64: 1773–1779. pmid:9572950

22. Levine AS, Fellers CR (1940) Action of acetic acid on food spoilage microorganisms. Journal of bacteriology 39: 499–515. pmid:16560309

23. Van Hofwegen G, Kuyper TW, Hoffland E, van den Broek JA, Becx GA (2009) Opening the Black Box: Deciphering Carbon and Nutrient Flow in Terra Preta. In: Woods WI, Teixeira WG, Lehmann J, Steiner C, WinklerPrins AMG., et al., editors. Amazonian Dark Earths: Wim Sombroek's Vision. Springer Science.

24. Glaser B, Haumaier L, Guggenberger G, Zech W (2001) The "Terra Preta" phenomenon: A model for sustainable agriculture in the humid tropics. Naturwissenschaften.

25. Smith NJ (1980) Anthrosols and Human Carrying Capacity in Amazonia. Annals of the Association of American Geographers 70. doi: 10.1111/j.1467-8306.1980.tb01332.x

26. Denevan WM (2002) Cultivated landscapes of native Amazonia and the Andes. Oxford and New York: Oxford University Press.

27. Schmidt MW, Skjemstad JO, Jäger C (2002) Carbon isotope geochemistry and nanomorphology of soil black carbon: Black chernozemic soils in central Europe originate from ancient biomass burning. Global Biogeochemical Cycles 16. doi: 10.1029/2002gb001939

28. Skjemstad JO, Clarke P, Taylor JA, Oades JM, McClure SG (1996) The chemistry and nature of protected carbon in soil. Australian Journal of Soil Research.

29. Scheffer F, Meyer B (1963) Berührungspunkte der archaeologsichen und bodenkundlichen Forschung. Neue Ausgrabungen und Forschungen in Niedersachsen 1.

30. Hartmann FA, West SE (1995) Antimicrobial Susceptibility Profiles of Multidrug-Resistant Salmonella Anatum Isolated from Horses. Journal of Veterinary Diagnostic Investigation 7: 159–161. pmid:7779955 doi: 10.1177/104063879500700128

31. Jeffers GT, Bruce JL, Wiedemann M (2001) Comparative genetic characterization of Listeria monocytogenes isolates from human and animal listeriosis cases. Microbiology 147: 1095–1104. pmid:11320113

32. Naylor SW, Low JC, Besser TE, Mahajan A, Gunn GJ, et al. (2003) Lymphoid Follicle-Dense Mucosa at the Terminal Rectum Is the Principal Site of Colonization of Enterohemorrhagic Escherichia coli O157:H7 in the Bovine Host. Infection and Immunity 71: 1505–1512. doi: 10.1128/IAI.71.3.1505-1512.2003. pmid:12595469

33. Songer GJ (2010) Clostridia as agents of zoonotic disease. Veterinary Microbiology.

34. Tavechio AT, Ghilardi ÂC, Peresi JT, Fuzihara TO, Yonamine EK, et al. (2002) Serotypes Isolated from Nonhuman Sources in São Paulo, Brazil, from 1996 through 2000. Journal of Food Protection 65.

35. Collins MD, Lawson PA, Farrow JE (1994) The Phylogeny of the Genus Clostridium: Proposal of Five New Genera and Eleven New Species Combinations. International Journal of Systematic Bacteriology 44: 812–826. pmid:7981107 doi: 10.1099/00207713-44-4-812

36. VDLUFA Methodenbuch Band III Die chemische Untersuchung von Futtermitteln (n.d.). VDLUFA—Verband Deutscher Landwirtschaftlicher Untersuchungs- und Forschungsanstalten e.V.

37. Linde K (1981) Hoch Immunogene Salmonella-Mutanten mit zwei unabhängig voneinander attenuierden Markern als potentielle Impfstoffe aus vermehrungsfähigen Bakterien. Zbl Bakt Hyg I Abt Orig.

38. Linde K, Beer J, Bondarenko V (1990) Stable Salmonella live vaccine strains with two or more attenuating mutations and any desired level of attenuation. Vaccine 8. doi: 10.1016/0264-410x(90)90058-t

39. Colwell RR (1979) Enumeration of specific populations by the most-probable-number (MPN) method. In: Costerton J., Colwell RR, editors. Native Aquatic Bacteria: Enumeration, Activity and Ecology. Philadelphia. pp. 56–61.

40. Alberto F, Broussolle V, Mason D., Carlin F, Peck M (2003) Variability in spore germination response by strains of proteolytic Clostridium botulinum types A, B and F. Letters in Applied Microbiology 36: 41–45. doi: 10.1046/j.1472-765X.2003.01260.x. pmid:12485340

41. Slepecky RA, Hemphill EH (1992) The Genus Bacillus—Nonmedical. In: Balows A, Trüpfer HG, Dworkin M, Harder W, Schleifer K-H, editors. The prokaryotes. New York and NY and Berlin and Heidelberg: Springer, Vol. 2. pp. 1663–1696.

42. Shehata AA, Sultan H, Hafez MH, Krüger M (2012) Safety and efficacy of a metabolic drift live attenuated Salmonella Gallinarum vaccine against fowl typhoid. Avian Diseases In-Press.

43. Bayne HN, Bayne GH, Garibaldi JA (1969) Heat Resistance of Salmonella: the Uniqueness of Salmonella senftenberg 775W. Applied and Environmental Microbiology 17: 78–82. pmid:5774764

44. .Ade-Kappelmann K (2008) Untersuchungen zur seuchenhygienischen Unbedenklichkeit von Gärresten aus Bioabfällen nach der von Gärresten aus Bioabfällen nach der: INAUGURAL-DISSERTATION Berlin: Freie Universität Berlin.

45. Martens W, Fink A, Philipp W, Weber A, Winter D, et al. (1998) Inactivation of viral and bacterial pathogens in large scale slurry treatment plants. Proceedings from Ramiran: 529–539.

46. Haacker K, Block HJ, Weissbach F (1983) Zur kolorimetrischen Milchsäurebestimmung in Silagen mit p-Hydroxydiphenyl. Archiv für Tierernaehrung 33. doi: 10.1055/s-0033-1360260. pmid:24715412

47. Richtlinien für die Prüfung chemischer Desinfektionsmittel der Deutschen Veterinärmedizinischen Gesellschaft e.V (2007) Giessen: DVG-Verlag.

48. Hrubant GR (1975) Changes in Microbieal Population During Fermenation of Feedlot Waste with Corn. Applied Microbiology and Biotechnology 30: 113–119.

49. Steiner C, Teixeira WG, Zech W (2009) The Effect of Charcoal in Banana (Musa Sp.) Planting Holes—An On-Farm Study in Central Amazonia, Brazil. In: Woods WI, Teixeira WG, Lehmann J, Steiner C, WinklerPrins AMG., et al., editors. Amazonian Dark Earths: Wim Sombroek's Vision. Springer Science. pp. 423–432.

50. Jones P, Martin M (2003) A review of the literature on the occurrence and survival of pathogens of animals and humans in green compostBanbury OX16 0AH: Institute for Animal Health.

51. Kjellberg Christensen K, Carlsb\aek M, Norgaard E, Warberg KH, Venelampi O, et al. (2002) Supervision of the sanitary quality of composting in the Nordic countries: Evaluation of 16 full-scale facilities. Copenhagen: Nordisk Council of Minister.

52. Lehmann E, Birsgal H (1960) Untersuchung von Futtermitteln mit Hilfe der Ameisensäuremethode zur Bestimmung von Zellstoff, Stärke und Protein. Zeitschrift für Pflanzenernährung, Düngung, Bodenkunde 89: 42–49. doi: 10.1002/jpln.19600890105. pmid:19349661

53. Factura H, Bettendorf T, Buzie C, Pieplow H, Reckin J, et al. (2010) Terra Preta Sanitation: re-discovered from an ancient Amazonian civilisation—integrating sanitation, bio-waste management and agriculture. Water Science & Technology, 61: 2673. doi: 10.2166/wst.2010.201

54. Linde K, Grosse-Herrenthey A, Heisig P, Schwarz S, Jacobsen ID, et al. (2012) Metabolic Drift (MD) Mutanten als potentielle Impfstämme Einzelbefunde oder ein für alle Erreger zutreffendes Attenuierungsprinzip Leipzig: Verlag der DVG Service GmbH.

55. Csonka LN (1989) Physiological and Genetic Response of Bacteria to

Osmotic Stress. Microbiological Reviews 53.

56. Eghball B, Power JF, Gilley JE, Doran JW (1997) Nutrient, carbon, and mass loss during composting of beef cattle feedlot manure. Journal of environmental quality 26: 189–193. doi: 10.2134/jeq1997.00472425002600010027x

57. Bernal MP, Sánchez-Monedero MA, Paredes C, Roig A (1998) Carbon mineralization from organic wastes at different composting stages during their incubation with soil. Agriculture, Ecosystems & Environment 69. doi: 10.1016/s0167-8809(98)00106-6

58. Back W (2008) Mikrobiologie der Lebensmittel, Getränke. Hamburg: Behrs-Verlag.

59. Boegel K, Mussgay M, Kunger L (1960) Isolierung und Charakterisierung eines Enterovirus des Rindes. Zentralblatt für Veterinärmedizin 7: 534–552. doi: 10.1111/j.1439-0442.1960.tb00270.x

60. Nazir J (2011) Persistence of H4N6, H5N1, and H6N8 avian influenza viruses, H1N1 human influenza virus, and two model viruses (NDV and ECBO) in various types of water, lake sediment, duck feces, and meat Giessen: VVB LaufersweilerVerlag.

61. Lund B, Jensen VF, Have P, Ahring B (1996) Inactivation of virus during anaerobic digestion of manure in laboratory scale biogas reactors. Antonie van Leeuwenhoek 69. doi: 10.1007/bf00641608

62. Biermann U, Herbst W, Schliesser T (1990) Untersuchungen zur Widerstandsfähigkeit von bovincm Enterovirus und Aujeszkyvirus in Rindergülle bei verschiedenen Lagerungstemperaturen. Berliner und Münchener Tierärztliche Wochenzeitschrift 103. pmid:9480080

63. Germans W (1954) Laboratioriumsuntersuchungen über die Resistenz der Eier des menschlichen Spulwurmes Ascaris Lumbricoides L. Zeitschrift für Parasitenkunde 16. doi: 10.1107/S0108767309007235. pmid:19349661

64. Seamster AP (1958) Developmental Studies Concerning the Eggs of Ascaris lubricoides var. suum. The American Midland Naturalist 43. doi: 10.2307/2421913

65. Johnson P, Dixon R, Ross A (1998) An in-vitro test for assessing the viability of Ascaris suum eggs exposed to various sewage treatment processes. International Journal for Parasitology 28: 627–633. doi: 10.1016/S0020-7519(97)00210-5. pmid:9602387

66. Boisvenue RJ (1990) Effects of aeration and temperature in in vitro and in vivo studies on developing and infective eggs of Ascaris suum. Journal

of the Helminthological Society of Washington 57.

67. Maya C, Ortiz M, Jiménez B (2010) Viability of Ascaris and other helminth genera non larval eggs in different conditions of temperature, lime (pH) and humidity. Water Science and Technology 62. doi: 10.2166/wst.2010.535

68. O'Donell CJ, Meyer KB, Schaefer FW III (1984) Survival of Parasite Eggs Upon Storage in Sludge. Applied and Environmental Microbiology 48.

69. Herrmann C, Heiermann M, Idler C (2011) Effects of ensiling, silage additives and storage period on methane formation of biogas crops. Bioresource Technology 102: 5153–5161. doi: 10.1016/j.biortech.2011.01.012. pmid:21334882

Chapter 9

SCREENING OF NON- SACCHAROMYCES CEREVISIAE STRAINS FOR TOLERANCE TO FORMIC ACID IN BIOETHANOL FERMENTATION

Cyprian E. Oshoma[1] , Darren Greetham[1] , Edward J. Louis[2] , Katherine A. Smart[3] , Trevor G. Phister[4], Chris Powell[1] , Chenyu Du[1,5]

[1] Bioenergy and Brewing Science Building, School of Biosciences, University of Nottingham, Sutton Bonington Campus, Loughborough, Leics, United Kingdom,

[2] Centre for Genetic Architecture of Complex Traits, University of Leicester, Leicester, United Kingdom,

[3] SAB Miller PLC, Surrey, United Kingdom,

[4] PepsiCo Int. Beaumont Park, Leycroft Road, Leicester, United Kingdom,

[5] School of Applied Sciences, University of Huddersfield, Queensgate, Huddersfield, United Kingdom

ABSTRACT

Formic acid is one of the major inhibitory compounds present in hydrolysates derived from lignocellulosic materials, the presence of which can significantly hamper the efficiency of converting available sugars into bioethanol. This study investigated the potential for screening formic acid tolerance in non-*Saccharomyces cerevisiae* yeast strains, which could be used for the development of advanced generation bioethanol processes. Spot plate and phenotypic microarray methods were used to screen the formic acid tolerance of 7 non-*Saccharomyces cerevisiae* yeasts. *S. kudriavzeii* IFO1802 and *S. arboricolus* 2.3319 displayed a higher formic acid tolerance when compared to other strains in the study. Strain *S. arboricolus* 2.3319 was selected for further investigation due to its genetic variability among the *Saccharomyces* species as related to *Saccharomyces cerevisiae* and availability of two sibling strains: *S.arboricolus* 2.3317 and 2.3318 in the lab. The tolerance of *S. arboricolus* strains (2.3317, 2.3318 and 2.3319) to formic acid was further investigated by lab-scale fermentation analysis, and compared with *S. cerevisiae* NCYC2592.

S. arboricolus 2.3319 demonstrated improved formic acid tolerance and a similar bioethanol synthesis capacity to *S. cerevisiae* NCYC2592, while *S. arboricolus* 2.3317 and 2.3318 exhibited an overall inferior performance. Metabolite analysis indicated that *S. arboricolus* strain 2.3319 accumulated comparatively high concentrations of glycerol and glycogen, which may have contributed to its ability to tolerate high levels of formic acid.

INTRODUCTION

The importance of identifying alternative energy sources has become necessary due to the continuous depletion of limited fossil fuel stock and for the creation of a safe and sustainable environment. Recently, attention has focused on renewable or alternative sources of energy, as a means of supplementing the inevitable shortage of world's energy supply [1]. In some developing countries, there is a need for alternative sources of energy, such as those derived from lignocellulosic biomass including herbaceous and woody plants, agricultural, forestry residues, municipal solid waste and industrial waste streams [2, 3]. These feedstocks are of particular interest as they do not compete with food production for agricultural resources [4].

Lignocellulosic plant residues containing up to 70% carbohydrate (as cellulose and hemicellulose) are prominent substrates for the advanced generation of bioethanol production. However, due to the recalcitrant nature of lignocellulosic biomass, pretreatment is necessary for the release of fermentable sugars. Pretreatment processing can be carried out in different ways including mechanical, steam explosion, ammonia fiber explosion, acid or alkaline pretreatment and biological pretreatment [1]. Furthermore, a combination of two or more of these processes can be employed with a view to producing synergetic effects.

Rapid and efficient fermentation of lignocellulosic hydrolysates is limited because, in addition to the release of monomeric sugars, a range of inhibitory compounds are generated during pretreatment and hydrolysis [5, 6, 7, 8]. These inhibitory compounds fall into specific groups such as weak acids, furan derivatives and phenolic compounds [9]. The types of toxic compounds generated, and their concentrations in lignocellulosic hydrolysates, depend on both the raw material and the operational conditions employed for hydrolysis [10]. Toxic compounds can act to stress fermentative organisms to a point beyond which the efficient utilization of sugars is possible, ultimately leading to reduced product formation [11].

Formic acid is one of the weak acid inhibitors present in lignocellulosic hydrolysates, with a typical concentration of approximately 1.4 g/L (30 mM) [8, 12]. The inhibitory effect of formic acid has been ascribed to both uncoupling and intracellular anion accumulation [13, 14] and the reduction of the uptake of aromatic amino acids [12]. The undissociated form of weak acids can diffuse from the fermentation medium across the plasma membrane [13, 14] and dissociate due to higher intracellular pH, thus decreasing the cytosolic pH. The decrease in intracellular pH is compensated by the plasma membrane ATPase, which pumps protons out of the cell at the expense of ATP hydrolysis. Consequently, less ATP is available for biomass formation. According to the intracellular anion accumulation theory, the anionic form of the acid is captured inside the cell and the undissociated acid will diffuse out of the cell until equilibrium is reached. Weak acids have also been shown to inhibit yeast growth by reducing the uptake of aromatic amino acids from the medium, probably as a consequence of strong inhibition of the enzyme permease [12]. Formic acid is more toxic to yeast strains than either acetic acid or levulinic acid [12, 15], due to a lower pK_a value (3.75 at 20°C) than acetic (4.75 at 25°C) and levulinic acid (4.66 at 25°C). Its undissociated form should be found in lower concentrations at the same internal pH, and consequently be less toxic to the cells. The increased toxicity of formic acid seems to be associated with a smaller molecule size, which may facilitate its diffusion through the plasma membrane and possibly its higher anion toxicity [16].

Yeasts, mostly strains of *Saccharomyces cerevisiae*, have been widely used for bioethanol production industrially, due to their high fermentative ability, ethanol tolerance and rapid growth under anaerobic conditions [17]. These yeasts, however, are susceptible to inhibitory compounds present in lignocellulose derived hydrolysates [18]. One possible solution is to detoxify the hydrolysate to remove the inhibitors, however, this creates additional costs and a potential loss of sugar [12]. An alternative approach and long-term solution to overcome this problem is to either screen for high inhibitor tolerant yeast strains or create genetically modified strains with desired tolerance properties. Research in these areas has focused on *S.cerevisiae* strains while exploitation of alternative species for improved inhibitor tolerance has been limited. Wimalasena *et al.* [19] recently screened *Saccharomyces* spp. (previously termed Saccharomyces sensu stricto) for their tolerance to osmosis, temperature, ethanol and inhibitors using phenotypic microarray analysis. The results indicated that some non-*S.cerevisiae* yeast strains could have promising properties to be used in lignocellulosic bioethanol fermentation.

In this study, the screening of yeast strains other than *S. cerevisiae* for high formic acid tolerance was conducted. Selected high tolerance strains were investigated for ethanol fermentation under formic acid stress, compared to a typical reference *S. cerevisiae* strain. It is anticipated that the selected strains, with an innate tolerance to inhibitors, could lead to improved bioethanol production from lignocellulose.

MATERIALS AND METHODS

Microorganisms

All the yeast strains used in this study are listed in Table 1. The strains were stored in glycerol stock at -80°C until required. The inoculum was prepared by taking a loop full of stock culture to 10 mL YPD (yeast extract 10 g/L, peptone 20 g/L and glucose 20 g/L) broth and incubating in 30 mL sterlin tube at 30°C, 120 rpm for 48 hours.

Table 1: Strain number and names of *Saccharomyces* spp. used in the study. doi:10.1371/journal.pone.0135626.t001

Strain	Culture Number	Organism
1	DBVPG6466	*Saccharomyces paradoxus*
2	IFO1816	*Saccharomyces mikatae*
3	CBS432	*Saccharomyces paradoxus*
4	DBVPG6299	*Saccharomyces bayanus*
5	2.3319	*Saccharomyces arboricolus*
6	IFO1802	*Saccharomyces kudriavzeii*
7	DBVPG6298	*Saccharomyces castelli*
8	NCYC2592	*Saccharomyces cerevisiae*
9	2.3317	*Saccharomyces arboricolus*
10	2.3318	*Saccharomyces arboricolus*

Spot plate analysis

Spot plate tests were performed according to Homann *et al.* [20] with modifications. YPD agar (YPD plus No. 1 agar 15 g/L) incorporating varying concentrations of formic acid (0, 25, 30, 35 or 40 mM) was used to screen for the tolerance level of the various strains. Yeast Nitrogen Base (YNB) agar composed of YNB 6.7 g/L, glucose 20 g/L, No.1 agar 15 g/L. YNB agar with addition formic acid at concentrations of 0, 10, 15 20 or 25 mM was employed for tolerance level screening. The yeast culture was diluted to an OD of 1.0 (an estimated cell number of 10^7 cells/mL) with distilled water. Then 5 µL samples of ten-fold serial dilution of the yeast cultures were spotted on YPD or YNB agar plates. The plates were incubated anaerobically at 30°C for 48 hours and growth differences were recorded photographically using a Bio-Rad-transilluminator (Bio-Rad, Cambridge, UK).

Phenotype Microarray (PM) analysis

Biolog medium comprising YNB 6.7 g/L, glucose 60 g/L, nutrient supplement 2.6 µl/L (48x NS solution g/L: adenine HCl 4.12, L-histidine HCl monohydrate 1.01, L-leucine 6.3, L-lysine HCl 4.38, L-methionine 9.6, L-tryptophan 2.45 and Uracil 1.61) and a proprietary stain known as dye D 0.2 µL (Biolog, USA). Final volume of 30 µL was made up using distilled water and transferred into various wells with different concentrations of formic acid (0, 10, 15, 20 or 25 mM). Wells also contained 75 µL IFY buffer (Biolog, USA), 3.8 µL of yeast (previously adjusted to 62% transmittance) and 11.2 µL distilled water. 96-well plates were loaded into Omnilog reader (Biolog, USA) and incubated at 30°C for 96 hours under anaerobic condition; data was recorded photographically at 15-minute intervals. The conversion of dye intensity was detected and transformed into a signal value that reflected cell metabolic activity and dye conversion. The signal data were compiled upon completion of the incubation and data exported from the Biolog software into Microsoft Excel. Each experiment was performed in triplicate.

Culture propagation for laboratory scale fermentations

S. arboricolus 2.3317, 2.3318, 2.3319 and *S. cerevisiae* NCYC2592 were employed in lab scale fermentations to investigate formic acid tolerance. A loop full of strain stock was aseptically inoculated into 10 mL of YPD broth and incubated at 30°C for 48 hours and at 120 rpm. The 10 mL culture was transferred into 100 ml of YPD broth and cultured for 48 hours at 30°C, 120 rpm, and the whole culture was finally transferred into 1 L of YPD and cultured for 48 hours at 30°C and 120 rpm. The 1 L culture was centrifuged at 5000 rpm at 4°C for 5 minutes. The supernatant was discarded and the wet pellet was used for inoculation.

Fermentation with addition of formic acid

YPD broth with the addition of formic acid at 0, 10, 20, 30, 40, 50 or 60 mM was used in the laboratory scale fermentations. The pH of the media was adjusted to 4.5 using phosphoric acid and/or NaOH under aseptic conditions. From the broth, 100 mL was transferred into mini fermentation vessels (FVs). The prepared 0.4 g (wet weight) of yeast pellet was aseptically transferred into each of the bottles. Then the bottle was sealed and equipped with a bubbling CO_2 outlet. All bottles were incubated at 30°C with shaking at 200 rpm for 24 hours. Samples were collected at specific time intervals to determine the total cell count, and concentrations of glucose, ethanol, glycerol and glycogen. All fermentations were carried out in triplicate.

Total cell number analysis

The total cell number was determined with a haemocytometer according to the method of Sami*et al.* [21]. Methylene blue 0.01% (w/v) was dissolved in sodium citrate 2% (w/v) solution. Yeast broth was diluted using sterile water. The cell suspension was mixed with methylene blue solution in a ratio of 1:1. The solution was examined microscopically and total cell number was counted.

HPLC analysis

Glucose, ethanol, and glycerol concentrations were determined using a JASCO HPLC system composed of a JASCO AS-2055 Intelligent Autosampler (JASCO, Essex, UK), and a JASCO PU-1580 Intelligent HPLC pump. The Rezex ROA organic acid H⁺ organic acid column (5μm, 7.8mm × 300mm, Phenomenex, Macclesfield, UK) was used and the mobile phase was 0.005 N H_2SO_4 with a flow rate 0.5 mL/minute.

Intracellular glycogen analysis

The procedure of Parrou and Francois [22] was employed for the determination of intracellular glycogen. Total cells of 1×10^9 cell/mL were obtained from the culture, centrifuge at 3,500 rpm for 5 min at 4°C and pellet was washed three times with distilled water. The washed cell pellet was lysed in 250 μL of sodium carbonate (0.25 M) incubated for 2 hours in a 95°C water bath with occasional stirring. To the cell suspension, 600 μL of sodium acetate (0.2 M) and 150 μL of acetic acid (1 M) were added respectively. From the cell suspension, 500 μL was transferred into a fresh Eppendorf tube and 10uL of α-amyloglucosidase (10 mg/mL; Sigma-Aldrich) was added and incubated at a 57°C for 12 hours in water bath. After overnight incubation, samples were centrifuged at 11,000×g for 2 minutes. The liberated glucose was quantified using a glucose assay kit (GOPOD; Megazyme). Analyses were carried out in triplicate and results expressed in glucose concentration as a function of cell number.

RESULTS

Strain screening using spot plate assay

Seven non-*S. cerevisiae* strains listed in Table 1 (strain number 1 to 7) were screened for their tolerance to formic acid alongside a typical lab *S. cerevisiae* strain (NCYC2592), which has been used in our lab previously for ethanol tolerance analysis [8]. In this study, the concentration of formic acid in the media ranged from 0 to 40 mM, which corresponds to the formic acid concentrations typically reported in pretreated biomass hydrolysates [8, 12]. The spot plate results demonstrated that all strains were able to grow on YPD and YNB under control conditions (Fig 1A and 1B and Table 2). Cell growth could be seen on all YPD plates when formic acid was present at concentrations of 25 mM or lower (data not shown). Although the presence of 35 mM formic acid prevented cell growth on most strains (Fig 1A), growth was observed for strains *S. paradoxus* DBVPG6466, *S. kudriavzeii* IFO1802, *S. arboricolus* 2.3319 and *S. cerevisiae* NCYC2592. At a concentration of 40 mM, no cell growth was observed for any of the strains analyzed (data not shown). In assays using YNB medium, all strains displayed a lower tolerance to formic acid. Strains *S. paradoxus* DBVPG6466, *S. kudriavzeii*IFO1802, *S. arboricolus* 2.3319 and *S. cerevisiae* NCYC2592 were only tolerant to 20 mM formic acid on YNB medium plates (Fig 1B, Table 2), suggesting that YPD as an enriched medium may have a higher buffering capacity than YNB, a minimal medium. The formic acid critical concentrations for these strains were summarized in Table 2.

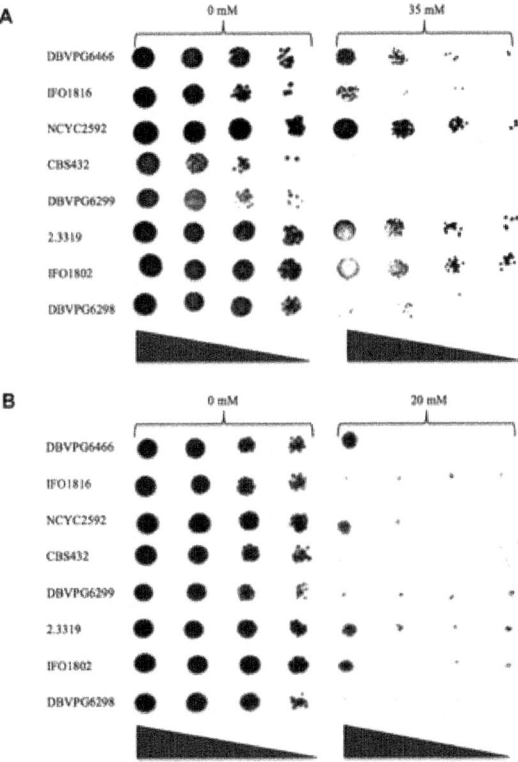

Figure 1: Tolerance to formic acid of *Saccharomyces* spp on solid medium using (A) YPD media with formic acid 0 and 35 mM; (B) YNB media with formic acid 0 and

20 mM. Aliquots 5 μL from tenfold serial dilution yeast cultures (initially suspended to an OD of 1.0 with an estimated cell number of 10^7 cells/mL) were spotted on the plates. Plates were incubated at 30°C under anaerobic conditions for 48 hours. The last concentration of formic acid that cell growth can be observed was defined as tolerance level. doi:10.1371/journal.pone.0135626.g001

Table 2: Summary of formic acid tolerance concentrations of *Saccharomyces* spp by spot plate analysis. doi:10.1371/journal.pone.0135626.t002

| Strain | Culture Number | Formic acid concentration (mM) | |
		YPD	YNB
1	DBVPG6466	35	20
2	IFO1816	30	15
3	CBS432	30	10
4	DBVPG6299	25	10
5	2.3319	35	20
6	IFO1802	35	20
7	DBVPG6298	30	15
8	NCYC 2592	35	20

Strain screening using Phenotypic Microarray

Phenotypic microarray was used to investigate the effect of formic acid on yeast metabolic output, defined here as redox signal intensity [23] (Fig 2). The presence of formic acid elicited a concentration-dependent reduction in redox signal intensity (Fig 2D). *S. cerevisiae* NCYC 2592, *S. paradoxus* DBVPG6466, *S. kudriavzeii* IFO1802 and *S. arboricolus* 2.3319 demonstrated their capacity to tolerate 10 and 15 mM formic acid as high redox signal intensity were shown in phenotypic microarray; while the sensitive strains to these concentrations were *S. paradoxus* CBS432 and *S. bayanus* DBVPG6299. Increase in formic acid to 20 mM exerted more inhibitory effects on all the yeast strains (Fig 2B) and reduced the glucose utilization as compared with the control. The redox signal intensity of strains *S.kudriavzeii* IFO1802, *S. arboricolus* 2.3319 and *S. cerevisiae* NCYC 2592 indicated that these stains tolerate 20 mM formic acid, although both lag phase and maximum redox signal intensity were affected (Fig 2C and 2D). At 25 mM formic acid, there was no metabolic output observed, showing none of the strains could tolerate 25 mM formic acid in phenotypic microarray experiments.

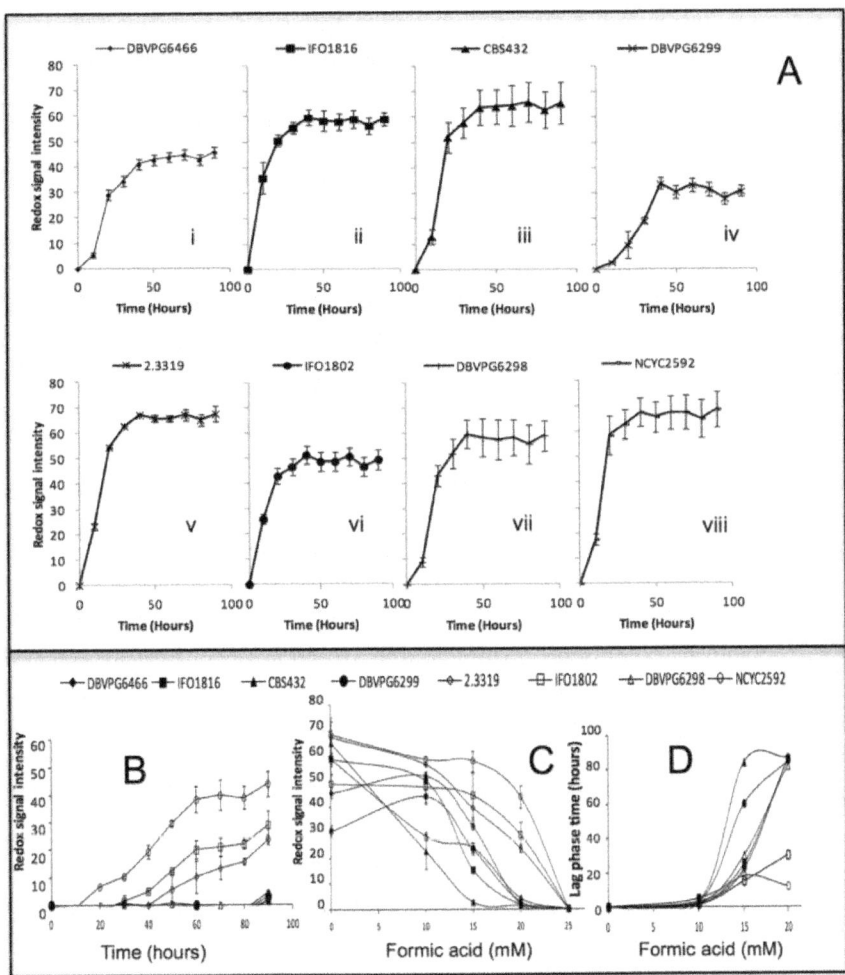

Figure 2: Metabolic output profiles of *Saccharomyces* spp on media containing formic acid (0–20 mM) incubated at 30°C under anaerobic conditions. (A) Control (0 mM) media without formic acid. (B) Formic acid 20 media. (C) Duration of lag phase in PM at formic acid 0,10,15 and 20 mM media. (D) Maximum redox signal intensity after 90 hours at formic acid 0, 10, 15, 20 and 25 mM media. Error bars represent the standard deviation of 3 replicates (Biolog unit = redox signal intensity). doi:10.1371/journal.pone.0135626.g002

The formic acid tolerance of *S. arboricolus* strains during fermentation

According to the spot plate and phenotypic microarray screening experiments, strain *S.kudriavzeii* IFO1802 and *S. arboricolus* 2.3319 showed higher formic acid tolerance than other strains screened in this study. *S. arboricolus* 2.3319 was chosen for further study based on the genome sequence variability among the *Saccharomyces* spp (formerly known as*Saccharomyces* sensu strict yeast) as related to *S. cerevisiae* and as novel *Saccharomyces*species isolated from Tree bark [19, 24, 25]; also the availability of two other *S. arboricolus*(2.3317 and 2.3318) for assessment of formic acid tolerance in our lab. It was reported that *S.arboricolus* 2.3317 and *S. arboricolus* 2.3318 shared 100% similarity based on their multigene analysis [26]. In these fermentation experiments, pH was controlled at 4.5 by buffering. Therefore, higher formic acid concentrations (up to 60 mM) were investigated. *S. arboricolus*2.3319 was firstly assessed for formic acid tolerance during fermentation and the results were compared with the reference strain *S. cerevisiae* NCYC2592 (Fig 3A and 3B). *S. arboricolus*2.3319 and *S. cerevisiae* NCYC2592 strains showed the highest cells number of 8.58 x 10^7cells/mL and 8.07 x 10^7 cells/mL respectively when cultured in the control medium (without formic acid addition). Increase in formic acid concentrations retarded yeast growth where at 60 mM cells number of 6.25± 0.03 × 10^7 cells/mL and 6.03± 0.10 x 10^7 cells/mL for *S. arboricolus*2.3319 and *S. cerevisiae* NCYC2592, were obtained respectively. In fermentations containing formic acid at 10–40 mM, *S. arboricolus* 2.3319 maintained over 90% of its relative cell growth in comparison to the control, while the relative cell growth of *S. cerevisiae* NCYC2592 dropped below 90% in fermentations in the presence of 20 mM or higher concentrations of formic acid. These results showed that *S. arboricolus* 2.3319 has higher formic acid tolerance than *S.cerevisiae* NCYC2592. Based on these results, *S. arboricolus* 2.3317 and 2.3318 were further assessed for the formic acid tolerance with the hypothesis that the strains from the same species may share similar weak acid tolerance properties. As shown in Fig 3C and 3D, strain*S. arboricolus* 2.3317 and 2.3318 exhibited good formic acid tolerance (relative cell growth is over 90%) at concentrations of 10 and 20 mM. However, when the formic acid concentration increased to 30 mM or higher, cell growth of strains *S. arboricolu* 2.3317 and 2.3318 was affected.

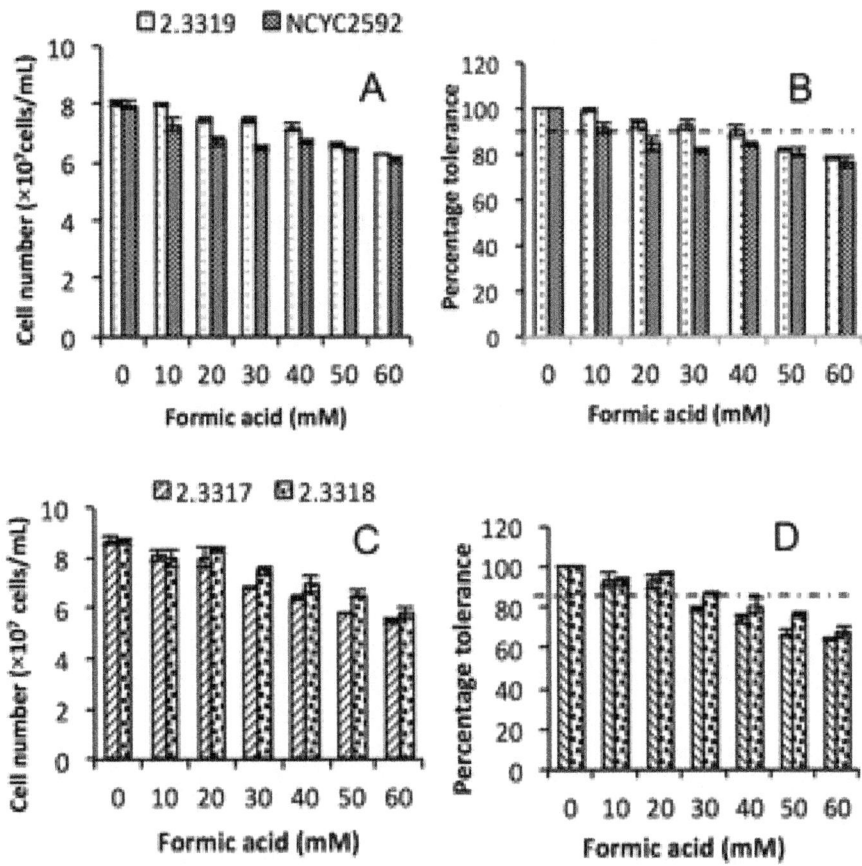

Figure 4: Growth profile (A and C) and percentage tolerance (B and D) of *Saccharomyces arboricolus* 2.3317, 2.3318 and 2.3319 and *Saccharomyces cerevisiae* NCYC2592 in the presence of formic acid (0–60 mM) using YPD medium. Values are the mean of three experiments and vertical error bars represent standard deviation. doi:10.1371/journal.pone.0135626.g003

During fermentations in the presence of 40 mM formic acid *S. arboricolus* 2.3318, 2.3319 and *S. cerevisiae* NCYC2592 all showed improved ethanol production (Table 3), however, further increase formic acid concentration to 50 or 60 mM led to a decrease in ethanol synthesis for all strains. *S. arboricolus* 2.3319 demonstrated similar ethanol fermentation capacity in terms of final ethanol concentration and ethanol yield as *S. cerevisiae* NCYC2592. This indicated that *S. arboricolus* strain 2.3319 could be a good candidate for bioethanol fermentations.

Table 3: Ethanol concentration and yield summary from *Saccharomyces* spp fermentation of glucose in the presence of formic acid. Data are the mean of triplicate experiments and standard deviation. doi:10.1371/journal.pone.0135626.t003

Formic acid (mM)	S. arboricolus 2.3317		S. arboricolus 2.3318		S. arboricolus 2.3319		S. cerevisiae NCYC2592	
	Ethanol concentration (g/l)	Ethanol yield	Ethanol concentration (g/l)	Ethanol yield	Ethanol concentration (g/l)	Ethanol Yield	Ethanol concentration (g/l)	Ethanol yield
0	8.54±0.15	0.43±0.01	8.92±0.46	0.45±0.02	9.61±0.09	0.48±0.00	9.80±0.05	0.49±0.00
10	8.48±0.67	0.42±0.03	8.97±0.16	0.45±0.01	9.54±0.08	0.48±0.00	9.50±0.16	0.48±0.01
20	8.48±0.20	0.42±0.01	8.92±0.80	0.45±0.09	9.48±0.22	0.47±0.01	9.62±0.11	0.48±0.01
30	8.38±0.45	0.42±0.02	9.20±0.28	0.46±0.01	9.95±0.47	0.50±0.02	9.92±0.42	0.50±0.02
40	8.53±0.10	0.43±0.01	9.03±0.33	0.45±0.02	9.91±0.12	0.50±0.01	10.20±0.36	0.51±0.02
50	8.50±0.04	0.42±0.00	8.81±0.44	0.44±0.02	9.89±0.11	0.49±0.01	10.04±0.31	0.50±0.02
60	8.41±0.03	0.42±0.00	8.96±0.05	0.45±0.00	9.13±0.45	0.47±0.02	8.89±0.11	0.44±0.01

In order to further explore the response of yeast to formic acid, glucose utilization, glycerol production and glycogen production of *S. arboricolus* 2.3319 and *S. cerevisiae* NCYC2592 were determined. There was no significant difference in glucose consumption between *S.arboricolus* 2.3319 (Fig 4A) and *S. cerevisiae* NCYC2592 at all concentrations (Fig 5A), although both strains consumed glucose faster in fermentations with a lower initial formic acid concentration. These results agreed with the cell growth curves observed in these fermentations (Figs 4B and 5B). Strains in fermentations with a higher formic acid concentration grew slower and ended at a relatively lower final cell concentration (Figs 3A, 3C,4B and 5B). Compared with *S. cerevisiae* NCYC2592, *S. arboricolus* 2.3319 consumed glucose faster at the 4-hour data point. Glycerol is generally considered to be associated with stress tolerance [27]. At all formic acid concentrations, *S. arboricolus* 2.3319 produced more glycerol than *S. cerevisiae* NCYC2592 (Fig 6A). In the fermentations with 40 mM formic acid, *S. arboricolus* 2.3319 produced 1.37 ± 0.16 g/L glycerol while *S. cerevisiae* NCYC2592 produced 1.02 ± 0.11 g/L glycerol. Compared within *S. arboricolus* strains, glycerol produced by *S. arboricolus* 2.3317 and *S. arboricolus* 2.3318 was similar to *S. arboricolus* 2.3319 (Fig 6).

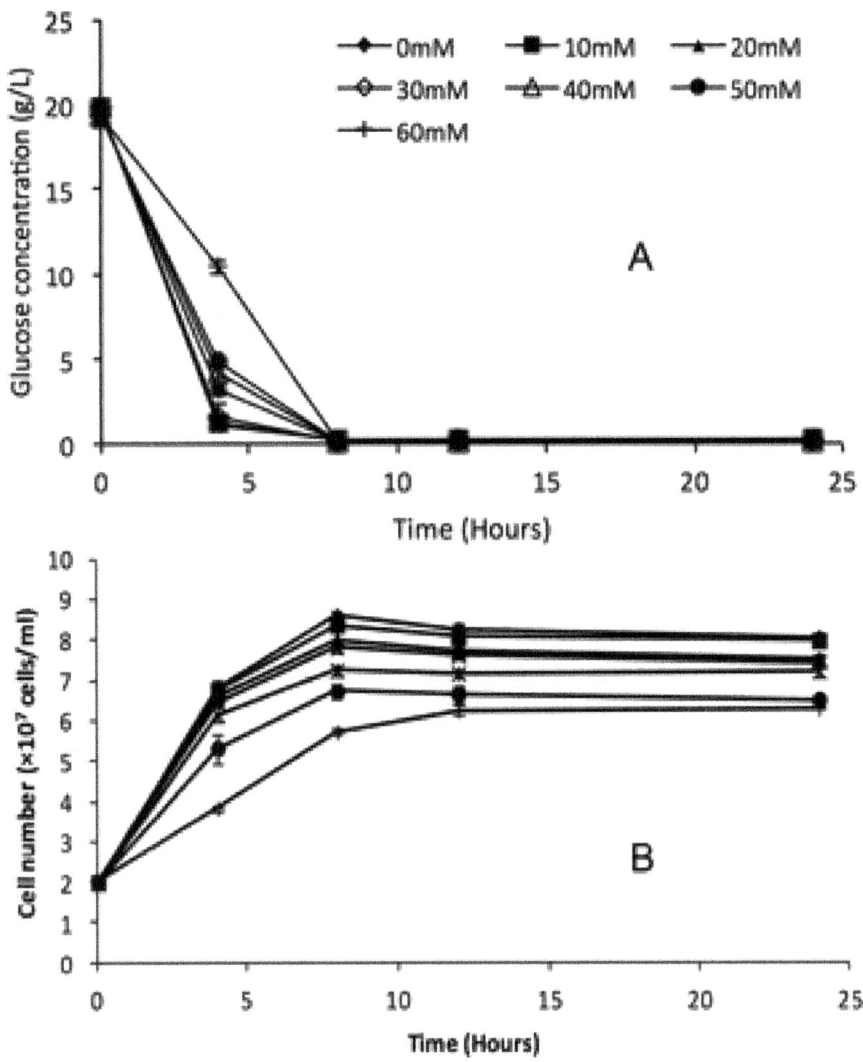

Figure 4: Time course profiles of glucose consumption (A) and cell growth curve (B) of *Saccharomyces arboricolus* 2.3319 in fermentation using YPD medium and formic acid (0–60 mM). Values are the mean of three experiments and vertical error bars represent standard deviation. doi:10.1371/journal.pone.0135626.g004

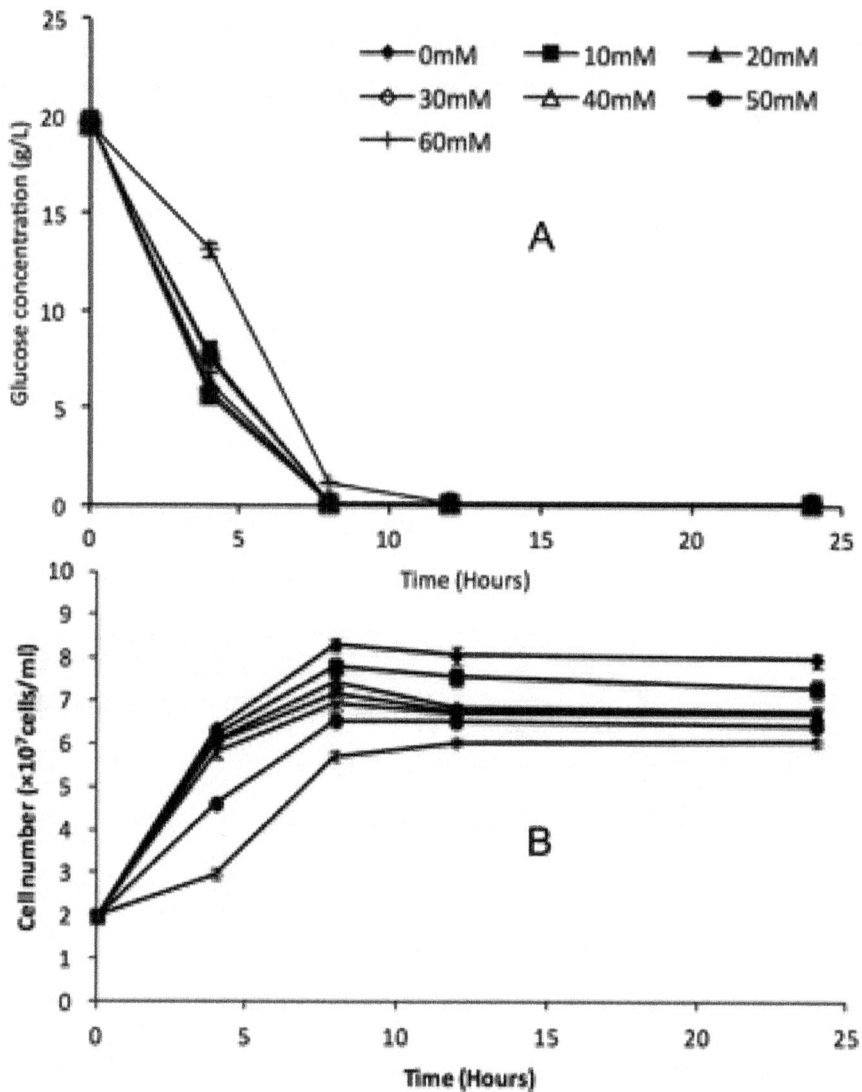

Figure 5: Time course profiles of glucose consumption (A) and cell growth curve (B) of *Saccharomyces cerevisiae* NCYC 2592 in fermentation using YPD medium and formic acid (0–60 mM). Values are the mean of three experiments and vertical error bars represent standard deviation. doi:10.1371/journal.pone.0135626.g005

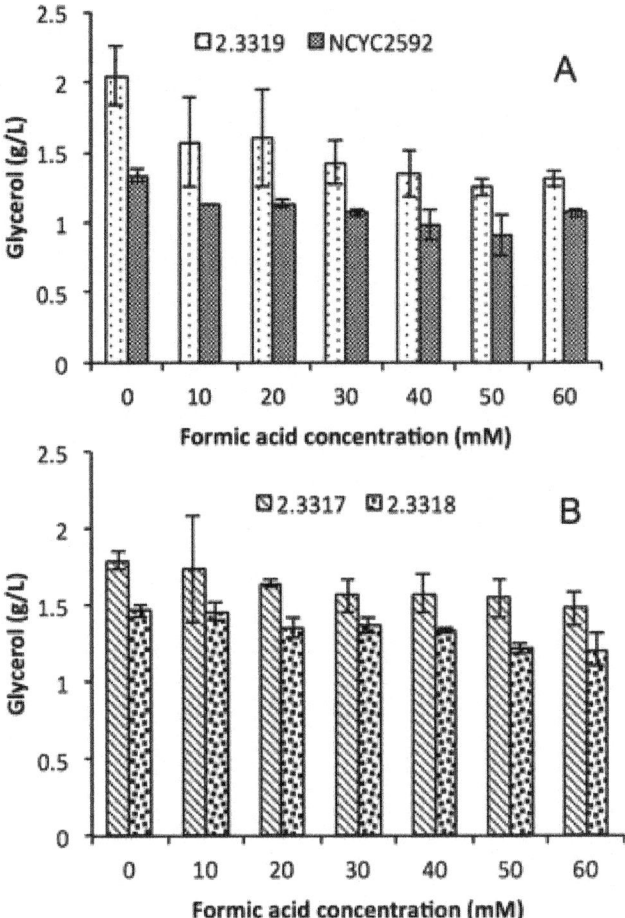

Figure 6: Level of glycerol produced by *Saccharomyces cerevisiae* NCYC 2592 and-*Saccharomyces arboricolus* 2.3319 (A), and *Saccharomyces arboricolus* 2.3317 and *Saccharomyces arboricolus* 2.3318 (B) in the presence of formic acid (0–60 mM) using YPD medium. Fermentation was carried out under anaerobic condition, incubated at 30°C and samples were analyzed after 24 hours. Values are the mean of three experiments and vertical error bars represent standard deviation. doi:10.1371/journal.pone.0135626.g006

Accumulation of intracellular glycogen in yeast cells exposed to formic acid stress during fermentation was also determined. Higher accumulation of glycogen in yeast cells may act as an energy reserve maintaining cell viability when stressed [28]. Fig 7 revealed that increase in formic acid concentration decreased intracellular glycogen accumulation in all strains. In comparison, *S. arboricolus* 2.3319 accumulated significantly higher concentrations of glycogen than *S. cerevisiae* NCYC2592 (Fig 7A), and *S. arboricolus* 2.3317 and *S. arboricolus* 2.3318 (Fig 7B) in formic acid media from 0 to 60 mM, with the maximum intracellular glycogen concentration of 100.64 ± 8.82 $\mu g/10^9$ cells. This higher accumulation of intracellular glycogen confirmed *S. arboricolus* 2.3319 to be more formic acid tolerant than *S. arboricolus* 2.3317, 2.3318 and *S. cerevisiae* NCYC2592.

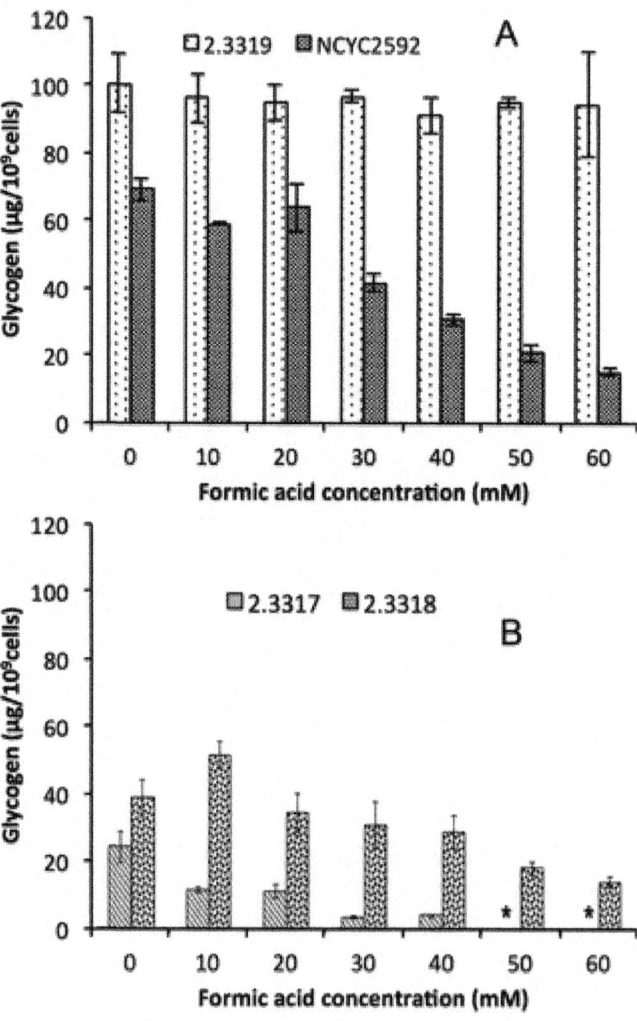

Figure 7: Intracellular glycogen accumulation by *Saccharomyces cerevisiae* NCYC 2592 and *Saccharomyces arboricolus* 2.3319 (A), and *Saccharomyces arboricolus*2.3317 and *Saccharomyces arboricolus* 2.3318 (B) in the presence of formic acid (0–60 mM) using YPD medium. Fermentation was carried out under anaerobic condition, incubated at 30°C and samples were analyzed after 24 hours. Values are the mean

of three experiments and vertical error bars represent standard deviation. *- Data are not available. doi:10.1371/journal.pone.0135626.g007

DISCUSSION

The hydrolysates of lignocellulosic substrates can contain toxic compounds which are released during pretreatment. These compounds can negatively affect the fermentation efficiency of *Saccharomyces cerevisiae* strains. Consequently, some form of strain- or process-adaptation is required to prevent the detrimental impact of inhibitors. Although the composition of toxic compounds differs among lignocellulosic biomass and pretreatment methods, it has been widely reported that weak acids such as formic acid are produced via a range of pre-treatment methods, including the use of dilute sulfuric acid and hot-compressed water treatments [12, 29]. Hasunuma *et al.* [30] reported that 10–20 mM concentrations of formic acid in the pretreated biomass hydrolysate were sufficient to hinder fermentation by*Saccharomyces cerevisiae*. Thus screening to identify yeast strains which are tolerant to formic acid may help to improve the efficiency of lignocellulosic ethanol production.

Yeast, particularly *Saccharomyces cerevisiae* strains had been screened for the tolerance of inhibitors in the lignocellulosic hydrolysate as well as osmotic, heat and acid stresses [31, 32]. However, screening of non-*Saccharomyces cerevisiae* strains for inhibitor tolerance, particularly formic acid has been limited. It has been shown that weak acids inhibit yeast cells through similar inhibitory mechanisms [33]. Therefore, screening of formic acid tolerance strains using spot plate and phenotypic microarray could identify yeasts with potential tolerance to other weak acids as well, especially as formic acid is considered to be more toxic than acetic acid and levulinic acid [12, 30]. The exploration of wild *Saccharomyces* sp. for inhibitor tolerance could lead to the discovery of novel yeast strains with distinct genetic background associated with formic acid tolerant. The resulting high tolerant strains can be mated with *Saccharomyces cerevisiae* to form hybrid diploids that will utilize hydrolysate for bioethanol production [19, 30].

In this report, we examined the formic acid tolerance of 7 non-*Saccharomyces cerevisiae*strains. YPD and YNB agar media were firstly employed with the addition of different concentrations of formic acid. With increasing formic acid concentration growth variation was observed amongst the strains. Strains *S. paradoxus* DBVPG6466, *S. kudriavzeii* IFO1802, *S.arboricolus* 2.3319 and *S. cerevisiae* NCYC2592 exhibited tolerance to 35 mM and 20 mM formic acid on YPD and YNB media respectively, while other strains did not grow. The tolerance levels of yeast strains to formic acid concentration in YPD was higher than tolerance in YNB because YNB as minimal medium affected cell growth [34]. The spot plate method has also been used by other researchers for the study of inhibitory effect on yeast strains caused by weak organic acids [35].

Screening of yeast strains for stress and inhibitor tolerance is an important step in lignocellulosic bioethanol production. The utilization of only spot plate analysis for the identification of tolerance strains is easy to operate and has less equipment dependent, but is slow and time consuming [8]. Phenotypic Microarray (PM) is an integrated system for high-throughput screening of microorganisms. Compared with spot plate method, PM is a quick, automatic and liquid-culture based technique that allows screening of a large number of strains at the same time under various conditions [8, 23]. However, it relies on the measurement of cell respiratory activity rather than cell growth [8, 23]. PM has recently been used to characterize yeast tolerance to various stresses, e.g. temperature, ethanol and inhibitors in the lignocellulosic hydrolysate [19]. Seven non-*Saccharomyces cerevisiae* strains were screened using the phenotypic microarray. The results demonstrated that *S. arboricolus* 2.3319 and *S.kudriavzeii* IFO1802 could tolerate 20 mM formic acid (Fig 2). This agreed with results obtained using YNB spot plate method, as PM uses a minimal medium as well. Strain *S. paradoxus*DBVPG6466 could not tolerate 20 mM formic acid in PM experiments, suggesting that strains may have different acid tolerance abilities between solid culture (spot plate) and liquid culture (PM). This indicated that PM is a good complement to the traditional spot plate screening method.

Based on the above results, *S. arboricolus* 2.3319 and two other *S. arboricolus* (2.3317 and 2.3318) strains were selected and tested for formic acid tolerance and bioethanol production at a laboratory scale. Compared with lab strain, *S. cerevisiae* NCYC2592, *S. arboricolus* 2.3319 maintained over 90% relative cell growth in the presence of formic acid (10 to 40 mM) (Fig 3). The reference strain, *S. cerevisiae* NCYC2592 also showed good formic acid tolerance, but the relative cell growth reduced to 90% or lower when formic acid concentration was 20 mM or higher. In all cases, the addition of formic acid decreased cell number (Fig 3). The exact reason for this reduction is unknown. A potential reason may be due to the diversion of energy (ATP) to pump out protons at the expense of cell biomass production [36]. Similar results were reported by Huang *et al.* [37] in the investigation of the inhibitory effect on *S. cerevisiae*. In comparison with *S. arboricolus* 2.3319, *S. arboricolus* 2.3317 and 2.3318 did not exhibited high formic acid tolerance, showing no correlation between species tolerance to formic acid. It was also established by Almeida *et al.* [31] who screened strains of *S. cerevisiae* and found out that inhibitor tolerance was strain specific and not species specific.

The formic acid has both positive and negative effects on bioethanol fermentation. At low or medium acid concentration (e.g. 30–40 mM), the presence of formic acid increased both ethanol yield and titer in both *S. arboricolus* 2.3319 and *S. cerevisiae* NCYC2592 strains (Table 3). A possible explanation is that glycolytic activity may be increased in order to produce more ATP required to pump protons out of the cells at the expense of cell biomass, while ethanol production was increased for ATP production [15, 30, 38]. This investigation supports the work of Teherzadeh *et al.* [39], who reported that during fermentation, exposure of *S. cerevisiae* to acetic acid (< 25 m) stimulated ATP production and increased the rate of ethanol production when compared to an unstressed control. Further increase in formic acid concentration to 60 mM led to drops in ethanol production (Table 3). The increased toxicity was associated with formic acid's high plasma membrane permeability [12, 30, 37]. But formic acid did not affect glucose consumption in either *S. arboricolus* or *S. cerevisiae* strains (Fig 4). Similar result was reported in yeast fermentations with acetic acid [34]. Compared with fermentations using *S.cerevisiae* NCYC2592, fermentations using *S. arboricolus* 2.3319 resulted in similar or higher ethanol yields, which were also very close to the theory glucose to ethanol yield of 0.51 (Table 3). These ethanol yields were also higher than or similar to several reports in fermentations using other high tolerance ethanol producing strains, e.g. *S. cerevisiae* YZ1 [31], *S. cerevisiae*Y-1528 [32] and an isolate of Bekonang [34].

In the fermentation process, *S. arboricolus* 2.3319 contains inherently higher glycerol concentrations under control and formic acid stress than *S. cerevisiae* NCYC2592 (Fig 6). This may be of benefit for *S. arboricolus* to tolerate formic acid than that of strain *S. cerevisiae*NCYC2592 though further experimentation would be required to establish if glycerol could be acting as a polyol to the cells [27]. The investigation agreed with the work of Tomas-Pejo *et al.* [40] that higher glycerol production resulted to strain tolerance to lignocellulosic hydrolysate inhibitors which is an indication of better cell growth. Lower amounts of glycerol were produced when formic acid addition was increased. The low glycerol formation may be attributed to the re-oxidation of NADH to NAD$^+$ for the ATP formation [41]. Taherzadeh *et al.* [42] also reported that the addition of acetate resulted in a decrease in glycerol production. Intracellular glycogen has been considered as an important reserved carbohydrate in the survival of yeasts when yeasts were exposed to stress [43, 44]. In this study, *S. arboricolus* 2.3319 accumulated more intracellular glycogen than *S. cerevisiae* NCYC2592 (Fig 7), which may help the strain to tolerate high formic acid concentrations. The intracellular glycogen may be playing a dual role according to Deshpande *et al.* [28], by providing energy and carbon skeleton required for cell growth as well as minimize leakage through plasma membrane by the stressful effect of formic acid. This agreed with Somani *et al.* [45] that higher content of glycogen favored the survival of yeast strain during environmental stress conditions.

CONCLUSION

This study reported a simple and fast method for screening non-*S. cerevisiae* strains for formic acid tolerance. In comparison to other non-*S. cerevisiae* strains, *S. arboricolus* 2.3319 was shown to be tolerant to formic acid using both spot plate and PM techniques, which was then confirmed by a series of fermentations. Fermentation experiments demonstrated that *S.arboricolus* 2.3319 produced similar amounts of ethanol to the reference strain *S. cerevisiae*NCYC2592, indicating its potential to be used as a novel bioethanol producer or as a source of gene donor to other higher ethanol producing strains that are inhibitor-sensitive. *S. arboricolus*2.3319 produced more glycerol and glycogen than the reference strain, which may enable its good formic acid tolerance ability.

ACKNOWLEDGMENTS

This study was supported by the Biotechnology and Biological Sciences Research Council (BBSRC), under the programme of Lignocellulosic Conversion to Ethanol (LACE) (BB/G01616X/1). The authors also thank the financial support provided by Educational Trust Fund from the University of Benin, Benin City, Nigeria for Cyprian Oshoma's PhD scholarship.

AUTHOR CONTRIBUTIONS

Conceived and designed the experiments: CO DG EL KS TP CP CD. Performed the experiments: CO DG. Analyzed the data: CO DG EL KS TP CP CD. Contributed reagents/materials/analysis tools: CO DG EL. Wrote the paper: CO DG EL KS TP CP CD.

REFERENCES

1.   Chandel AK, Chan EC, Rudravaram R, Narasu ML, Rao LV, Ravindra P. Economics and environmental impact of bioethanol production technologies: an appraisal. Biotechnol Mol Biol Rev. 2007: 2: 14–32.

2.   .Fujita Y, Takahashi S, Ueda M, Tanaka A, Okada H, Motikawa Y, et al. Direct and efficient production of ethanol from cellulosic material with a strain displaying cellulolytic enzyme. Appl Environ Microbiol. 2002: 68(10): 5135–5141. doi: 10.1128/aem.68.10.5136-5141.2002

3. Van Wyk JPH. Biotechnology and the utilization of biowaste as a resource for bioproduct development. Trends Biotechnol. 2001: 19: 172–177. pmid:11301129 doi: 10.1016/s0167-7799(01)01601-8

4. Sun Y, Cheng J. Hydrolysis of lignocellulosic materials for ethanol production: a review. Biores Technol. 2002: 83: 1–11. doi: 10.1016/s0960-8524(01)00212-7

5. Barber A.R, Hansson H, Pamment NB. Acetaldehyde stimulation of the growth of *Saccharomyces cerevisiae* in the presence of inhibitors found in lignocellulose-to-ethanol fermentations. J Indus Microbiol Biotechnol. 2000: 25: 104–108. doi: 10.1038/sj.jim.7000031

6. Palmqvist E, Grage H, Meinander NQ, Hahn-Hagerdal B. Main and interaction effects of acetic acid, furfural and hydroxybenzoic acid on growth and ethanol productivity of yeasts. Biotechnol Bioeng. 1999: 63: 46–55. pmid:10099580 doi: 10.1002/(sici)1097-0290(19990405)63:1<46::aid-bit5>3.3.co;2-a

7. Palmqvist E, Hahn-Hagerdal B. Fermentation of lignocellulosic hydrolysates. I: Inhibition and detoxification. Biores Technol. 2000a: 74: 17–24. doi: 10.1016/s0960-8524(99)00160-1

8. Greetham D, Wimalasena T, Kerruish DWM, Brandley S, Ibbett R, Linforth RL, et al. Development of a phenotypic assay for characterization of ethanologenic yeast strain sensitivity to inhibitors released from lignocellulosic feedstocks. J. Ind. Microbiol. Biotechnol. 2014 doi: 10.1007/s10295-014-1431-6

9. Sun W, Tao W. Comparison of cell growth and ethanol productivity on different pretreatment of rice straw hemicellulose hydrolysate by using *Candida shehatae* CICC 1766. Afri J Microbiol Research. 2010: 4(11): 1105–109.

10. Taherzadeh MJ, Niklasson C, Liden G. On-line control of fed-batch fermentation of dilute-acid hydrolyzates. Biotechnol Bioeng. 2000:69:330–338. pmid:10861413 doi: 10.1002/1097-0290(20000805)69:3<330::aid-bit11>3.0.co;2-q

11. Palmqvist E, Hahn-Hagerdal B. Fermentation of lignocellulosic hydrolysates. II: Inhibitors and mechanisms of inhibition. Biores Technol. 2000b: 74: 25–33. doi: 10.1016/s0960-8524(99)00161-3

12. Almeida JRM, Modig T, Peterson A, Hahn-Hagerdal B, Liden G, Gorwa-Grauslund MF. Increase tolerance and conversion of inhibitors in lignocellulosic hydrolysates by *Saccharomyces cerevisiae. J* Chem Technol Biotechnol. 2007: 82(4): 340–349. doi: 10.1002/jctb.1676

13. Pampulha ME, Loureiro-Dias MC. Energetics of the effect of acetic acid

on growth of*Saccharomyces cerevisiae*. FEMS Microbiol Letters. 2000: 184: 69–72. doi: 10.1111/j.1574-6968.2000.tb08992.x

14. Verduyn C, Postma E, Scheffers WA, Van Dijken JP. Effect of benzoic acid on metabolic fluxes in yeasts: a continuous culture study on the regulation of respiration and alcoholic fermentation. Yeast. 1992: 8:501–517. pmid:1523884 doi: 10.1002/yea.320080703

15. Hasunuma T, Sanda T, Yamada R, Yoshimura K, Ishii J, Kondo A. Metabolic pathway engineering based on metabolomics confers acetic and formic acid tolerance to a recombinant xylose-fermenting strain of *Saccharomyces cerevisiae*. Microbial Cell Factories. 2011a: 10:1–13. doi: 10.1186/1475-2859-10-2

16. Larsson S, Palmqvist E, Hahn-Hagerdal B, Tengborg C, Stenberg K, Zacchi G, et al. The generation of fermentation inhibitors during dilute acid hydrolysis of softwood. Enzyme Microbial Technol. 1999: 24:151–159. doi: 10.1016/s0141-0229(98)00101-x

17. Ivanova V., Petrova P. and Hristov J. Application in the ethanol fermentation of immobilized yeast cells in matrix of alginate/magnetic nanoparticles, on chitosan-magnetite microparticles and cellulose-coated magnetic nanoparticles. International Review of Chemical Engineering 2011: 3(2): 289–299.

18. .Martin C, Fernandez T, Garcia R, Carrillo E, Marcet M, Galbe M, et al. Preparation of hydrolysates from tobacco stalks and ethanolic fermentation by *Saccharomyces cerevisiae*. World J Microbiol Biotechnol. 2002: 18: 857–862.

19. Wimalasena TT, Greetham D, Marvin ME, Liti G, Chandelia Y, Hart A, et al. Phenotypic characterization of *Saccharomyces* spp. yeast for tolerance to stresses encountered during fermentation of lignocellulosic residues to produce bioethanol. Microb Cell Factories. 2014: 13: 47. doi: 10.1186/1475-2859-13-47

20. Homann OR, Cai H, Becker JM, Lindquist SL. Harnessing natural diversity to probe metabolic pathways. PLoS genetics. 2005: 1:715–729. doi: 10.1371/journal.pgen.0010080.eor

21. Sami M, Ikeda M, Yabuuchi S. Evaluation of the alkaline methylene blue staining method for yeast activity determination. J Ferment Bioeng. 1994: 78: 212–216. doi: 10.1016/0922-338x(94)90292-5

22. Parrou JL, Francois JA. A simplified procedure for a rapid and reliable assay of both glycogen and trehalose in whole yeast cells. Analyt Biochem. 1997: 248: 186–188. pmid:9177741 doi: 10.1006/abio.1997.2138

23. Bochner BR. Global phenotypic characterization of bacteria. FEMS Microbiol Rev. 2009: 33: 191–2005. doi: 10.1111/j.1574-6976.2008.00149.x. pmid:19054113

24. Liti G, Ba AN, Blythe M, Muller CA, Bergstrom A, Cubillos FA, et al. High quality de novo sequencing and assembly of the *Saccharomyces arboricolus* genome. BMC Geno. 2013: 14: 69. doi: 10.1186/1471-2164-14-69

25. Wang S, Bai F. *Saccharomyces arboricolus* sp. nov., a yeast species from tree bark. Int J System Evol Microbiol. 2008: 58: 510–514. doi: 10.1099/ijs.0.65331-0

26. Naumov GI, Naumova ES, Masneuf-Pomarede I. Genetic identification of new biological species *Saccharomyces arboricolus* Wang et Bai. Ant Van Leeuwen. 2010: 98: 1–7. doi: 10.1007/s10482-010-9441-5

27. .Walker G M. Yeast Physiology and Biotechnology. John Wiley and Sons, Chichester, New York. 1998: 350pp.

28. Deshpande PS, Sankh SN, Arvindekar AU. Study of two pools of glycogen in*Saccharomyces cerevisiae* and their role in fermentation performance. J. Institute Brew. 2011: 117: 113–119. doi: 10.1002/j.2050-0416.2011.tb00451.x

29. van Maris A J A, Abbott DA, Bellissimi E, van den Brink J, Kuyper M, Luttik MAH, et al. Alcoholic fermentation of carbon sources in biomass hydrolysates by*Saccharomyces cerevisiae*: current status. Antonie Van Leeuwenhock. 2006: 90:391 418. pmid:17033882 doi: 10.1007/s10482-006-9085-7

30. Hasunuma T, Sung K, Sanda T, Yoshimura K, Matsuda F, Kondo A. Efficient fermentation of xylose to ethanol at high formic acid concentrations by metabolically engineered *Saccharomyces cerevisiae*. Appl Microbiol Biotechnol. 2011b: 90:997–1004. doi: 10.1007/s00253-011-3085-x

31. Almeida JRM, Karhumaa K, bengtsson O, Gorwa-Grauslund MF. Screening of*Saccharomyces cerevisiae* strains with respect to anaerobic growth in non-detoxified lignocellulose hydrolysate. Biores Technol. 2009: 100: 3674–3677. doi: 10.1016/j.biortech.2009.02.057

32. Zheng D, Wu X, Tao X, Wang P, Li P, Chi X, et al. Screening and construction of*Saccharomyces cerevisiae* strains with improved multi-tolerance and bioethanol fermentation performance. Biores Technol. 2011: 102: 3020–3027. doi: 10.1016/j.biortech.2010.09.122

33. Hasunuma T, Kondo A. Development of yeast cell factories for consolidated bioprocessing of lignocellulose to bioethanol through

cell surface engineering. Biotechnol Adv. 2012: 30: 1207–1218. doi: 10.1016/j.biotechadv.2011.10.011. pmid:22085593

34. Keating JD, Panganiban C, Mansfield SD. Tolerance and adaptation of ethanologenic yeasts to lignocellulosic inhibitory compounds. Biotechnol Bioeng. 2006: 93: 1196–1206. pmid:16470880 doi: 10.1002/bit.20838

35. Buaer BE, Rossington D, Mollapour M, Mamnun Y, Kuchler K, Piper P. Weak organic acid stress inhibits aromatic amino acid uptake by yeast, causing a strong influence of amino acid auxotrophies on the phenotypes of membrane transporter mutants. Eur J Biochem. 2003: 270: 3189–3195. pmid:12869194 doi: 10.1046/j.1432-1033.2003.03701.x

36. Wikandari R, Millati R, Syamsiyah S, Muriana R, Ayuningsih Y. Effect of furfural, hydroxymethlfurfural and acetic acid on indigeneous microbial isolate for bioethanol production. Agric J. 2010: 5: 105–109. doi: 10.3923/aj.2010.105.109

37. Huang H, Guo X, Li D, Liu M, Wu J, Ren H. Identification of crucial yeast inhibitors in bioethanol and improvement of fermentation at high pH and high total solids. Biores Technol. 2011: 102:7486–7493. doi: 10.1016/j.biortech.2011.05.008

38. Casey E, Sedlak M, Ho NWY, Mosier NS. Effect of acetic acid and pH on the cofermentation of glucose and xylose to ethanol by genetically engineered strain of*Saccharomyces cerevisiae*. FEMS Yeast Res. 2010: 10: 385–393. doi: 10.1111/j.1567-1364.2010.00623.x. pmid:20402796

39. Taherzadeh MJ, Niklasson C, Liden G. Acetic acid friend or foe in anaerobic batch conversion of glucose to ethanol by *Saccharomyces cerevisiae*? Chem Eng Sci. 1997: 52: 2553–2659. doi: 10.1016/s0009-2509(97)00080-8

40. Tomas-Pejo E, Ballesteros M, Oliva JM, Olsson L. Adaptation of the xylose fermenting yeast *Saccharomyces cerevisiae* F12 for improving ethanol production in different fed-batch SSF processes. J. Ind. Microbiol. Biotechnol. 2010: 37: 1211–1220. doi: 10.1007/s10295-010-0768-8. pmid:20585830

41. Teherzadeh MJ, Gustaffson L, Niklasson C. Physiological effects of 5-hydroxymethyfurfural on *Saccharomyces cerevisiae*. Appl Microbiol Biotechnol. 2000: 63: 701–708.

42. Teherzadeh MJ, Gustaffson L, Niklasson C, Liden G. Conversion of furfural in aerobic and anaerobic batch fermentation of glucose by *Saccharomyces cerevisiae*. J Biosci Bioeng. 1998: 87: 169–174. doi: 10.1016/s1389-1723(99)89007-0

43. Francois J, Parrou JL. Reserve carbohydrates metabolism in the yeast*Saccharomyces cerevisiae*. FEMS. 2001: 25: 125–145. doi: 10.1111/ j.1574-6976.2001.tb00574.x

44. Samokhvalov VA, Mel'nikov GV, Ignatov VV. The role of trehalose and glycogen in the survival of aging *Saccharomyces cerevisiae* cells. Microbiol. 2004: 73: 378–382. doi: 10.1023/b:mici.0000036979.96786.78

45. Somani A, Bealin-Kelly F, Axcell B, Smart KA. Impact of storage temperature on lager brewing yeast viability, glycogen, trehalose and fatty acid content. J Am Soc Brew Chem. 2012: 70(2): 123–130. doi: 10.1016/j.cervis.2013.09.017

Chapter 10

EFFECT OF FERMENTATION PROCESS ON NUTRITIONAL COMPOSITION AND AFLATOXINS CONCENTRATION OF DOKLU, A FERMENTED MAIZE BASED FOOD

Marina C. N. Assohoun[1,2], Théodore N. Djeni[1*], Marina Koussémon-Camara[1], Kouakou Brou[2]

[1]Department of Biotechnology and Food Microbiology, University Nangui Abrogoua, Abidjan, Côte d'Ivoire; [2]Department of Nutrition and Food Security, University Nangui Abrogoua, Abidjan, Côte d'Ivoire.

ABSTRACT

Investigations were carried out to determine the influence of spontaneous fermentations as achieved at household level on the nutrients composition and aflatoxins concentration of maize during the processing into doklu, a fermented maize food product consumed in Côte d'Ivoire with legumes, soup and fried fish. Results showed that maize grains contained aflatoxin B1, G1 and G2 and that during fermentation all physicochemical parameters significantly (P ≤ 0.05) decreased except moisture and total titratable acidity contents which were significantly (P ≤ 0.05) increased. Fermentation also caused significant reduction in the concentration of total aflatoxins (72%), with the most important reduction in aflatoxin B1 (80%) after the soaking of maize grains. However, no aflatoxin was detected after 24 hours of fermentation until the final product was obtained. Despite the losses in some nutritional compounds, the fermented product, doklu, was found to have appreciable nutritional quality.

INTRODUCTION

Traditional cereal foods play an important role in the diet of the people of Africa particularly in cereal producing zones. Flour from various cereals is one of the main raw materials used in the production of popular food products with high acceptability, good storage characteristics and affordable cost [1]. These cereals are very widely utilized as food in African countries and account for as much as 77% of total caloric consumption [2]. Indeed, Cereals have a relatively better mineral profile but the availability of these minerals to human system is low [3]. Phytic acid present in considerable amount in cereal grains [4] may be partly responsible for the low digestibility of starch [5], protein [6] and low bioavailability of minerals [7,8]. A majority of traditional cereal based foods consumed in Africa are processed by natural fermentation and are particularly important as weaning foods for infants and as dietary staples for adults [9,10]. Fermentation of food grains is known to be an effective method of improving the starch and protein digestibility [11] and bioavailability of minerals [4]. Fermentation also brings down the level of antinutrients like phytic acid and polyphenols [7,12].

In Côte d'Ivoire, the indigenous fermented foods are numerous and varied. Doklu is one of these traditional fermented foods produced mainly in the southern parts of country at household level and for family consumption only. It is a snack made from maize flour and eaten at any time of the day. The people often appreciate doklu for its sour taste due to fermentation. In the preparation of doklu, cleaned and washed whole maize grains are steeped in water for 1 or 2 days, milled, mixed into a dough and left to undergo a spontaneous fermentation for 2 - 3 days by desirable microbes in the environment. One portion of the fermented dough is firstly precooked andthen shaped into balls, wrapped in maize husks and boiled for about 3 h (**Figure 1**).

Besides improving the digestibility, bioavailability of minerals and reducing the level of antinutrients, fermentation may also change the level of nutrients in the food grains. Indeed, during manufacture of these fermented cereal foods, nutrients including protein and minerals are lost from the grains thereby affecting nutritional quality adversely, as reported by [13].

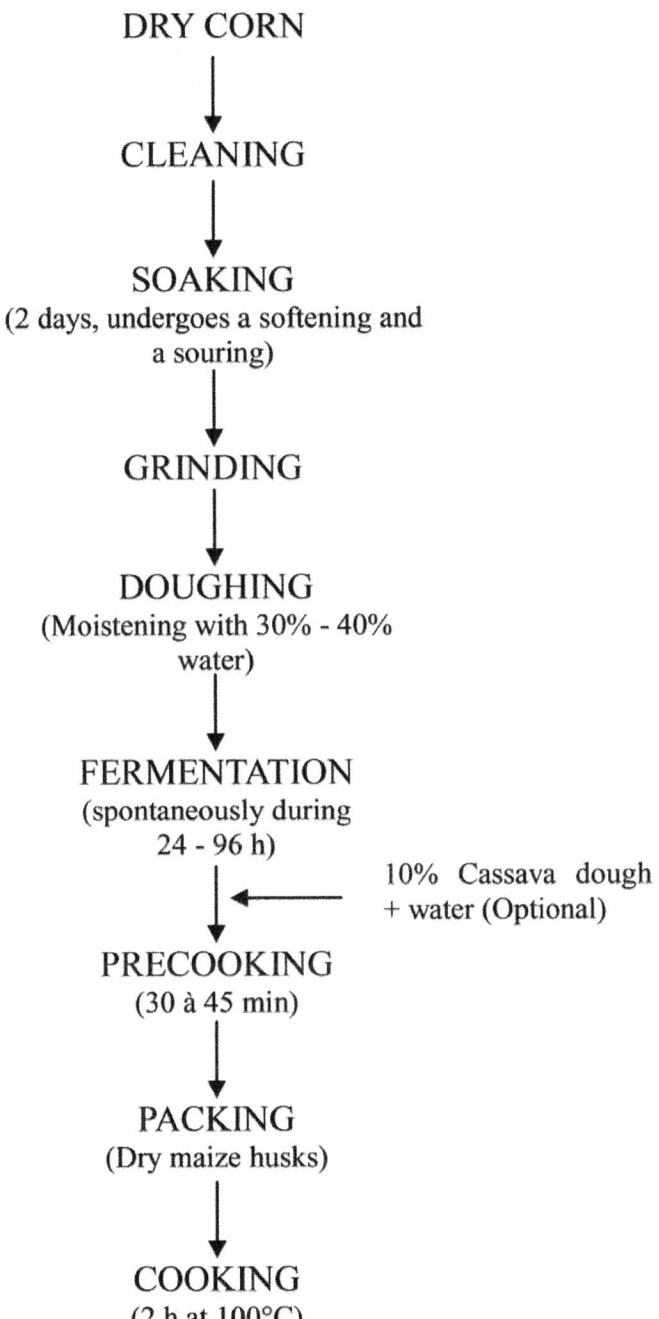

Figure 1: Flow diagram for production of doklu.

This paper reports the effect of spontaneous fermentation as applied at household level on the nutrients composition and aflatoxins rate of maize during the processing into doklu.

MATERIALS AND METHODS

Sample Collection

The different samples (250 g) for this study were collected in sterile containers from 6 processors at different stages of processing (maize grains, milled flours, fermented dough and finished product doklu) in Abidjan (the economic capital of Côte d'Ivoire). All collected samples were immediately transported in an icebox directly to the laboratory for analyses.

Proximate Composition

Forty grams of samples were ground in 300 ml of distilled water in a porcelain mortar and then centrifuged at 4000 tours/min for 30 min. The pH was determined on 50 ml of the supernatant using a pH-meter (P107 Consort). Total titratable acidity was determined by titrating 30 ml of supernatant used for pH determination against 0.1 M NaOH using phenolphthalein as indicator. The total titratable acidity was calculated as percentage of lactic acid.

Organic acids of samples were before extracted and then analyzed by high performance liquid chromatography using an ion exclusion ORH-801 column (300 mm × 6.5 mm) (Interchrom, France) as achieved by [14]. Running conditions were: mobile phase, H_2SO_4 40 mmol·L^{-1}; flow rate, 0.8 ml·min^{-1}; wave length, 210 nm; room temperature (25°C). The separated components were detected with an UV spectrophotometric detector (SPD-6A, Shimadzu Corporation, Japan).

Total sugars content were determined by the phenol sulphuric acid method according to Dubois et al. [15] and the values were expressed in g/100g of fresh dough. Total carbohydrates were determined according to the method of [16]. The contents of protein, fat, ash and moisture were determined according to the methods described in [17]. The determination of the energy value was done by calculation according to the method proposed by [18], with specific coefficients of starchy foods:

Energy value (kcal) = (3.87 × Proteins) + (4.12 × Total carbohydrates) + (8.37 × Fats).

Mineral Analyses

The method described by the Association of Official Analytical Chemists [19] was used for mineral analysis. The samples were ashed at 550°C. The ash was boiled with 10 ml of 20% hydrochloric acid in a beaker and then filtered into a 100 ml standard flask. This was made up to the mark with deionized water. The minerals were determined from the resulting solution. Sodium [Na] and Potassium [K] were determined using the standard flame emission photometer. NaCl and KCl were used as the standards [19]. Phosphorus [P] was determined calorimetrically using the spectronic 20 [Gallenkamp, UK] with KH_2PO_4 as the standard. Calcium [Ca] was determined using Atomic Absorption Spectrophotometer [AAS Model SP9]. All values were expressed in mg/ 100g.

Aflatoxin Extraction from Maize Samples and HPLC Analysis

Aflatoxins were extracted after the method of [20], with slight modifications. Twenty (20) g of maize samples collected throughout the doklu process were extracted with 100 ml acetonitrile: potassium chloride (90:10, v/v). The extract was filtered and the resultant filtrate further purified with a Sepak cleanup column (Merck). Three aliquots (200 μl) of the purified extract were transferred into vials. The solvent was evaporated under nitrogen gas and the samples were stored at 4°C. The dried samples were dissolved in 1 ml acetonitrile: potassium chloride (1:1, v/v) and filtered. An aliquot of the filtrate (300 μl) was injected into a Shimadzu HPLC system (Shimadzu, model CRB-GA, Kyoto, Japan). Separation was carried out isocratically using H_3PO_4 (0.33 M): acetonitrile: propanol-2 (650:400:50, v/v/v) as the mobile phase. The flow rate was maintained at 0.5 ml/min) and the fluorometric was used. Identification of aflatoxins (B1, B2, G1, and G2) in each sample was achieved by comparison with retention times of standard peaks. A series of each aflatoxin standards were used to construct a calibration curve. The equation obtained from the calibration curve was used to calculate the concentration of aflatoxins in each sample.

Statistical Analysis

The effect of processing steps on the concentration nutrient and aflatoxins in maize samples was analyzed using the analysis of variance (ANOVA) with the use of post hoc tests. Tests with P-values less than 0.05 were considered to be statistically significant.

RESULTS AND DISCUSSION

Acid Production in Doklu Fermentation

The types and amounts of the main acids produced during the process of doklu production are shown in **Table 1**. Two acids were produced in detectable amounts, namely, lactic and acetic acids. Lactic acid rate was already higher in the maize flour, due to the 2 days soaking stage underwent by maize grains. In fact, according to reference [21], lactic acid bacteria dominate during the soaking stage of the traditional process. And as a result, a significant increase of organic acids takes place. The concentrations of these organic acids increased continuously during the fermentation and reached a maximum total of about 1.8 g/100g lactic and 0.5 g/100g acetic acids the second day of fermentation. After this period, a decrease was observed.The amounts of acids produced in this fermentation are comparable to amounts reported for many other traditional fermented foods. Reference [22] reported 0.78% lactic acid in ogi, 0.6% in kaffir beer and 1% in bussa, three spontaneous fermented cereal based products.

The pH dropped significantly during the fermentation from 5.4 in the grains to 2.2 at 48 h contrarily to the total titratable acidity which amount increased to 0.6% at the same time (**Table 1**). These are relatively quick variations if compared to similar fermentations, and indicates a relatively high fermentation rate. According to references [23-25], the organic acids released (e.g. lactic, acetic, propionic and butyric acids), as by-products during lactic acid fermentation, lower the pH to levels of 3 to 4 with a titratable acidity of about 0.6%. The undissociated forms of the acetic and lactic acids at low pH exhibit inhibitory activities against a wide range of pathogens. This improves food safety by restricting the growth and survival, in fermented cereal beverages, of spoilage organisms and some pathogenic organisms such as Shigella, Salmonella and E. coli [26,27]. Fermented maize dough for doklu production with pH value below 3 could have undoubtedly inhibited the growth of such organisms if they were present.

Macronutrient and Mineral Composition

The macronutrient and mineral composition of samples collected during doklu fermentation are presented in Tables 2 and 3 respectively. The moisture content values of the samples ranged between 14.2 ± 0.1 in the grains and

Table 1: Effect of fermentation stages on pH, total titratable acidity and organic acids of maize during its processing into doklu.

Sampling steps	Products	Parameters			
		pH	TTA (%)	Lactic acid (g/100g)	Acetic acid (g/100g)
Maize	Grains	5.4 ± 0.1^a	0.02 ± 0.001^a	nd	nd
Grinding	Flour	3.2 ± 0.3^b	0.3 ± 0.03^b	0.12 ± 0.03^a	0.04 ± 0.002^a
Fermentation	Dough$_{24h}$	2.4 ± 0.2^{cd}	0.6 ± 0.01^c	0.7 ± 0.002^b	0.07 ± 0.01^b
	Dough$_{48h}$	2.2 ± 0.1^d	0.6 ± 0.01^c	1.8 ± 0.1^c	0.5 ± 0.01^c
	Dough$_{72h}$	2.7 ± 0.04^c	0.4 ± 0.002^b	1.6 ± 0.2^d	0.2 ± 0.01^d
Cooking	Doklu	nd	nd	1.3 ± 0.2^e	0.1 ± 0.04^d

nd: not detected, data were analyzed on triplicates. Mean values with the same superscript in a column are not significantly different (P > 0.05). TTA: total titratable acidity.

Table 2: Effect of fermentation stages on proximate composition of maize during its processing into doklu.

Sampling steps	Products	Parameters						
		Moisture (%)	Proteins (%)	Fatty matters (%)	Carbo hydrates (%)	Total sugars (%)	Ash (%)	Energy value (Kcal)
Maize	Grains	14.2 ± 0.1^a	8.2 ± 0.7^a	0.64 ± 0.01^a	60.5 ± 3.0^a	0.64 ± 0.01^a	1.85 ± 0.07^a	280.56^a
Grinding	Flour	16.2 ± 0.3^b	8.1 ± 0.2^a	0.61 ± 0.07^a	60.1 ± 3.3^a	0.62 ± 0.03^a	1.95 ± 0.07^b	278.29^d
Fermentation	Dough$_{24h}$	20.5 ± 0.1^c	7.3 ± 0.2^{bc}	0.18 ± 0.07^c	60.2 ± 3.01^a	0.24 ± 0.01^b	1.95 ± 0.07^b	271.62^b
	Dough$_{48h}$	20.7 ± 0.2^c	7.1 ± 0.1^c	0.18 ± 0.07^c	60.2 ± 2.2^a	0.19 ± 0.01^b	1.80 ± 0.07^a	270.82^b
	Dough$_{72h}$	20.4 ± 0.2^c	7.1 ± 0.3^c	0.18 ± 0.07^c	60.1 ± 2.1^a	0.16 ± 0.02^c	1.84 ± 0.14^a	270.42^b
Cooking	Doklu	20.7 ± 0.1^c	6.9 ± 0.1^{cd}	0.20 ± 0.07^c	60.2 ± 3.4^a	0.10 ± 0.01^d	1.66 ± 0.2^c	270.2^b

Data were analyzed on triplicates, mean values with the same superscript in a column are not significantly different (P > 0.05).

Table 3: Effect of fermentation stages on minerals composition of maize during its processing into doklu.

Sampling steps	Products	Minerals (%)			
		Ca	P	K	Na
Maize	Grains	0.05 ± 0.003^a	0.17 ± 0.07^a	0.02 ± 0.003^a	0.01 ± 0.007^a
Grinding	Flour	0.05 ± 0.003^a	0.17 ± 0.01^a	0.02 ± 0.003^a	0.01 ± 0.007^a
Fermentation	Dough$_{24h}$	0.03 ± 0.01^a	0.14 ± 0.07^a	0.02 ± 0.003^a	0.01 ± 0.007^a
	Dough$_{48h}$	0.03 ± 0.01^a	0.14 ± 0.05^a	0.02 ± 0.003^a	0.01 ± 0.007^a
	Dough$_{72h}$	0.03 ± 0.003^a	0.14 ± 0.07^a	0.02 ± 0.003^a	0.01 ± 0.007^a
Cooking	Doklu	0.03 ± 0.003^a	0.14 ± 0.01^a	0.02 ± 0.003^a	0.01 ± 0.007^a

Data were analyzed on triplicates, mean values with the same superscript in a column are not significantly different (P > 0.05).

20.7% ± 0.1% for the final product. This variation in the moisture content is due to the fact that an amount of water is added to the maize flour to make the dough. However, the value obtained for the final product is different of those stated by [28] for kenkey, a similar maize food from Ghana. Moreover, scientific investigation has reported that low moisture content in food samples increased the storage periods of the food products [29]; while high moisture content in foods encourage microbial growth; hence, food spoilage [30].

The proteins, fatty matters and total soluble sugars contents were respectively of 8.2 ± 0.7, 0.64 ± 0.01 and $0.64\% \pm 0.01\%$ in maize grains (**Table 2**). These values were significantly reduced during the process of doklu production. As a result of fermentation significant reducetion in crude protein and soluble sugars contents of the food may be attributed in one hand to an increase in protein catabolism by the fermenting microorganisms which leads to the escape of the by-product of metabolic deamination, i.e. ammonia and in other hand to the utilization of sugars as a carbon source. The results are similar to those reported by [31] who observed a reduction in protein content of fermented cereal-legume food mixtures by the action of bacteria and yeasts. Reference [32] also noticed a significant reduction in the protein content of pearl millet when it was fermented with L. acidophilus. According to Reference [33], fermentation may slightly alter the proximate composition substrates. A slight increase in the percentage of protein can be noted. This increase reflects the decrease of other constituents which the microorganisms might have consumed for growth. The decrease of soluble sugars and fatty matters in maize dough during doklu fermentation suggested that the fermenting microorganisms had used them as an energy source. Reduction of fat content was previously mentioned by [34] in their study on the fermentation of pear millet.

Ash content of maize seeds was 1.85% and this value was significantly affected during the various stage of processing (**Table 2**), whilst mineral contents and energy value remained unaffected by the fermentation process (**Table 3**). A similar trend was observed during the fermentation of cassava during fufu production by [35]. These authors showed that the fermentation process caused an increase in the concentration of calcium (+12%) in cassava but reductions in the levels of potassium, sodium, manganese, iron, copper, zinc and phosphorus. Data have not been published on changes in the caloric content of food as a result of fermentation processes. Generally only small changes would be expected. In processes such as tempeh production, which are aerobic, the fermentation period is too short to allow large decreases in the total lipids, carbohydrate, or protein components of the food [36].

Effect on Aflatoxins Rate

Figure 2 shows the high-performance liquid chromatogram of aflatoxins B1, G1 and G2 extracted from maize grains samples. As it could be seen on the figure, aflatoxin B2 was not detected in maize grains and consequently during all the process of doklu production.

The total amount of aflatoxins detected in the grains was 4.59 ± 0.03 µg/kg consisting in 2.52 ± 0.01 µg/kg of aflatoxin B1, 2.52 ± 0.01 µg/kg of aflatoxin G1 and 0.33 ± 0.02 µg/kg aflatoxin G2 (**Table 4**). Maize and maize products are known to be susceptible to contamination by fungi that produce secondary metabolites such as aflatoxins [37]. Aflatoxins have been described as extremely toxic and carcinogenic compounds, which appear to be ubiquitous in the environment [38]. The incidence and level of aflatoxin contamination in various food commodities have been monitored worldwide [19] and continues to be of great concern. Aflatoxins, particularly aflatoxin B1 (AFB1), are considered to be the most important of the mycotoxins due to their high toxicity and they are still of major concern to the feed industry and farmers as many raw materials which are used as components of animal feeds are prone to contamination [39]. Their adverse effects involves their mutagenic, carcinogenic (especially to kidneys and liver), teratogenic and oestrogen immunosuppressive effects. Aflatoxin B1 is one of the strongest carcinogens and it was included by WHO and the International Agency for Research on Cancer in the list of carcinogenic substances Group I, i.e. substances with confirmed carcinogenic effect in humans [40,41]. However, during processing of maize into doklu, important decreases in aflatoxins rates were observed (**Table 4**). After the 2 days soaking, the amount of aflatoxins was reduced to 1.3 ± 0.04 µg/kg (about 72%) for total aflatoxins and to 0.51 ± 0.02 µg/kg (about 80%) for aflatoxin B1. At the fermentation steps no aflatoxin was detected, involving that all the mycotoxins degraded. This study has shown that natural fermentation of maize dough for doklu production can substantially reduce the amount of aflatoxins contamining the raw material. The toxicity of the product was significantly reduced after the soaking and the first time fermentation period with progressive decrease in the pH. This is in agreement with other studies, which clearly show that lactic acid bacteria (Lactobacillus strains) efficiently remove aflatoxin B1 from the culture solution [42,43]. It has been suggested that removal of toxins is through noncovalent binding of mutagens by fractions of the cell wall skeleton of lactic acid bacteria [44]. However, other alternative mechanism of aflatoxin B1 removal has been reported, in which LAB fermentation opens up the aflatoxin B1 lactone ring resulting in its complete detoxification [45]. The lower pH of the media could also have contributed to the removal of toxins from the media as other studies have shown that treatment of LAB pellets with hydrochloric acid significantly enhanced the binding ability of the bacteria [36].

Conclusion

It may be concluded from this study that natural fermentation during the processing of maize grains into doklu resulted in reduction of some nutritional parameters, but a high increase in acidity, an important characteristic for the product safety was observed. Natural fermentation also leads to a total elimination of aflatoxins in the product. Despite the losses in some nutritional compounds, the fermented product, doklu, was found to have appreciable nutritional quality.

ACKNOWLEDGEMENTS

This research was made possible through funds provided by the International Foundation for Science, Sweden

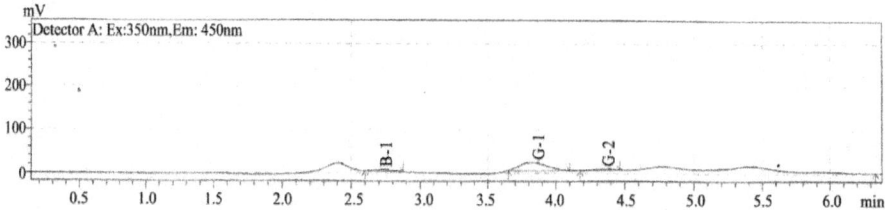

Figure 2: Typical high-performance liquid chromatogram of extract of maize grains samples showing picks of aflatoxins B1, G1, G2 and their retention times.

Table 4: Effect of fermentation stages on aflatoxins content of maize during its processing into doklu

		Aflatoxins (µg/kg)				
Sampling steps	Products	AFB1	AFB2	AFG1	AFG2	Total
Maize	Grains	2.52 ± 0.01^a	nd	2.52 ± 0.01^a	0.33 ± 0.02	4.59 ± 0.03^a
Grinding	Flour	0.51 ± 0.02^b	nd	0.80 ± 0.02^b	nd	1.3 ± 0.04^b
Fermentation	Dough₂₄ₕ	nd	nd	nd	nd	nd
	Dough₄₈ₕ	nd	nd	nd	nd	nd
	Dough₇₂ₕ	nd	nd	nd	nd	nd
Cooking	Doklu	nd	nd	nd	nd	nd

nd: not detected, data were analyzed on triplicates, mean values with the same superscript in a column are not significantly different (P > 0.05).

(E/4955-1). The authors gratefully acknowledge M. SORO and M. OUATTARA from LANADA for their assistance in HLPC analysis for aflatoxins quantification.

REFERENCES

1. T. I. Mbata, M. J. Ikenebomeh and J. C. Alaneme, "Studies on the Microbiological, Nutrient Composition and Antinutritional Contents of Fermented Maize Flour Fortified with Bambara Groundnut (Vigna subterranean L)," African Journal of Food Science, Vol. 3, No. 6, 2009, pp. 165-171.

2. O. O. Taiwo, "Physical and Nutritive Properties of Fermented Cereal Foods," African Journal of Food Science, Vol. 3, No. 2, 2009, pp. 23-27.

3. S. Mahajan and B. M. Chauhan, "Phytic Acid and Extractable Phosphorus of Pearl Millet Flour as Affected by Natural Lactic Acid Fermentation," Journal of the Science of Food and Agriculture, Vol. 41, No. 4, 1987, pp. 381-382.http://dx.doi.org/10.1002/jsfa.2740410410

4. B. M. Chauhan, N. Suneja and C. M. Bhat, "Nutritional Value and Fatty Acid Composition of Some High Yielding Varieties of bajra," Bulletin of Grain Technology, Vol. 21, 1986, pp. 441-442.

5. N. Dhankher and B. M. Chauhan, "Effect of Temperature and Period of Fermentation on the Protein and Starch Digestibility (in Vitro) of rabadi—A Pearl Millet Fermented Food," Journal of Food Science, Vol. 52, No. 2, 1987, pp. 489-490.http://dx.doi.org/10.1111/j.1365-2621.1987.tb06648.x

6. N. Dhankher and B. M. Chauhan, "Effect of Temperature and Fermentation Time on Phytic Acid and Polyphenol Content of rabadi—A Fermented Pearl Millet Food," Journal of Food Science, Vol. 52, No. 3, 1987, pp. 828- 829. http://dx.doi.org/10.1111/j.1365-2621.1987.tb06739.x

7. K. B. Nolan and P. A. Duffin, "Effect of Phytate on Mineral Bioavailability. In Vitro Studies on Mg^{2+}, Ca^{2+}, Fe^{3+} and Zn^{2+} in the Presence of Phytate," Journal of the Science of Food and Agriculture, Vol. 40, No. 1, 1987, pp. 79-83.http://dx.doi.org/10.1002/jsfa.2740400110

8. Q. Liu, E. Donner, Y. Yin, R. L. Huang and M. Z. Fan, "The Physicochemical Properties and in Vitro Digestibility of Selected Cereals, Tubers and Legumes Grown in China," Food Chemistry, Vol. 99, No. 3, 2006, pp. 470- 477.http://dx.doi.org/10.1016/j.foodchem.2005.08.008

9. E. N. T. Akobundu and F. H. Hoskins, "Protein Losses in Traditional agidi Paste," Journal of Food Science, Vol. 47, No. 5, 1982, pp. 1728-1729. http://dx.doi.org/10.1111/j.1365-2621.1982.tb05021.x

10. V. Umoh and M. J. Fields, "Fermentation of Corn for Nigerian Agidi," Journal of Food Science, Vol. 46, No. 3, 1981, pp. 903-905. http://dx.doi.

org/10.1111/j.1365-2621.1981.tb15376.x

11. M. A. M. Ali, A. H. El-Tinay and Abdalla, "Effect of Fermentation on in Vitro Protein Digestibility of Pearl Millet," Food Chemistry, Vol. 80, No. 1, 2003, pp. 51-54.http://dx.doi.org/10.1016/S0308-8146(02)00234-0

12. A. Sharma and N. Khetarpaul, "Effect of Fermentation on Phytic Acid Content and in Vitro Digestibility of Starch and Protein of Rice-Black Gram Dhal—Whey Blends," Journal of Food Science and Technology, Vol. 34, 1997, pp. 20-22.

13. E. R. Aminigo and J. O. Akingbala, "Nutritive Composition and Sensory Properties of Ogi Fortified with Okra Seed Meal," Journal of Applied Science and Environment Management, Vol. 8, No. 2, 2004, pp. 23-28.

14. N. T. Djeni, K. F. N'guessan, D. M. Toka, K. A. Kouame and K. M. Dje, "Quality of attieke (a Fermented Cassava Product) from the Three Main Processing Zones in Côte d'Ivoire," Food Research International, Vol. 44, No. 1, 2011, pp. 410-416.http://dx.doi.org/10.1016/j.foodres.2010.09.032

15. M. Dubois, K. A. Gilles, J. K. Hamilton, P. A. Rebers and F. Smith, "Colorimetric Method for Determinations of Sugars and Related Substances," Analytical Chemistry, Vol. 28, No. 3, 1956, pp. 350-356. http://dx.doi.org/10.1021/ac60111a017

16. FAO, "Food Energy—Methods of Analysis and Conversion Factors," Report of a Technical Workshop, Rome, 3-6 December 2002, FAO Food and Nutrition Paper 77, 93 p.

17. AOAC, "Official Methods of Analysis," 16th Edition, Vol. 1-2, Association of Official Analytical Chemists, Arlington, 1998.

18. FAO, "Composition des Aliments en Principes Nutritifs Calorigènes et Calcul des Valeurs Energétiques Utiles," Rapport du Comité Chargé de l'Etude des Aliments et des Facteurs de la Division de la Nutrition de la FAO, Washington DC, 1947, 30 p.

19. AOAC, "Official Methods of Analysis of AOAC International," 18th Edition, Association of Analytical Communities Gaithersburg, 2005.

20. A. E. Ce'spedes and G. J. Diaz, "Analysis of Aflatoxins in Poultry and Pig Feeds and Feedstuffs Used in Columbia," Journal AOAC International, Vol. 80, No. 6, 1997, pp. 1215-1219.

21. R. K. Mulyowidarso, G. H. Fleet and K. A. Buckle, "Changes in the Concentration of Organic Acids during the Soaking of Soybeans for tempe Production," International Journal of Food Science and Technology, Vol. 26, No. 6, 1991, pp. 607-614.http://dx.doi.org/10.1111/j.1365-2621.1991.tb02006.x

22. S. A. Odunfa and S. Adeyde, "Microbiological Changes during the Traditional Production of ogibaba, a West African Fermented Sorghum Gruel," Journal of Cereal Science, Vol. 3, No. 2, 1985, pp. 173-180. http://dx.doi.org/10.1016/S0733-5210(85)80027-8

23. W. H. Holzapfel, "Appropriate Starter Culture Technologies for Small-Scale Fermentation in Developing Countries," International Journal of Food Microbiology, Vol. 75, No. 3, 2002, pp. 197-212. http://dx.doi.org/10.1016/S0168-1605(01)00707-3

24. M. O. Edema and A. I. Sanni, "Functional Properties of Selected Starter Cultures for Sour Maize Bread," Food Microbiology, Vol. 25, No. 4, 2008, pp. 616-625.http://dx.doi.org/10.1016/j.fm.2007.12.006

25. P. Mensah, "Fermentation—The Key to Food Safety Assurance in Africa?" Food Control, Vol. 8, No. 5-6, 1997, pp. 271-278. http://dx.doi.org/10.1016/S0956-7135(97)00020-0

26. A. Blandino, M. E. Al-Aseeri, S. S. Pandiella, D. Cantero and C. Webb, "Cereal-Based Fermented Foods and Beverages," Food Research International, Vol. 36, No. 6, 2003, pp. 527-543. http://dx.doi.org/10.1016/S0963-9969(03)00009-7

27. N. B. Omar, H. Abriouel, R. Lucas, M. Martinez-Ca- ñamero, J. Guyot and A. Gálvez, "Isolation of Bacteriocinogenic Lactobacillus plantarum Strains from ben saalga, a Traditional Fermented Gruel from Burkina Faso," International Journal of Food Microbiology, Vol. 112, No. 1, 2006, pp. 44-50.http://dx.doi.org/10.1016/j.ijfoodmicro.2006.06.014

28. A. Annan-Prah and J. A. Agyeman, "Nutrient Content and Survival of Selected Pathogenic Bacteria in kenkey Used as a Weaning Food in Ghana," Acta Tropica, Vol. 65, No. 1, 1997, pp. 33-42. http://dx.doi.org/10.1016/S0001-706X(97)00650-5

29. Y. E. Alozie, M. A. Iyam, O. Lawal, U. Udofia and I. F. Ani, "Utilization of Bambara Ground Flour Blends in Bread Production," Journal of Food Technology, Vol. 7, No. 4, 2009, pp. 111-114.

30. V. J. Temple, E. J. Badamosi, O. Ladeji and M. Solomon, "Proximate Chemical Composition of Three Locally Formulated Complementary Foods in West African," Journal of Biological Sciences, Vol. 5, 1996, pp. 134-143.

31. Binita, N. Khetarpaul and R. Kumar, "Development, Acceptability and Nutritional Composition of Food Blends Fermented with Probiotic Organisms," Annals of Biology, Vol. 12, 1996, pp. 127-133.

32. A. Sharma, "Fermentative Improvement of Pearl Millet and Utilization of the Fermented Product," Ph.D. Thesis Dissertation, Haryana Agricultural University, Hisar, 1994.

33. H. L. Wang, "Nutritional Quality of Fermented Foods," In: C. W. Hesseltine and H. L. Wang, Eds., Mycologia Memoir No. 11, Indigenous Fermented Food of Non-Western Origin, 1968, pp. 289-301.

34. N. Khetarpaul and B. M. Chauhan, "Effect of Fermentation on Protein, Fat, Minerals and Thiamine Content of Pearl Millet," Plant Foods for Human Nutrition, Vol. 39, No. 2, 1989, pp. 169-177. http://dx.doi.org/10.1007/BF01091897

35. O. B. Oyewole and S. A. Odunfa, "Effects of Fermentation on the Carbohydrate, Mineral, and Protein Contents of Cassava during 'Fufu' Production," Journal of Food Composition and Analysis, Vol. 2, No. 2, 1989, pp. 170- 176. http://dx.doi.org/10.1016/0889-1575(89)90078-1

36. R. F. McFeeters, "Effects of Fermentation on the Nutritional Properties of Food," In: E. Karmas and R. S. Harris, Eds., Nutritional Evaluation of Food Processing, Van Nostrand Reinhold Co., New York, 1988, pp. 423-446. http://dx.doi.org/10.1007/978-94-011-7030-7_16

37. M. P. Mokoena, P. K. Chelule and N. Gqaleni, "The Toxicity and Decreased Concentration of Aflatoxin B1 in Natural Lactic Acid Fermented Maize Meal," Journal of Applied Microbiology, Vol. 100, No. 4, 2006, pp. 773-777. http://dx.doi.org/10.1111/j.1365-2672.2006.02881.x

38. T. E. Massey, R. K. Steewart, J. M. Daniels and L. Liu, "Biochemical and Molecular Aspects of Mammalian Susceptibility to Aflatoxin B1 Carcinogenicity," Proceedings of the Society for Experimental Biology and Medicine, Vol. 208, No. 3, 1995, pp. 213-227.http://dx.doi.org/10.3181/00379727-208-43852A

39. K. Ślizewska and S. Smulikowska, "Detoxification of Aflatoxin B1 and Change in Microflora Pattern by Probiotic in Vitro Fermentation of Broiler Feed," Journal of Animal and Feed Sciences, Vol. 20, 2011, pp. 300-309.

40. JECFA, "Safety Evaluation of Certain Food Additives and Contaminants Aflatoxins," WHO Food Additives Series, Vol. 40, 1998, pp. 897-913.

41. FAO/WHO/UNEP, "Mycotoxin Prevention and Decontamination, HACCP and Its Mycotoxin Control Potential: An Evaluation of Ochratoxin A in Coffee Production," Third Joint FAO/WHO/UNEP International Conference on Mycotoxins, 1999, pp. 1-13.

42. H. el-Nezami, P. Kankaanpaa, S. Salminen and J. Ahokas, "Physicochemical Alterations Enhance the Ability of Dairy Strains of

Lactic Acid Bacteria to Remove Aflatoxin from Contaminated Media," Journal of Food Protection, Vol. 61, No. 4, 1998, pp. 466-468.

43. C. A. Haskard, H. S. El-Nezami, P. E. Kankaanpaa, S. Salminen and J. T. Ahokas, "Surface Binding of Aflatoxin B_1 by Lactic Acid Bacteria," Applied and Environmental Microbiology, Vol. 67, No. 7, 2001, pp. 3086- 3091. http://dx.doi.org/10.1128/AEM.67.7.3086-3091.2001

44. X. B. Zhang and Y. Ohta, "Binding of Mutagens by Fractions of the Cell Wall Skeleton of Lactic Acid Bacteria on Mutagens," Journal of Dairy Science, Vol. 74, No. 5, 1991, pp. 1477-1481. http://dx.doi.org/10.3168/jds.S0022-0302(91)78306-9

45. N. J. R. Nout, "Fermented Foods and Food Safety," Food Research International, Vol. 27, No. 3, 1994, pp. 291-298. http://dx.doi.org/10.1016/0963-9969(94)90097-3

Chapter 11

LARGE SCALE EXPERIMENTS ON THE INVESTIGATION OF THE EFFECT OF HIGH CONCENTRATIONS OF AFLATOXIN B1 ON THE FERMENTATION OF DIFFERENT WINES

Cs. Csutorás[1], K. Rácz[2], G. Z. Nagy[1], O. Hudák[1], L. Rácz[1]

[1] Institute of Food Science, Egerfood Regional Knowledge Center, Eszterhazy Karoly University, Eger, Hungary

[2] Eger Crown Winehouse Ltd., Kerecsend, Hungary

ABSTRACT

The change of aflatoxin B1 (AFB1) content during must fermentation processes in different white, rosé and red musts was investigated, using selected yeast strains of Saccharomyces cerevisiae as starter cultures. Levels of AFB1 in must and lees were determined by high-performance liquid chromatography (HPLC) combined with diode array detection (DAD). Reductions of the AFB1 content between 77% - 97% were recorded after 90 days must fermentations in the model systems, while the relative adsorption level of AFB1 in lees was around ~0.63 in case of white wines, ~0.41 in case of rosé wines and ~0.23 in case of red wines. The results show that even extremely high AFB1 levels do not affect the fermentation process and the life-circle of yeast strains. The concentration of AFB1 in wine can be controlled by using appropriate yeast strains during the alcoholic fermentation.

INTRODUCTION

Aflatoxins produced by Aspergillus and Penicillum strains can be found mainly in cereals, cheese, peanuts, corn, almond, fruits and a wide group of other foods and feeds [1,2]. The main problem with these secondary metabolites is that they have mutagenic, toxic, carcinogenic and immunosuppressive effects on living organisms and we cannot destroy them easily, only by harsh heat-treatment [3]. Their production depends on environmental factors, such as temperature, humidity and other storage conditions [4,5]. Among 18 different types of aflatoxins, the B1 (AFB1) is the most widespread food contaminant [6,7]. Ochratoxin A (OTA) is a thoroughly investigated mycotoxin that is responsible mainly for toxin contaminations of wines [8,9]. Grapes can be contaminated with a wide variety of moulds including Aspergillus and Penicillium genera. Among Aspergillus genus, black aspergilli seemed to be the most common contaminant of grapes namely Aspergillus niger aggregates and Aspergillus carbonarius the most known OTA-producing species. Reports by several authors showed that grape and its derived products such as dried vine fruits [10], grape juices and wines [9,11,12] were highly contaminated by OTA. However until now only a few number of publications deal with the possibility of contamination of wines with aflatoxins. There is a few available information on the occurrence of the AFB1 on grapes and its derived products since the studies conducted in Mediterranean countries revealed a very low occurrence of Aspergillus flavus in the vineyards [13-15]. However in these studies no quantification of AFB1 in grapes or its products was performed. It is certain, that AFB1 can be produced by certain strains of the fungi Aspergillus flavus and Aspergillus parasiticus that are grown on grapes in the vineyards. These fungi can colonize and contaminate the grapes before grape-gathering and during storage, if water is allowed to exceed critical values for mould growth [16,17]. At grape-crushing they might contaminate the must thereby wines too. The minimal amount of the AFB1 allowed in plant foodstuffs is 2 μg/kg. A thorough study would be important to reveal the real risk of AFB1 contamination of grapes and wines. Different analytical methods are known for the determination of aflatoxins in a variety of matrices, like thin layer chromatography (TLC) [18,19], overpressure layer chromatography (OPLC) [20], high-pressure liquid chromatography (HPLC) [18,21], enzyme-linked immunosorbent assay (ELISA) [22] and optical waveguide lightmode spectroscopy (OWLS) [4]. Sripatomswat and Thasnakorn [18] were extracted fermented rice, soybean sauce, peanut butter, soy sauce, Thai red and white wine, and rice sugar wine and tested for aflatoxins by TLC and HPLC at 350 nm. Adányi et al. [4] applied OWLS technique for the analysis of ochratoxins and aflatoxins in both competitive and direct immunoassays. The sensitive detection range of the competitive detection method was between 0.5 and 10 ng/mL. Rasch et al. [3] accomplished the qualitative and quantitative analysis of AFB1 in different wine and beer samples by one- and two-photon-induced fluorescence (OPIF and TPIF) techniques. The limit of detection were 31 ng/mL in white wine, 43 ng/mL in rose wine and 62 ng/mL in beer at 720 nm, using OPIF. A better limit of detection can be achieved by TPIF, where the limit of detection at 360 nm wavelength were 6 ng/mL in white wine, 20 ng/mL in rose wine and 46 ng/mL in beer samples. Previous studies showed that

AFB1 can be detected in wines from southern regions, like Mediterranean vineyards [16], but the proofs of a real risk of aflatoxin contaminations in wines have not been enough yet. Further studies are necessary to be prepared for a possible aflatoxin contamination in grape and its derived products, like wines. Our previous study focused on the investigation of OTA during wine fermentation processes [23] and showed that the fermentation process and life-circle of the yeast strains were not affected even in the presence of extremely high levels of OTA. The aim of our present work was to investigate the changes of AFB1 during the fermentation of red, white and rose wines that were artificially contaminated with different amounts of the mycotoxin. We focused in the paper on the investigation of the fermentation process, whether the fermentation is influenced or not in real must samples, that were inoculated with high concentrations of AFB1. Our experiments intended to build models for the effect of AFB1 contamination on the growth of yeast strains and to investigate the alterations of the levels of AFB1 during the fermentation process, to be able to estimate a possible risk of contamination. Thus previous studies carried out only micro-scale laboratory experiments, in this paper we focused on industrial scale investigations modelling real fermentation processes for the best.

MATERIALS AND METHODS

Reagents and Standards

Samples

Red, rose and white wine musts were purchased from Eger Crown Winehouse Ltd., whose cellars were freely used for the experiments. Yeast strains of Saccharomyces cerevisiae type "Fermol Premier Cru" were purchased from AEB Biochemical Inc. (San Francisco, USA).

Reagents

Solvents used for HPLC were ethanol 96% of gradient grade for liquid chromatography and methanol, dichloromethane and acetonitrile for liquid chromatography, all purchased from Merck. In all analytical steps, deionized water generated by a Milli-Q P Ultra-Pure Water System from Millipore (Billerica, MA, USA) was used. Benzene-free AFB1 reference standard material was delivered by Sigma-Aldrich Chemical Ltd. (Schnelldorf, Germany)

Method and Sample Preparation

The stemmed, crushed and pressed red, rose and white wine musts were sulfurized (60 mg/L potassium- pyrosulfite) similarly to real wine making procedures. The musts were inoculated with yeast strains of Saccharomyces cerevisiae (200mg/L). The inoculated musts were artificially contaminated with AFB1 at 0.5; 1.0; 2.0; 4.0 µg/mL in triplicate, and stored in 16 L glass balloons in wine cellar at a permanent temperature of 12°C. After 20 and 90 days samples were taken from the wines and after 90 days also from lees in order to determine the concentration of AFB1. HPLC analyses needed appropriate sample pre-treatment as follows: 8 mg sodium chloride was added to 100 mL must or lees sample then it was extracted with acetonitrile (4 × 30 mL). The combined solutions were centrifuged for 5 minutes at 3800 rpm (Centrifuge 5810R, Eppendorf, Germany), then the clear solutions were evaporated by rotary evaporator (Laborota 4001, Heidolph, Germany). Before the HPLC measurements the samples were dissolved in 2 mL acetonitrile and filtered using a 0.2 µm filter

Instrumentation for the Analysis of AFB1

Analysis of AFB1: Quantitative analysis of must and wine samples has been carried out by a Shimadzu LCMS-2010EV instrument using a LC-20AB binary pump, SIL-20A sample changer, CTO-20A column oven and a SPD-M20A photodiode-array (PDA) detector. An Agilent Zorbax SB-C18 (4.6 × 250 mm, 5 µm) column was used for the chromatographic separations in an isochratic eluent setup with water/acetonitrile/methanol solvent system (3/1/1 ratio), the eluent flow rate was 1 mL/min and 10 µL sample was injected at the beginning of the chromatographic run. The signal of AFB1 was detected at 362 nm wavelength with a retention time of 18 min. 1 - 5 µg/mL standard AFB1 solutions were prepared and injected under the same experimental conditions and the signal integral values were used to calculate the linear calibration curve which was further used to calculate the AFB1 content of the wine and must samples. The following main analytical performance data were observed: linear range: 0.03 - 10 µg/mL, recovery: 96%, quantification limit: 0.02 µg/mL, precision: ±0.01 µg/mL.

RESULTS AND DISCUSSION

Our investigations focused on the experimental modelling of the change of AFB1 level during alcoholic fermentation of musts, applying a model system, which was similar to real conditions. Three different pre-treatments were applied of the must at the beginning of the experiments (selected yeast strains of Saccharomyces cerevisiae and sulfur were added in the first case, selected yeast strains of Saccharomyces cerevisiae but no sulfur were added to the must in the second case, and no additives were applied in the third part of experiments—in the latter case only wild yeasts can be worked). The results of the experiments are summarized in Table 1 and Table 2. The most important observation of our investigations was that the fermentation process was not affected by the addition of high levels of AFB1. This means a potential risk in the case of the appearance of AFB1 producing mold species on grape, therefore our experiments appeal to researchers to study AFB1 in the wine product chain thoroughly in the near future. Our second observation is in correlation with our previous findings in the case of another mycotoxin, OTA [23]. As it can be seen from Table 1, the AFB1 contents in white, rose and red wines significantly decreased during the fermentation process in all concentration ranges. In white wines the concentration of AFB1 decreased after 90 days fermentation from 0.5 to 0.07 µg/mL, from 1 to 0.16 µg/mL, from 2 to 0.29 µg/mL and from 4 to 0.55 µg/mL. These findings imply a mean reduction percentage of 91% considering the initial concentrations. In rose wines the concentration of AFB1 decreased after 90 days fermentation from 0.5 to 0.05 µg/mL, from 1 to 0.06 µg/mL, from 2 to 0.07 µg/mL and from 4 to 0.14 µg/mL, which means 95% reduction considering the initial concentrations. In red wines the concentration of AFB1 decreased from 0.5 to 0.08 µg/mL, from 1 to 0.08 µg/mL, from 2 to

Table 1: AFB1 level in musts during the fermentation process (Pre-treatment 1 = yeast and sulfur added; 2 = yeast but no sulfur added; 3 = no yeast and no sulfur added).

Wine	Pre-treatment method	AFB₁ in wine and must during fermentation (µg/mL)			Reduction rate
		0 day	20 days	90 days	
chardonnay	1	0.49	0.21	0.07	86%
chardonnay	2	0.48	0.24	0.08	84%
chardonnay	3	0.47	0.21	0.09	82%
chardonnay	1	1.01	0.48	0.16	84%
chardonnay	2	0.96	0.34	0.22	78%
chardonnay	3	0.95	0.31	0.23	77%
chardonnay	1	1.93	0.90	0.26	87%
chardonnay	2	1.92	0.46	0.29	86%
chardonnay	3	1.91	0.45	0.31	85%
chardonnay	1	3.85	1.91	0.65	84%
chardonnay	2	3.86	0.95	0.60	85%
chardonnay	3	3.84	0.68	0.55	86%
bluefrank rose	1	0.48	0.09	0.05	90%
bluefrank rose	1	0.95	0.14	0.06	94%
bluefrank rose	1	1.90	0.30	0.07	97%
bluefrank rose	1	3.82	0.59	0.14	97%
cabernet sauvignon	1	0.47	0.15	0.08	84%
cabernet sauvignon	1	0.97	0.26	0.08	96%
cabernet sauvignon	1	1.94	0.44	0.21	90%
cabernet sauvignon	1	3.88	0.76	0.18	96%

*Reduction rate (%): 100-(100× Measured concentration of AFB1 in wine after 90 days /Added amount of AFB1).

Table 2: AFB1 level in lees at the end of the fermentation process (Pre-treatment 1 = yeast and sulfur added; 2 = yeast but no sulfur added; 3 = no yeast and no sulfur added).

Wine	Pre-treatment method	AFB1 in lees after fermentation (µg/mL)	Relative adsorption values*
chardonnay	1	0.29	0.59
chardonnay	2	0.16	0.33
chardonnay	3	0.26	0.51
chardonnay	1	0.38	0.38
chardonnay	2	0.85	0.86
chardonnay	3	0.74	0.74
chardonnay	1	1.37	0.69
chardonnay	2	1.21	0.61
chardonnay	3	0.74	0.37
chardonnay	1	2.75	0.69
chardonnay	2	2.10	0.53
chardonnay	3	2.41	0.60
bluefrank rose	1	0.26	0.52
bluefrank rose	1	0.42	0.42
bluefrank rose	1	0.42	0.21
bluefrank rose	1	2.00	0.50
cabernet sauvignon	1	0.12	0.24
cabernet sauvignon	1	0.18	0.18
cabernet sauvignon	1	0.44	0.22
cabernet sauvignon	1	1.03	0.26

*Relative adsorption: Measured concentration of AFB1 in lees/Added amount of AFB1.

0.21 µg/mL and from 4 to 0.18 µg/mL after 90 days of fermentation, which means 90% reduction ratio. The pre-treatment procedures had no effect on the overall decrease of the AFB1 content of the samples. Even in the absence of added starter yeasts the natural yeast cultures of the grape and cellar could start the fermentation process and therefore no significant differences were observed in the reduction percentages. The greatest reduction of AFB1 concentration in the must was achieved in the case of rose wine (95%), followed by white wine (91%). On the other hand red wine showed the lowest AFB1 reduction rates with an average reduction percentage of 90. This value represents still a significant reduction rate of AFB1 level during fermentation. Experiments were also done on a different way, whose results are not indicated in Table 1. In these experiments musts were sulfurized and no yeasts were applied. In this case the fermentation of wine was not started, that can be explained with the destroying effect of sulfur on the natural yeast species of musts. These experiments were finished after 20 days, because the fermentation process did not start. But the results were surprising because no change of AFB1 levels were obtained. These findings indicate that the mechanism of the reduction of AFB1 concentrations in musts should be similar to the supposed mechanism for OTA which was thoroughly investigated in contrast with AFB1 [23-25]. The mechanism involved physical adsorption process by the cell-walls of the microorganisms similarly to other mycotoxins [26]. The results that are summarized in Table 2 harmonize with the literature data, namely mycotoxins were accumulated in lees. The chemistry and the molecular basis of mycotoxin binding was previously thoroughly examined, according to the present knowledge from literature data the reduction mechanism may consist of physical binding of the toxin to the cell wall proteins of the applied yeast [27,28]. According to our experimental data a similar model can be supposed for the explanation of the reduction of AFB1 level during wine fermentation processes, namely a physical adsorption of the toxin to the yeast cells can be occurred to obtain significant reduction of its concentration in must. We can conclude that a natural clarification process occurs during wine fermentation process which results in an average 90% reduction of mycotoxin concentration during the ripening process.

CONCLUSION

Saccharomyces cerevisiae, the most important yeast strain involved in wineries, can reduce AFB1 level in wine must during the fermentation process even by 77% - 97%. The reduction rate changed depending on the variety of must, but no significant differences were observed in AFB1 removal with regard to the initial AFB1 concentrations. The experiments which were carried out without adding starter yeasts after sulfurization delivered indirect evidences about the mechanism of the reduction of the AFB1 level. According to our results we can suppose that the reduction can be connected to the yeast cells, presumably to an adsorption process. In the developed model system we used AFB1 in extremely high concentrations to investigate the toxin's effect on yeast cells and on changes of the amount of toxin during the fermentation. These changes can be extrapolated onto lower AFB1 concentrations according to our previous findings in the case of OTA [23]. The results of our experiments demonstrated that even at a contamination level of 10 µg/L, the AFB1 concentration can be reduced under 2 µg/L, which is the European limit value for AFB1 in alcoholic beverages.

ACKNOWLEDGEMENTS

This work was supported by a GOP-1.3.1. project of New Szechenyi Plan, titled "Elaboration of an efficient wine technology for the elimination of mycotoxins".

REFERENCES

1.   Czerwiecki, L., Wilczynska, G. and Kwiecien, A. (2006) Mycotoxins in Several Polish Food Products in 2004-2005. Mycotoxin Research, 22,

159-162. http://dx.doi.org/10.1007/BF02959269

2. Moss, M.O. (1989) Mycotoxins of Aspergillus and Other Filamentous Fungi. Journal of Applied Bacteriology Symposium Supplement, 67, 69-81. http://dx.doi.org/10.1111/j.1365-2672.1989.tb03771.x

3. Rasch, C., Böttcher, M. and Kumke, M. (2010) Determination of Aflatoxin B1 in Alcoholic Beverages: Comparison of One and Two-Photon-Induced Fluorescence. Analytical and Bioanalytical Chemistry, 397, 87-92. http://dx.doi.org/10.1007/s00216-010-3530-1

4. Adányi, N., Levkovets, I.A., Rodriguez-Gil, S., Ronald, A., Váradi, M. and Szendrő, I. (2007) Development of Immunosensor Based on OWLS Technique for Determining Aflatoxin B1 and Ochratoxin A. Biosensors and Bioelectronics, 22, 797-802. http://dx.doi.org/10.1016/j.bios.2006.02.015

5. Ramos, A. J., Labernia, N., Marin, S., Sanchis, V. and Magan, N. (1998) Effect of Water Activity and Temperature on Growth and Ochratoxin Production by Three Strains of Aspergillus ochraceus on a Barley Extract Medium and on Barley Grains. International Journal of Food Microbiology, 44, 133-140. http://dx.doi.org/10.1016/S0168-1605(98)00131-7

6. Cervino, C., Knopp, D., Weller, M.G. and Niessner, R. (2007) Novel Aflatoxin Derivatives and Protein Conjugates. Molecules, 12, 641-653. http://dx.doi.org/10.3390/12030641

7. Yu, J., Cleveland, T.E., Nierman, W.C. and Bennett, J.W. (2005) Aspergillus flavus Genomics: Gateway to Human and Animal Health, Food Safety, and Corp Resistance to Diseases. Revista Iberoamericana Micologica, 22, 194-202. http://dx.doi.org/10.1016/S1130-1406(05)70043-7

8. Spadaro, D., Loré, A., Garibaldi, A. and Gullino, M.L. (2010) Occurrence of Ochratoxin A before Bottling in DOC and DOCG Wines Produced in Piedmont (Northern Italy). Food Control, 21, 1294-1297. http://dx.doi.org/10.1016/j.foodcont.2010.02.017

9. Zimmerli, B. and Dick, R. (1996) Ochratoxin A in Table Wine and Grape Juice: Occurrence and Risk Assessment. Food Additives and Contaminants, 13, 655-668. http://dx.doi.org/10.1080/02652039609374451

10. MacDonald, S., Wilson, P., Barmes, K., Damant, A., Massey, R., Mortby, E. and Shepherd, M.J. (1999) Ochratoxin A in Dried Fruit: Method Development and Survey. Food Additives and Contaminants, 16, 253-260. http://dx.doi.org/10.1080/026520399284019

11. Otteneder, H. and Majerus, P. (2000) Occurrence of Ochratoxin A in Wines: Influence of the Type of Wine and Its Geographical

Origin. Food Additives and Contaminants, 17, 793-798. http://dx.doi. org/10.1080/026520300415345

12. Visconti, A., Pascale, M. and Centonze, G. (1999) Determination of Ochratoxin A in Wine by Means of Immunoaffinity Column Clean-Up and High-Performance Liquid Chromatography. Journal of Chromatography, 864, 89-101. http://dx.doi.org/10.1016/S0021-9673(99)00996-6

13. Martinez-Culebras, P.V. and Ramon, D. (2007) An ITS-RFLP Method to Identify Black Aspergillus Isolates Responsible for OTA Contamination in Grapes and Wine. International Journal of Food Microbiology, 2, 147-153. http://dx.doi.org/10.1016/j.ijfoodmicro.2006.06.023

14. Medina, A., Mateo, R., Lopez-Ocana, L., Valle-Algarra, F.M. and Jimenez, M. (2005) Study of Spanish Grape Mycobiota and ochratoxin A Production by Isolates of Aspergillus tubingensis and Other Members of Aspergillus Section Nigri. Applied Environmental Microbiology, 71, 4696-4702. http://dx.doi.org/10.1128/AEM.71.8.4696-4702.2005

15. Melki Ben Fredj, S., Chebil, S., Lebrihi, A., Ghorbel, A. and Mliki, A. (2007) Occurrence of Pathogenic Fungal Species in Tunisian Vineyards. International Journal of Food Microbiology, 3, 245-250. http://dx.doi. org/10.1016/j.ijfoodmicro.2006.07.022

16. Khoury, A.E.L., Rizk, T., Lteif, R., Azoury, H., Delia, M.L. and Lebrihi, A. (2008) Fungal Contamination and Aflatoxin B1 and Ochratoxin A in Lebanese Wine-Grapes and Musts. Food Chemical Toxicology, 6, 2244-2250. http://dx.doi.org/10.1016/j.fct.2008.02.026

17. Pietri, A., Rastelli, S. and Bertuzzi, T. (2010) Ochratoxin A and Aflatoxins in Liquorice Products. Toxins, 2, 758-770. http://dx.doi.org/10.3390/toxins2040758

18. Sripatomswat, N. and Thasnakorn, P. (1981) Survey of Aflatoxin-Producing Fungi in Certain Fermented Foods and Beverages in Thailand. Mycopathologia, 73, 83-88. http://dx.doi.org/10.1007/BF00562595

19. Var, I., Kabak, B. and Gök, F. (2007) Survey of Aflatoxin B1 in Helva, a Traditional Turkish Food, by TLC. Food Control, 18, 59-62. http://dx.doi.org/10.1016/j.foodcont.2005.08.008

20. Móricz, Á.M., Fáter, Zs., Otta, K.H., Tyihák, E. and Mincsovics, E. (2007) Overpressured Layer Chromatographic Determination of Aflatoxin B1, B2, G1 and G2 in Red Paprika. Microchemical Journal, 85, 140-144. http://dx.doi.org/10.1016/j.microc.2006.03.007

21. Calleri, E., Marrubini, G., Brusotti, G., Massolini, G. and Caccialanza, G. (2007) Development and Integration of an Immunoaffinity Monolithic Disk for the On-Line Solid-Phase Extraction and HPLC Determination

with Fluorescence Detection of Aflatoxin B1 in Aqueous Solutions. Journal of Pharmaceutical and Biomedical Analysis, 44, 396-403. http://dx.doi.org/10.1016/j.jpba.2007.01.030

22. Pei, S.C., Zhang, Y.Y., Eremin, S.A. and Lee, W.J. (2009) Detection of Aflatoxin B1 in Milk Products from China by ELISA Using Monoclonal Antibodies. Food Control, 20, 1080-1085. http://dx.doi.org/10.1016/j.foodcont.2009.02.004

23. Csutorás, C., Rácz, L., Rácz, K., Fűtő, P., Forgó, P. and Kiss, A. (2013) Monitoring of Ochratoxin A during the Fermentation of Different Wines by Applying High Toxin Concentrations. Microchemical Journal, 107, 182-184. http://dx.doi.org/10.1016/j.microc.2012.07.001

24. Bejaoui, H., Mathieu, F., Taillandier, P. and Lebrihi, A. (2004) Ochratoxin A Removal in Synthetic Medium and Natural Grape Juices by Selected Oenological Saccharomyces Strains. Journal of Applied Microbiology, 97, 1038-1044. http://dx.doi.org/10.1111/j.1365-2672.2004.02385.x

25. Meca, G., Blaiotta, G. and Ritieni, A. (2010) Reduction of Ochratoxin A during the Fermentation of Italian Red Wine Moscato. Food Control, 21, 579-583. http://dx.doi.org/10.1016/j.foodcont.2009.08.008

26. Bueno, D.J., Casale, C.H., Pizzolitto, R.P., Salvano, M.A. and Oliver, G. (2007) Physical Adsorption of Aflatoxin B1 by Lactic Acid Bacteria and Saccharomyces cerevisiae: A Theoretical Model. Journal of Food Protection, 70, 2148- 2154.

27. Devegowda, G., Raju, M.V.L.N. and Swamy, H.V.L.N. (1998) Mycotoxins: Novel Solutions for Their Counteraction. Feedstuffs, 70, 12-15.

28. Ringot, D., Lerzy, B., Bonhoure, J.P., Auclair, E., Oriol, E. and Larondelle, Y. (2005) Effect of Temperature on in Vitro Ochratoxin A Biosorption onto Yeast Cell Derivatives. Process Biochemistry, 40, 3008-3016. http://dx.doi.org/10.1016/j.procbio.2005.02.006

Chapter 12

LACTIC ACID FERMENTATION OF PEPPERS

Maria Rosa Alberto[1,2], Maria Francisca Perera[1], Mario Eduardo Arena[1,2]

[1]Facultad de Bioquímica, Química y Farmacia, Universidad Nacional de Tucumán (UNT), Tucumán, Argentina;

[2] Centro Científico Tecnológico Tucumán—Consejo Nacional de Investigaciones Científicas y Tecnológicas (CCT-CONICET), Tucumán, Argentina.

ABSTRACT

Different peppers fermentations (Capsicum annum, grossum variety) were assayed: spontaneous, native microflora supplemented individually with Lactobacillus plantarum N8, Leuconostoc mesentereroides L. or Pediococcus pentosaceus 12p and by pure or combined cultures of these lactic acid bacteria (LAB). In order to eliminate the native flora, different kinds of heat treatment were assayed. The treatment selected was heating in autoclaved after research 3/4 atmosphere and to turn off. Fermentations were carried out at 22°C and 30°C and the culture media contained 2% or 0.2% glucose and 4% NaCl. Sugar consumption, pH reduction and acid production were higher at 30°C than at 22°C. At both temperatures, spontaneous fermentation showed a slower rate reduction in pH than inoculated samples. Diminution in pH in presence of 2% glucose was faster than at 0.2%, but minimum pH was in both case lower than 3.0. Maximum growth was reached between 2 and 5 days of fermentation in all the samples assayed. After 30 days of incubation in presence of 2% glucose the survival of LAB was nearly 5 log ufc/ml. The survival was higher at the lower temperature assayed for both glucose concentrations. Organoleptic properties of peppers fermented with a mixed culture of Leuconostoc mesenteroides and Pediococcus pentosaceus were found best by a human panel. This sample has a relation lactic acid/acetic acid of nearly 3 in the conditions assayed

INTRODUCTION

There are different forms to conserve food. One of them consists of increasing the acidity, which can be obtained artificially through addition of weak acids, or naturally by fermentation, obtaining free additive products. Fermentation can be developed spontaneously by the native microflora or after inoculation with lactic acid bacteria (LAB). In many cases, the fermentation is led by the indigenous flora and varies regarding substrate, temperature and storage conditions; consequently, the final product has variable sensorial properties. The use of starter cultures would be an appropriate approach for the control and optimization of the fermentation process in order to minimize variations in the organoleptic quality and microbiological stability. LAB are responsible for the fermentation of many vegetables and this process contributes to flavour, texture and aroma characteristics of the food. Additionally it guaranties a hygienically conservation and commercial stability. Lactic acid fermentation requires no or very little energy in the form of heat, allows the preservation of fresh vegetables or vegetables process minimally [1] and it improves the digestibility and nutritional value of the food [2]. The demand of fermented products has experienced an important increase in recent years, as consumers recognize that fermentation plays an important and beneficial role in human nutrition, health and nourishing safety [3]. Peppers are consumed mature or immature, raw or in conserves or pickles. The information available on fermentation of peppers is little. According to data provided by the National Centre of Studies and Agricultural Investigations of Cuba (C.E.N.A.I.C.) peppers can be fermented by the native microflora. Peppers represent an important crop in the northwest of Argentina, but the product is not available all year round, and for this reason a presservation process is necessary. Considering the technological importance of controlled fermentation of vegetables for the industry, different heat treatments and fermentation processes were assessed, in order to obtain an adequate product. The aim was to select a suitable starter culture in order to conduct an appropriate fermentation of Argentine peppers and to obtain a controlled process and a product of stable quality through time

MATERIALS AND METHODS

Organisms

Lactobacillus plantarum N8 [4], Leuconostoc mesentereroides L. [5] and Pediococcus pentosaceus 12p [6] were isolated from orange, tomato and grape, respectively. The bacteria were pre-cultured in MRS [7] broth supplemented with 15% (v/v) tomato juice and incubated at 30°C.

Peppers

Mature peppers (Capsicum annum variety grossum) were obtained from Salta province, Argentina, and carefully selected, without blows, apparent damages or microbeological alterations. The peppers were washed with abundant water and cut in fine strips. The seeds were eliminated and the peppers were processed within 48 hours of cultivation.

Heating Procedures

In order to eliminate the native flora without changing sensory properties, different heating techniques were assessed. Peppers were placed in 250 ml of a sterile solution of glucose and NaCl with or without inoculation with LAB (Lactobacillus plantarum N8). Heating techniques assayed were: heating the samples in autoclaved with fluent steam during 5 min.; heating in autoclaved after research 3/4 atmosphere and to turn off, and heating in autoclaved during 3 min. after research 3/4 atmosphere. Peppers (40 g) were subjected to heat treatments in a solution of 5% glucose and 4% NaCl (250 ml). After each treatment they were incubated at 30°C for one week. In order to evaluate the best technique: cell counts (cfu/ ml), pH and organoleptic characteristics such as colour of the peppers and consistency and colour of the solution were tested

Fermentation

Fermentation was carried out under previously laboratory-optimized conditions, at 22°C and at 30°C. The peppers (40 g) were incubated in 235 ml sterile solution containing (g/l): glucose (2 and 20) and NaCl (4); initial pH was 5.0. Each glucose concentration and temperature was therefore assayed with the 12 samples. Without heating: Natural Fermentation with the native flora (NF), NF plus Lactobacillus plantarum N8; NF plus Leuconostoc mesenteroides L.; NF plus Pediococcus pentosaceus 12p. Samples with heating: without inoculation (Control); with pure cultures of Lactobacillus plantarum N8; Leuconostoc mesenteroides L. or Pediococcus pentosaceus 12p; with mixed cultures of two pure cultures (Lactobacillus plantarum N8 and Leuconostoc mesenteroides L.; Lactobacillus plantarum N8 and Pediococcus pentosaceus 12p or Leuconostoc mesenteroides L. and Pediococcus pentosaceus 12p); and the mixed cultures of the three strains cited. In order to conserve the fermentation atmosphere of each sample different flasks were used for each assay (0, 1, 2, 5, 10, 20 and 30 days), because once the flasks were opened the samples could not continue being incubated due to the entrance of oxygen and the risk of loss of the atmosphere generated by the fermentation process.

Starter Culture

For the preparation of the starter culture, microorganisms grown in MRS were centrifuged at 30,000 g during 10 min., washed with sterile distilled water, centrifuged again and resuspended in a solution of glucose and NaCl, fitting an OD_{560} between 0.9 and 1 (10^7 cfu/ml). In the mixed cultures proportions were 1:1 and 1:1:1. The bacteria were inoculated in experimental media at a total cell concentration of 1 - 2 × 107 cfu/ml. Samples were taken after 0, 1, 2, 5, 10, 20 and 30 days incubation for growth measurement and stored frozen (−18°C) for subsequent analyses.

Growth Measurement

Bacterial growth was determined spectrophotometrically by measurement of optical density at 560 nm and by direct counting of cells on MRS agar supplemented with 15% (v/v) tomato juice, pH 6.0

Analytical Determinations

The pH was determined with a pH-meter equipped with a glass electrode, which was calibrated against standard buffer solutions (Anedra) at pH 4.0 and 7.0. Glucose and fructose were analysed by HPLC [8]. Organic acids were determined by HPLC analysis. Sample proteins were eliminated: 0.5 ml of a 6% trichloroacetic acid solution was added to 0.5 ml of the sample. The mixture was stirred on a vortex during 3 min. and then centrifuged during 5 min at 30,000 g. The pellet was discarded and the supernatant was membranefiltered (0.45 µ). The solvent used for separation was 0.01 N sulphuric acid. The samples were filtered using a sterile membrane of 0.45 µ stirrer. HPLC was performed with Gilson equipment with an infrared detector and in tegrator (Hewlett Packard, HP 3396 Series II). An ORH- 801 column for organic acids was used, containing a matrix of 300 × 6.5 mm, packed with a polymer of cationic interchange in its hydrogenated form. The column was operated at 22°C with a flow speed of 0.500 ml/min.

Sensorial Determinations

Organoleptic characteristics were evaluated by a group of selected people using the double blind test. The group was integrated by 10 people of both sexes (6 men and 4 women) and different age (23 - 45 years old). The human testers evaluated the peppers fermented under the conditions assayed according to their visual aspect, flavour and aroma. The parameters were selected according to those proposed by Seseña et al. [9] for the tasting of fermented eggplants.

Conservation of the Fermented Peppers

The fermented peppers were conserved during three months at room temperature in the same fermentation medium or in commercial vinegar (5% acetic acid) with 2% NaCl. In the case of commercial vinegar, the peppers were washed with distilled sterile water after 30 days of fermentation and they were placed with the vinegar in sterile bottles.

Spoilage Microorganism

Possible spoilage of the pickles was assayed for the following microorganisms: yeasts, Clostridium botulinum and enterobacteria, using Sabouraud, SPS agar and Mac Conkey, respectively.

RESULTS AND DISCUSSION

Heat Effect on the Organoleptic Characteristic of Peppers

Table 1 shows the effect of heating techniques on organoleptic properties before fermentation. The results were similar for inoculated and noninoculated samples and indicate that all heat treatments affect colour and consistency of peppers. In absence of heat (control) or in presence of fluent steam the bacteria (wild or inoculated) can grow and a decrease in pH was observed after 7 days of incubation. In the control media, with or without inoculation, the pH decreased two units, whereas the pH decreased only 0.3 units after seven days in samples treated with fluent steam. Consequently, fluent steam was inappropriate as a bactericidal procedure. The lowest alteration in the sensory properties occurred when the products were put under fluent steam and when they were heating in autoclaved until research 3/4 atmosphere and turn off immediately. The last procedure has the advantage that inactive the native flora and produces fewer organoleptic modifications than the same treatment during 3 min. Therefore, the technique applied in this study to study the effect of the bacterial inoculums in the vegetable fermentation was heating peppers in solution at 3/4 atmosphere in autoclave and extinguished.

Cell Growth

Table 2 shows maximum development of the microorganisms under the different fermentation conditions. The starters were inoculated at a concentration 100 times higher than the native flora, according to procedures proposed by Gardner et al. [10] and in agreement with Seseña et al. [9], who used lactic acid bacteria starters to carry out the fermentation of vegetables at a concentration of 107 cfu/ml. Maximum values of viable cells were obtained between the second and fifth day of fermentation; as of this time the number of viable cells began diminishing or remained stable. This is common in diverse vegetable fermentation processes, such as cucumbers and cabbage for the elaboration of sauerkraut [11]. For samples without heating procedure at both glucose concentrations, in general highest growth was observed at 22°C. This effect could be due to adaptation of the native microflora to growth at room temperature. However, inoculated samples with heat treatment showed higher growth at 30°C than at 22°C. In these conditions, at glucose concentration of 2% maximum development was higher than at 0.2%, nevertheless the 10-fold higher glucose concentration did not produce a proportional increase in the cell number. At both glucose concentrations, survival at room temperature (22°C) was higher than at 30°C, with the exception of NF samples at 2% glucose, in which survival was higher at 30°C (Table 3). After 30 days incubation at 30°C, lowest microbial survival was observed after heat treatment and inoculated with a pure culture of Leuconostoc mesenteroides or in a combination with one or two LAB at both glucose concentrations (Table 3). In the controlled fermentations and inoculated with Leuconostoc mesenteroides L. the lower survival can be explained by weak resistance to the low pH. The results agree with those reported by Gardner et al. [10] for carrots, onions and cabbages

Table 1: Effect of heating techniques on sensorial properties

Sample treatment	Solution colour	Pepper colour	Pepper consistency[a]
Control	Transparent	Red intense	++++
Fluent steam	Yellow	Red orange	+++
Autoclaved at 3/4 atmosphere and extinguished	Yellow	Red orange	+++
Autoclaved at 3/4 atmosphere during 3 minutes	Orange	Orange	+

[a]Consistence intensity with respect to an untreated sample (++++).

Table 2: Maximum growth of microorganisms at different temperatures in presence of 0.2% and 2% glucose.

Starter culture	Glucose			
	0.2%		2%	
	22°C	30°C	22°C	30°C
Natural Fermentation (NF)	8.09 ± 0.04[a]	7.57 ± 0.04	7.91 ± 0.05	6.99 ± 0.04
NF + L. plantarum	8.11 ± 0.03	7.63 ± 0.05	8.26 ± 0.02	8.19 ± 0.04
NF + Lc. mesenteroides	8.01 ± 0.04	7.67 ± 0.06	8.09 ± 0.05	8.19 ± 0.05
NF + P. pentosaceus	8.07 ± 0.05	7.92 ± 0.04	8.23 ± 0.06	8.09 ± 0.05
L. plantarum	7.91± 0.05	8.04 ± 0.03	7.96 ± 0.05	8.19 ± 0.05
Lc. mesenteroides	7.95 ± 0.04	8.03 ± 0.02	8.34 ± 0.01	8.35 ± 0.02
P. pentosaceus	7.92 ± 0.02	7.95 ± 0.01	8.09 ± 0.05	8.34 ± 0.05
L. plantarum + P. pentosaceus	7.50 ± 0.05	7.89 ± 0.05	8.25 ± 0.02	8.29 ± 0.01
L. plantarum + Lc. mesenteroides	7.80 ± 0.02	7.86 ± 0.02	8.09 ± 0.05	8.37 ± 0.05
Lc. mesenteroides + P. pentosaceus	7.80 ± 0.02	7.93 ± 0.05	8.29 ± 0.03	8.29 ± 0.02
L. plantarum + Lc. mesenteroides + P. pentosaceus	7.50 ± 0.06	7.95 ± 0.03	8.33 ± 0.02	8.35 ± 0.01

[a]Data are expressed in Log cfu/ml. Initial concentration 1.00×10^7 cells/ml, with the exception of Natural Fermentation in which cases the initial concentration was 3.12×10^5 cells/ml.

Table 3: Survival of microorganisms at different temperatures after 30 days of incubation in presence of 0.2% and 2% glucose at 22°C and 30°C.

Starter culture	Glucose			
	0.2%		2%	
	22°C	30°C	22°C	30°C
Natural Fermentation (NF)	5.47 ± 0.02[a]	3.38 ± 0.03	4.17 ± 0.01	5.40 ± 0.05
NF + L. plantarum	5.34 ± 0.03	4.00 ± 0.05	5.30 ± 0.05	5.70 ± 0.03
NF + Lc. mesenteroides	5.17 ± 0.02	3.92 ± 0.04	5.40 ± 0.04	5.70 ± 0.05
NF + P. pentosaceus	4.69 ± 0.04	3.84 ± 0.05	5.20 ± 0.02	5.60 ± 0.05
L. plantarum	4.53 ± 0.01	1.75 ± 0.01	5.60 ± 0.04	4.70 ± 0.04
Lc. mesenteroides	4.30 ± 0.03	1.00 ± 0.01	5.40 ± 0.02	4.50 ± 0.05
P. pentosaceus	5.30 ± 0.05	1.87 ± 0.01	5.70 ± 0.03	4.80 ± 0.03
L. plantarum + P. pentosaceus	5.84 ± 0.01	3.70 ± 0.02	5.80 ± 0.05	4.80 ± 0.05
L. plantarum + Lc. mesenteroides	5.82 ± 0.06	1.50 ± 0.05	5.45 ± 0.02	4.70 ± 0.05
Lc. mesenteroides + P. pentosaceus	5.90 ± 0.05	1.30 ± 0.02	5.60 ± 0.05	4.60 ± 0.02
L. plantarum + Lc. mesenteroides + P. pentosaceus	5.47 ± 0.04	1.20 ± 0.05	5.40 ± 0.04	4.70 ± 0.01

[a]Data are expressed in Log cfu/ml. Initial concentration 1.00×10^7 cells/ml, with the exception of Natural Fermentation in which cases the initial concentration was 3.12×10^5 cells/ml

Analytical Determinations in Culture Media

After 30 days of incubation production of lactic and acetic acid and consumption of glucose and fructose were determined under the different fermentation conditions (Tables 4 and 5). Initial glucose was higher for peppers subjected to thermal treatment. This increase was due to the diffusion of the sugar from the vegetable to the solution or the liberation of glucose from sucrose (data not shown). In addition, fructose was not added to the media, but it was detected in the culture media, perhaps due to liberation from the peppers. The microorganisms used in the pepper fermentations consumed as much glucose as fructose. In general, fructose and glucose consumption and acid production were higher at 30°C than at 22°C. Glucose consumption was faster in the natural fermentations supplemented with pure cultures (NF + Lactobacillus plantarum; NF + Leuconostoc mesenteroides; and in NF + Pediococcus pentosaceus) than with the others fermentations including the NF (data not shown). In all natural fermentations glucose was totally consumed after 20 days of incubation. The smallest amount of glucose was consumed by Lactobacillus plantarum as pure culture. The mixed LAB cultures used glucose faster than pure cultures (data not shown). Not all the glucose consumed was recovered as final fermentation products; maybe, it was used for to the production of biomass and cellular maintenance. The percentage of recovery of carbon in the final products determined oscillates between 58 and 99%. Highest recovery was found with 20 g/l of glucose at room temperature. Acetic acid was only formed in fermentations in the presence of Leuconostoc mesenteroides as starter culture at both glucose concentrations and in the natural fermentations in the presence of 20 g/l glucose (Tables 4 and 5). This indicates the presence of heterofermentatives microorganisms in the natural flora of the pepper. The relationship lactic acid/acetic acid in the natural fermentations did not remain constant under the different conditions; this demonstrates the variability of the natural flora of the vegetables, and therefore the inability of obtaining a product of stable quality and a reproducible process when the fermentation is spontaneous. The amount of free sugar appears to be important for the development of heterofermentatives microorganisms. In fermentations carried out by a pure culture of Leuconostoc mesenteroides the relationship lactic acid/acetic acid was somewhat higher than 1. In fermentations carried out by cultures of homofermentative LAB (Lactobacillus plantarum and Pediococcus pentosaceus) the production of lactic acid was high. In fermentations carried out by homo/heterofermentative mixed cultures the relationship lactic acid/acetic acid was about 3, whereas in fermentations carried out by a mixed culture of the three strains, the relationship lactic acid/acetic acid was nearly 5. Spyropoulou et al. [3] informed that in the fermentation of olives the production of lactic acid was 5 times higher when the initial glucose concentration increased from 1 to 10 g/l. Lactic acid production in our study with fermented peppers was between 8 and 10 times higher,

Table 4: Sugar consumption and organic acid production after 30 days pepper fermentations with 2 g/l of glucose

Temperature	Samples	Glucose consumption	Fructose consumption	Lactic acid formation	Acetic acid formation	Lactic acid/ acetic acid
	Natural Fermentation (NF)	11.11[a]	0.61	10.30	0.00	-
	NF + L. plantarum	11.11	0.59	16.81	0.00	-
	NF + Lc. mesenteroides	11.11	0.66	14.11	6.66	2.11
	NF + P. pentosaceus	11.11	0.64	17.78	0.00	-
	L. plantarum	12.40	0.94	18.40	0.00	-
22°C	Lc. mesenteroides	13.05	0.94	11.33	9.50	1.19
	P. pentosaceus	13.72	1.05	20.10	0.00	-
	L. plantarum + P. pentosaceus	13.57	1.23	21.78	0.00	-
	L. plantarum + Lc. mesenteroides	13.28	1.20	15.31	4.37	3.51
	Lc. mesenteroides + P. pentosaceus	13.22	1.18	14.00	4.50	3.11
	Lc. mesenteroides + P. pentosaceus + L. plantarum	13.33	0.88	15.55	3.10	5.02
	Natural Fermentation (NF)	11.11	0.88	13.22	1.00	13.22
	NF + L. plantarum	11.11	0.95	13.55	0.84	16.13
	NF + Lc. mesenteroides	11.11	0.88	15.40	7.56	2.04
	NF + P. pentosaceus	11.11	0.98	14.78	2.78	5.31
	L. plantarum	12.78	1.94	19.20	0.00	-
30°C	Lc. mesenteroides	13.11	2.38	11.60	9.60	1.21
	P. pentosaceus	13.89	2.38	21.10	0.00	-
	L. plantarum + P. pentosaceus	13.99	2.29	22.67	0.00	-
	L. plantarum + Lc. mesenteroides	13.77	2.33	14.78	4.88	3.02
	Lc. mesenteroides + P. pentosaceus	13.61	2.38	15.00	5.30	2.83
	Lc. mesenteroides + P. pentosaceus + L. plantarum	13.44	2.27	16.67	3.52	4.73

[a]Data are expressed in mmol/l. Initial values: glucose 11.11 mmol/l and fructose 0.88 mmol/l in media without heat treatment; glucose 14.44 mmol/l and fructose 2.38 mmol/l in media with heat treatment. Initial value of lactic and acetic acids 0.00 mmol/l. Relative Standard deviation (RSD) ≤ 2%.

Table 5: Sugar consumption and organic acid production in pepper fermentations with 20 g/l of glucose in 30 days.

Temperature	Samples	Glucose consumption	Fructose consumption	Lactic acid formation	Acetic acid formation	Lactic acid/ acetic acid
	Natural Fermentation (NF)	111.11[a]	0.55	68.88	8.30	8.29
	NF + L. plantarum	111.11	0.58	155.78	6.23	25.00
	NF + Lc. mesenteroides	111.11	0.77	144.40	49.33	2.93
	NF + P. pentosaceus	111.11	0.60	147.00	7.32	20.08
	L. plantarum	83.55	0.61	166.67	0.00	-
22°C	Lc. mesenteroides	100.00	1.38	133.33	98.36	1.35
	P. pentosaceus	87.22	0.50	157.78	0.00	-
	L. plantarum + P. pentosaceus	104.00	0.57	169.00	0.00	-
	L. plantarum + Lc. mesenteroides	111.98	0.87	139.00	53.90	2.57
	Lc. mesenteroides + P. pentosaceus	113.80	1.61	142.00	46.60	3.05
	Lc. mesenteroides + P. pentosaceus + L. plantarum	94.40	0.50	143.77	27.33	5.26
	Natural Fermentation (NF)	111.11	1.10	79.68	5.30	15.03
	NF + L. plantarum	111.11	1.12	176.67	4.90	36.06
	NF + Lc. mesenteroides	111.11	0.20	165.40	56.33	2.94
	NF + P. pentosaceus	111.11	1.08	171.23	5.12	33.44
	L. plantarum	101.10	1.94	170.62	0.00	-
30°C	Lc. mesenteroides	116.00	2.11	143.50	100.35	1.43
	P. pentosaceus	111.80	1.22	167.78	0.00	-
	L. plantarum + P. pentosaceus	112.01	1.15	178.00	0.00	-
	L. plantarum + Lc. mesenteroides	115.15	1.89	148.99	56.87	2.61
	Lc. mesenteroides + P. pentosaceus	114.16	2.11	152.00	48.60	3.12
	Lc. mesenteroides + P. pentosaceus + L. plantarum	116.06	1.88	145.53	30.33	4.80

[a]Data are expressed in mmol/l. Initial values: glucose 111.11 mmol/l and fructose 1.10 mmol/l in media without heat treatment, glucose 116.66 mmol/l and fructose 2.10 mmol/l in media with heat treatment. Initial value of lactic and acetic acids 0.00 mmol/l. Relative Standard deviation (RSD) ≤ 2%.

when the glucose concentration increased from 2 to 20 g/l. Optimum relation between lactic acid and acetic acid in the production of sauerkraut is between 3.5 and 5.0 [10,12]

pH Variations

From an initial pH of 5.0, reduction in pH in the spontaneous fermentation was slower than in inoculated samples and the final pH was higher at both temperatures. At 30°C in media with 2 g/l glucose, after one day incubation in the inoculated samples, the pH deceased nearly 1.5 units, with the exception the sample inoculated with of Lactobacillus plantarum, in this case the pH values diminished 1.0 unit. In NF the pH diminution was 0.4 units. At 22°C in all the cases the diminution was lower than at 30°C, and less than a unit. At 30°C, after 2 days of incubation average pH was 3.5. At room temperature (22°C), the same value was reached after 5 days. These results could be related to a faster consumption of glucose and fructose produced in the fermentations carried out at 30°C than at 22°C. Final pH at 22°C was reached between 10 and 20 days. In the experiment with 20 g/l of glucose, microbial growth was higher and the decrease in pH was faster than that in media supplemented with 2 g/l of glucose. Acid production in pepper fermentation depended on the initial glucose concentration. Consequently, fermentation carried out at 30°C and with 20 g/l of glucose and using starter cultures confers more microbiological stability to the product, because the rapid decrease in pH compared to fermentations at 22°C or at 0.2 g/l or by spontaneous fermentations. Nevertheless, fermentation of peppers with a lower glucose concentration allowed a reduction in the sugar used and therefore lowers cost and could reduce the development of NF heterofermentatives (no formation of acetic acid)

Organoleptic Evaluation

Organoleptic evaluation of the peppers fermented under the different conditions, revealed that those fermented by a mixed culture of Leuconostoc mesenteroides and Pediococcus pentosaceus at both temperatures were considered the best. At least 70% of the members of the tasting panel agreed and no significant difference was observed between either temperatures. However, the tasting panel found the peppers fermented at lower temperature slightly sweeter, which is probably due to the elevated concentration of residual sugars in the fermentation at 22°C (Table 6)

Table 6: Organoleptic evaluation of the fermented peppers

OLFACTORY EXAMINATION		1	2	3	4	5	6	7	8	9	10	11
	Exaggerated											
	Powerful						X				X	
Scent intensity	Sufficient	X	X	X	X	X		X	X	X		X
	Weak											
	Nonexistent											
VISUAL EXAMINATION												
	+											
	++						X		X			
Colour intensity	+++	X	X	X	X		X	X		X		X
	++++										X	
FLAVOUR EXAMINATION												
	Excessive						X	X	X	X		X
Acidity	Balanced				X	X					X	
	Insufficient	X	X	X								
	Smooth			X			X	X	X	X	X	
Texture	Moderate	X	X		X	X						X
	Rough											
	Soft	X	X	X	X							
Consistency	Interval					X					X	X
	Hard						X	X	X			X
FINAL EXAMINATION												
Highest conformity											X	

1: Natural Fermentation (NF); 2: NF + *L. plantarum* +; 3: NF + *Lc. mesenteroides*; 4: NF + *P. pentosaceus*; 5: *L. plantarum*, 6: *Lc. mesenteroides*, 7: *P. pentosaceus*; 8: *L. plantarum* + *P. pentosaceus*; 9: *L. plantarum* + *Lc. mesenteroides*; 10: *Lc. mesenteroides* + *P. pentosaceus*; and 11: *L. plantarum* + *Lc. mesenteroides* + *P. pentosaceus*.

The fact that a combination of one heterofermentative, Leuconostoc mesenteroides L., and one homofermentative, Pediococcus pentosaceus 12p, was found the best and selected as a starter by the panel members, is in agreement with results reported previously for other vegetables. The relation lactic acid/acetic acid for the starter of pepper fermentation, selected by the tasting panel (Leuconostoc mesenteroides L.—Pediococcus pentosaceus 12p), was between 2.8 and 3.1. In the mixed culture, the acetic acid production by heterofermentative microorganisms contributed to reach a balanced acidity in the taste examination. Gardner et al. [10] used pure and mixed starters of three lactic acid bacteria: Lactobacillus plantarum NK 312, Pediococcus acidilactici AFERM 772 and Leuconostoc mesenteroides BLAC, to lead the fermentation of juice from vegetable mixtures (onion, carrot, beet and cabbage) and selected as most suitable starter to lead the process to the constituted by the three lactic acid bacteria.

Conservation of Fermented Peppers

Before each organoleptic evaluation samples were tested in order to determine the presence of spoilage microorganisms. Clostridum or enterobacteria could not be detected in the fermentation media, probably because of the low pH obtained after a short period of time and the production of CO_2 that can eliminate O_2 from the fermentation atmosphere. A superficial layer with a creamy white colour could be observed at 22°C in the natural fermentation after 5 days and in the nonheating and inoculated fermentations after 10 days. The layer was examined optically and identified microscopically as yeasts. At 30°C in the nonheating and inoculated fermentations presence of yeasts was not detected by plating out on Sabouraud medium. LAB under adequate growth conditions seem to inhibit the development of yeasts. Our results agree with those observed previously by Gardner et al. [10], who informed that when inoculating vegetable juices with a mixture of three LAB (Lactobacillus plantarum NK 312, Pediococcus acidilactici AFERM 772 and Leuconostoc mesenteroides BLAC) to carry out the fermentation, the yeasts growth was inhibited. According to Bayrock and Ingledew [13], inhibition of yeasts can be due to their competition for nutrients with lactic acid bacteria and not to the production of lactic acid.

Bonestroo et al. [14] outlined that the inhibition of spoilage yeasts in fermented salads is probably due to a combination of the formation of lactic acid and CO_2 and the reduction in concentration of residual oxygen. The large presence of health-promoting compounds and the sensory features of pepper fruits may encourage food processing that aims at preserving functional compounds and agreeable sensory characteristics for extended shelf-life, possibly at room temperature [15-18]. Di Cagno et al. [17] demonstrated that fermentation by autochthonous and selected lactic acid bacteria strains (Lactobacillus plantarum, Lactobacillus curvatus and Weissella confusa), combined with heat treatment, allowed the manufacture and storage at room temperature (30 days) of safe red and yellow peppers with sensory attributes similar to raw fruits. The microbial and sensory features of peppers stored with sunflower seeds oil were almost similar to those stored without suspending liquid. In this study, two possible methods for pepper conservation were assayed after 30 days of fermentation: one in the some fermentation media and the other remove the peppers to the fermentation media and conserve them in commercial vinegar. The peppers conserved in the same media of fermentation during 3 months, did not present significant modifycations in the sensorial property. The survival of LAB after 3 months stayed in the order to 104 cfu/ml, suggesting the possibility of use pepper fermentation as source of probiotic. This could be given additional value to the fermentation of vegetables, and consider it as a functional food. Peppers conserved by the same period of time in vinegar, were excessively acid and had bleached completely after two months of storage, probably due to destabilization of the peppers caused by the acidity. LAB, after 3 months, were not detected. Growth of pathogenic microorganisms was not detected under either conservation condition. However, it is more convenient to conserve fermented vegetables in their own fermentation media, because they do not negatively modify the sensorial properties and the survival of LAB after 3 months of storage are high. Peppers in commercial conserves lack skin, whereas in this peppers fermentation, the peppers were not peeled, since the fermentation process improves the digestibility of the skin. From the results it is possible to suggest that fermentations be controlled at 30°C, because the fast drop in pH gives greater microbiological stability to the product. Fermented vegetables are generally not pasteurized and do not have artificial preservatives. The use of LAB cultures with adequate technological conditions could acidify the media quickly and thus diminishing the possibility of their deterioration during storage. LAB compete with other microorganisms for nutrients and space, so their growth eliminates undesirable microorganisms. Biopreservation is mainly due to the synthesis of a wide variety of antagonistic primary and secondary metabolites including organic acids, carbon dioxide [19]. Lactic acid fermentation undoubtedly represents the easiest and the most suitable way for increasing the daily consumption of fresh-like vegetables and fruits [18]. Moreover BAL consumption can exert beneficial effects on health level. In this work we have demonstrated the possibility of obtaining fermented peppers with excellent organoleptic qualities without inclusion of artificial additives. Controlled lactic acid fermentation of Capsicum annum could be an interesting technological procedure to conserve this product.

ACKNOWLEDGEMENTS

This work was supported by grants from Consejo Nacional de Investigaciones Científicas y Técnicas (CONICET), Consejo de Investigaciones de la Universidad Nacional de Tucumán (CIUNT).

REFERENCES

1. K. H. Steinkraus, "Lactic Acid Fermentation in Production of Foods from Vegetables, Cereals and Legumes," Antonie Van Leewenhoek, Vol. 49, No. 3, 1983, pp. 337- 348. http://dx.doi.org/10.1007/BF00399508

2. E. Caplice and G. F. Fittzgerald, "Food Fermentations: Role of Microorganisms in Food Production and Preservation," International Journal of Food Microbiology, Vol. 50, No. 1-2, 1999, pp.131-149. http://dx.doi.org/10.1016/S0168-1605(99)00082-3

3. K. E. Spyropoulou, N. G. Chorianopoulos, P. N. Skandamis and G. J. Nychas, "Survival de Escherichia coli O157: H7 during the Fermentation of Spanish Style Green Olives (Conserveola Variety) Supplemented with Different Carbon Sources," International Journal of Food Microbiology, Vol. 66, 2001, pp. 3-11. http://dx.doi.org/10.1016/S0168-1605(00)00510-9.

4. M. E. Arena, F. M. Saguir and M. C. Manca de Nadra, "Inhibition of Growth of Lactobacillus plantarum Isolated from Citrus Fruits in the Presence of Organics Acids," Microbiology-Aliments-Nutrition, Vol. 14, No. 3, 1996, pp. 219-226.

5. S. A. Sajur, F. M. Saguir and M. C. Manca de Nadra, "Effect of Dominant Specie of Lactic Acid Bacteria from Tomato on Natural Microflora Development in Tomato Purée," Food Control, Vol. 18, No. 5, 2007, pp. 594-600. http://dx.doi.org/10.1016/j.foodcont.2006.02.006

6. A. M. Strasser de Saad and M. C. Manca de Nadra, "Isolation and Identification of Lactic Acid Bacteria from Cafayate (Argentina) Wines," Microbiology—Aliments —Nutrition, Vol. 5, 1987, pp. 45-49.

7. J. C. De Man, M. Rogosa and M. R. Sharpe, "A Medium for the Cultivation of Lactobacilli," Journal of Applied Bacteriology, Vol. 23, No. 1, 1960, pp. 130-135. http://dx.doi.org/10.1111/j.1365-2672.1960.tb00188.x

8. A. M. Wilson, T. M. Work, A. A. Bushway and R. J. Bushway, "HPLC Determination of Fructose, Glucose, and Sucrose in Potatoes," Journal of Food Science, Vol. 46, No. 1, 1981, pp. 300-301. http://dx.doi.org/10.1111/j.1365-2621.1981.tb14589.x

9. S. Seseña, I. Sanchez Hurtado, M. A. González Vinas and L. Palop,

"Contribution of Starter Culture to the Sensory Characteristics of Fermented Almagro Eggplants," International Journal of Food Microbiology, Vol. 67, No. 3, 2001, pp. 197-205. http://dx.doi.org/10.1016/S0168-1605(01)00442-1

10. N. J. Gardner, T. Savard, P. Obermeier, G. Caldwell and C. P. Champagne, "Selection and Characterization of Mixed Starter Cultures for Lactic Acid Fermentation of Carrot, Cabbage, Beet and Onion Vegetable Mixtures," International Journal of Food Microbiology, Vol. 64, No. 3, 2001, pp. 261-275. http://dx.doi.org/10.1016/S0168-1605(00)00461-X

11. W. C. Frazir and D. C. Westhoff, "Contaminación, Conservación y Alteración de las Hortalizas y de las Frutas," Food Microbiology, McGraw Hill, New York, 1993, pp. 259-288.

12. [M. R. Adams and M. O. Mosa, "Alimentos Fermentados y Alimentos Microbianos," Microbiología de los Alimentos, Zaragoza,1995.

13. D. P. Bayrock and W. M. Ingledew, "Inhibition of Yeast by Lactic Acid Bacteria in Continuous Culture: Nutrient Depletion and/or Acid Toxicity?" Journal of Industrial Microbiology and Biotechnology, Vol. 31, No. 8, 2004, pp. 362-368. http://dx.doi.org/10.1007/s10295-004-0156-3

14. M. H. Bonestroo, J. C. Wit, B. J. Kusters and F. M. Rombouts, "Inhibition of the Growth of Yeasts in Fermented Salads," International Journal of Food Microbiology, Vol. 17, No. 4, 1993, pp. 311-320. http://dx.doi.org/10.1016/0168-1605(93)90201-Q

15. J. M. Navarro, P. Flores, C. Garrido and V. Martinez, "Changes in the Contents of Antioxidants Compounds inPepper Fruits at Different Ripening Stages, as Affected by Salinity," Food Chemistry, Vol. 96, No. 1, 2006, pp. 66-73. http://dx.doi.org/10.1016/j.foodchem.2005.01.057

16. S. M. Castro, J. A. Saraiva, J. A. Lopes-da-Silva, I. Delgadillo, A. Van Loey, C. Smout and M. Hendrickx, "Effect of Thermal Blanching and High Pressure Treatments on Sweet Green and Red Bell Peppers Fruits (Capsicum annum L.)," Food Chemistry, Vol. 107, No. 4, 2008, pp. 1436-1449. http://dx.doi.org/10.1016/j.foodchem.2007.09.074

17. R. Di Cagno, R. F. Surico, G. Minervini and M. De Angelis, "Use of Autochthonous Starters to Ferment Red and Yellow Peppers (Capsicum annum L.) to Be Stored at Room Temperature," International Journal of Food Microbiology, Vol. 130, No. 2, 2009, pp. 108-116. http://dx.doi.org/10.1016/j.ijfoodmicro.2009.01.019

18. R. Di Cagno, R. Coda, M. De Angelis and M. Gobbetti, "Exploitation of Vegetables and Fruits through Lactic Acid Fermentation," Food

Microbiology, Vol. 33, No. 1, 2013, pp. 1-10. http://dx.doi.org/10.1016/j. fm.2012.09.003

19. L. Fan and L. Truelstrup Hansen, "Fermentation and Biopreservation of Plant Based Foods with Lactic Acid Bacteria," In: Y. H. Hui, Ed., Handbook of Plant Based Fermented Food and Beverage Technology, 2nd Edition,CRC Press, Boca Raton, pp. 35-48.

Chapter 13

APPLICATION OF ARTIFICIAL NEURAL NETWORKS TO FOOD AND FERMENTATION TECHNOLOGY

Madhukar Bhotmange and Pratima Shastri

Laxminarayan Institute of Technology, Rashtrasant Tukadoji Maharaj Nagpur University, Nagpur 440033. India

INTRODUCTION

Every system is controlled by certain parameters and works at its best for a certain combination of the values of these parameters Input parameters of the system are defined as the independent variables or causes, which affect the values of output parameters commonly identified as effects. The relationship in many case is typically nonlinear, and complex. Different input parameters –apart from their individual influences – may affect the output parameter in synergistic or antagonistic way. The knowledge of cause-and-effect relationships is important in the solution of problems in all fields of endeavor. In the simplest of cases, these relationships may take on a linear form, while in others, highly nonlinear and complex, relationships may be appropriate. Some relationships are static, while others involve dynamic or time varying elements. A complex system like thermal processing requires maximum destruction of undesirable microorganisms with minimum loss of freshness, taste, texture and flavor as the outputs, with time temperature, can size, etc. as extrinsic causes, along with the composition, viscosity, and thermal properties of food material as intrinsic causes. Product development happens to be an equally complex system where level and proportion of ingredients are the inputs, which determine the sensory parameters, cost and marketability. Modeling of bioprocesses for engineering applications is equally challenging task, due to their complx nonlinear dynamic behaviour. The conditions of best functioning are called optimum operating / functioning conditions. Large number of experiments need to be performed under certain set of conditions, for obtaining these optimum parameters. Still, the results at selected data points need not necessarily represent the optimum functioning of a process, specially for typical nonlinear systems. Performing permutations and combination with experimental parameters till the optimum combination of parameters is achieved is not only time consuming and laborious, but also contributes to increased expenses, hazard possibility and error incorporations. In such situation, several structured and unstructured models can be developed from the available data, and the possible outputs can be successfully predicted at any combination of values, within the frame work. Artificial Neural Network (ANN) is one such tool for prediction of outputs for nonlinear systems at various combinations. The process is based on learning of the network with the experimental values, thus knowing the system behavior, & then predicting the output values of the desired set of parametric combinations. Food science and technology represents a potential area for application of ANN. Critical review by Huang et al. (2007) discusses the basic theory of the ANN technology and its applications in food science, providing food scientists and the research community an overview of the current research and future trend of the applications of ANN technology in the field.

WHAT IS NEURAL NETWORK?

Mother nature's most complex creation, the human brain has evolved over million of years and has very complex and powerful architecture. It consists of large number of nerve cells called neurons. The axon or output path of a neuron splits up and connects to dendrites or input paths of other neurons through a junction known as a synapse (Fig.1) The transmission across this junction is chemical in nature, and the amount of signal transferred depends on the amount of chemicals (Acetylchloline) released by the axon and in turn received by the dendrites. This synaptic strength is modified when the brain learns. Each neuron will have of the order of 10,000 dendrites through which they accept inputs.

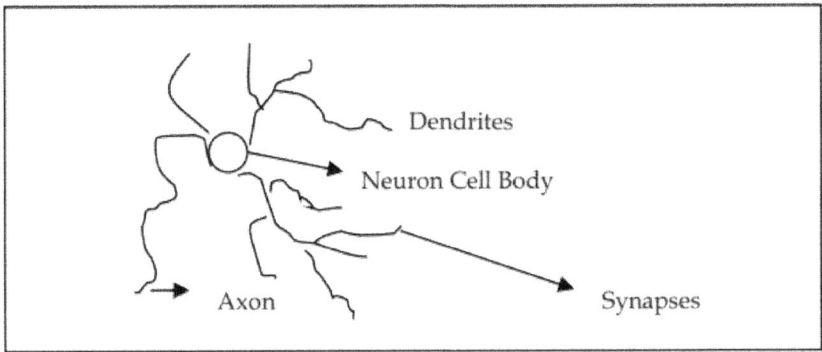

Figure 1: Biological Neuron

Artificial Neural Network (ANN)

An artificial neural network (ANN) is a data processing system based on the structure of the biological neural simulation by learning from the data generated experimentally or using validated models.

Some terms required to be defined for ANN users are:

- ANN: A neural network is a processing device, either an algorithm, or actual hardware, whose inspired by the design in and functioning of animal brains and components thereof. It is computer program designed to simulate the brain neurons.

- Processing element: In an ANN, the unit analogous to the biological neuron is a processing elements (PE). Each PE has many inputs and outputs. The network consists of many units or neurons, each possibly having a small amount of local memory. The unit by undirectional communication channels "connections" which carry numeric data. The units operate only on their local data and on the inputs they receive connection.

- Connection weight: The output path of a processing element is connected to input paths of other PEs through connection weights, analogous to the synaptic strength of neural connections.

- Input, output and hidden layers: A network consists of a sequence of layers with connections between successive layers. Data to the network is presented at input layer and the response of the network to the given data is produced in the output layer. There may be several layers between these two principal layers, which are called hidden layers.

- Training: Most neural networks have some sort of "training" rule whereby the weights of connection are adjusted on the basis of presented patterns. In other words, neural network patterns "learn from example".

- Error: It is defined as the total sum of the difference between desired output and output produced by the network for the set of inputs.

- Learning rate: A learning rule, which changes the connection weights of the network in response to the example inputs and desired output to those inputs. The training of neural network model is similar to the way humans or animals are trained by reinforcement technique, where certain synapses that connect the neurons selectively get strengthened leading to increase in the gain.

- Recall: Recall refer to how the network processes a data set presented at its input layer and produces a response at the output layer. The weights are not changed during the recall process.

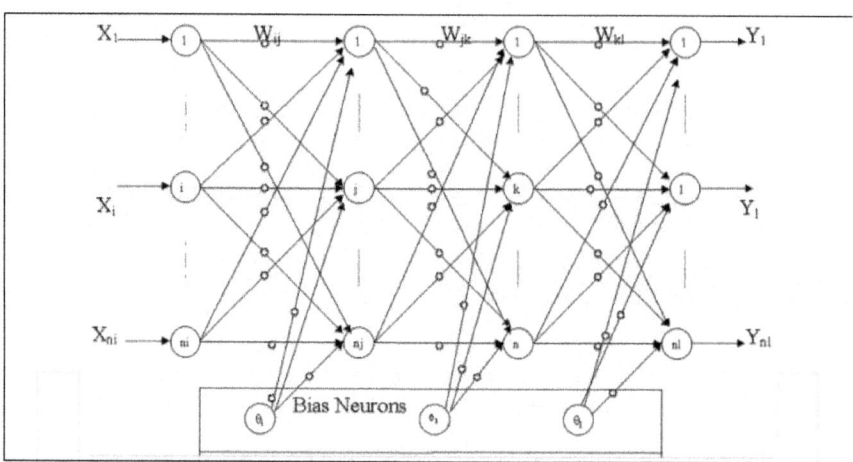

Figure 2: Artificial Neural Network : A Multilayer Perceptron

Derived from their biological counterparts, ANNs are based on the concept that a highly inter-connected system of simple processing elements can learn complex inter relationships between independent and dependent variables. ANNs offer an attractive approach to the black-box modeling of highly complex, nonlinear systems having a large number of inputs and out puts in the form of massively connected parallel structures. It has three-layered system, an input layer, and intermediate layer called hidden layer, and an output layer (Fig.2). Each layer contains a number of neurons. The number of neurons in the input layer equals the number on inputs to the neural network while the number of neurons in the output layer equals the number outputs in the system.

Although numerous guidelines have been proposed for selecting the number of units in the hidden layer, they do not work in all situations, and the number is often determined heuristically. Each neuron is connected to all the neurons in the next layer by means of a "connection weight". The output from neurons can be calculated by suitable "transform equations" provided the inputs and the connection weights are known. The sequence of neural network modeling is to assume a set of weights initially, compute the outputs and the predict error, and then adjust the weights according to an error minimization technique until the prediction error falls to an acceptable level. This activity of finding optimal weight is called network training. Once the network is so trained, the black –box model is ready, and may be used to predict outputs for a set of new inputs, not originally part of those used in training.

Types of ANN

Back Propogation Network (BPN)

Back Propogation Network has been extensively studied, theoretically, and has been the most successful. The BPN is usually built from a three layered system consisting of input, hidden, and output layers. An equation in the hidden layers (transfer function) determines whether inputs are sufficient to produce an output (Hornik et al 1989). There are several kinds of transfer functions, e.g. threshold or sigmoid functions. In training a NN, the values predicted by the net work are compared to experimental values using the delta rule, an equation which minimizes error between experimental values and net work predicted values. The errors are then back propagated to hidden and input layers to adjust weights. This is repeated many times until errors between predicted and experimental values are minimized. General reviews, and references of NN procedure are discussed by Eberhart and Dobbins (1990) .

General Regression Neural Network (GRNN)

General Regression Neural Network are memory based feed forward networks meaning that all the training samples are stored in the network. It possess a special property that they do not require iterative training.

NEURAL NETWORK VS STATISTICAL REGRESSION

In statistical regression, the parameters or constants of the equation are determined for a given mathematical equation, which relates the inputs to the output(s), so that the difference between the desired output and the output of the equation for the set of inputs is a minimum. Here the type and nature of the equation relating the inputs with the output has to be initially formulated clearly. Neural Network (NN) doesn't require such explicit relationship between the inputs and the output(s). In Neural network parameter values cannot be extracted after the simulation. In statistics the analysis is limited to a certain number of possible interactions. However, more terms can be examined for interaction and included in Neural Network. By allowing more data to be analyzed at the same time, more complex and subtle interactions can be determined. Fuzzy and not so clear data sets can also be analyzed and their interaction studied with Neural Network, whereas statistical regression analysis will fail in such situation. It can perform better than statistical regression analysis for prediction, modeling & optimization even if the data is noisy and incomplete. It is also ideally suited when the inputs are qualitative in nature and when the inputs or the output cannot be represented as mathematical terms (Pandharipande, 2004). Unlike other modeling such as expert system, an ANN can use more than two parameters to predict two or more parameters. In addition, ANN differs from traditional methods due to their ability to learn about the system to be modeled without a prior knowledge of the process parameter. ANN results are straight forward and do not need any transformations. ANN is amongst various intelligent modeling methods which are able to solve a very important problem –processing of unstructured ,scarce and incomplete numerical information about nonlinear and non-stationary systems , as well as biotechnological processes (Vassileva et al, 2000).. ANN has the ability for relearning according to new data. and it is possible to add new observations at any time. Unlike ANN, when new observations are added to the data set in PCR, principal components have to be calculated before regression analysis is applied (Vallejo-Cordoba et al, 1995)

APPLICATIONS OF ANN IN FOOD TECHNOLOGY

Artificial Neural Networks (ANNs) have been applied in almost every aspect of food science over the past two decades, although most applications are in the development stage. ANNs are useful tools for food safety and quality analyses, which include modeling of microbial growth and from this predicting food safety, interpreting spectroscopic data, and predicting physical, chemical, functional and sensory properties of various food products during processing and distribution. ANNs hold a great deal of promise for modeling complex tasks in process control and simulation and in applications of machine perception including machine vision and electronic nose for food safety and quality control.

ANN for prediction of food quality, properties and shelf life

Quality of food is complex term, and is assessed by suitable combination of physical, chemical and organoleptic tests. Physical / chemical parameters- though convenient to measure - do not always have straightforward correlations with the sensory evaluation results. However, frequent sensory evaluation is restricted due to the availability of trained judges, and proper ambience. Several investigators have attempted to apply ANN models for prediction of food properties, and changes during processing and storage of foods. Zhang and Chen (1997) introduced a method of food sensory evaluation employing artificial neural networks. The process of food sensory evaluation can be viewed as a multiinput and multi-output (MIMO) system in which food composition serves as the input and human food evaluation as the output. It has proved to be very difficult to establish a mathematical model of this system; however, a series of samples have been obtained through experiments, each of which comprises input and output data. On the basis of these sample data, the back-propagation algorithm (BP algorithm) is applied to "train" a threelayer feed-forward network. The result is a neural network that can successfully imitate the food sensory evaluation of the evaluation panel. This method can also be applied in other fields such as food composition optimizing, new product development and market evaluation and investigation. Lopez et al (1999) have applied ANN for identification of registered designation of origin areas of portugese cheese defined by microbial phenotypes and artificial neural networks. The human sense of smell is the faculty which has very important role to play in industries such as beverages, food and perfumes. Studies have been carried out to construct an instrument that mimics the remarkable capabilities of the human olfactory system (Gardner et al 1990). The instrument or electronic nose consists of a computer-controlled multi-sensor array, which exhibits a differential response to a range of vapors and odors. The authors report on a novel application of artificial neural networks (ANNS) to the processing of data gathered from the integrated sensor array or electronic nose. This technique offers several advantages, such as adaptability, fault tolerance, and potential for hardware implementation over conventional data processing techniques. Results of the classification of the signal spectra measured from several alcohols are reported and they show considerable promise for the future application of ANNs within the field of sensor array processing. Electronic/ artificial nose, developed as systems for the automated detection and classification of odors, vapors, and gases is generally composed of a chemical sensing system (e.g., sensor array or spectrometer) and a pattern recognition system (e.g., artificial neural network). Electronic noses for the automated identification of volatile chemicals for environmental, medical and food industry applications are being developed A similar report on application of electronic nose for classification of pig fat has been reported by Carrapsio et al. (2001). Fatty acid analysis is frequently performed in fat and other raw materials to classify them according to their fatty acid composition, but the need to carry out online determinations has generated a growing interest in more rapid options. This research was done to evaluate the ability of a polymer-sensor based electronic nose to classify Iberian pig fat samples with different fatty acid compositions. Significant correlations were found between individual fatty acids and sensor responses, proving that sensor response data were not fortuitously sorted. Significant correlations also appeared between some sensors and water activity, which was considered during the sample classification. Two supervised pattern recognition techniques were attempted to process the sensor responses: 85.5% of the samples were correctly classified by discriminant analysis, but the percentage increased to 97.8% using a one-hidden layer back-propagation artificial neural network. An artificial olfactory system based on Gas Sensor Array and Back-Propagation Neural Network is constructed to determine the individual gas concentrations of gas mixture (CO and H_2) with high accuracy. Back-Propagation (BP) neural network algorism has been designed using MATLAB neural network toolbox, and an effective study to enhance the parameters of the neural network, including pre-processing techniques and early stopping method is presented in this paper. It is showed that the method of BP artificial neural improves the selectivity and sensitivity of semiconductor gas sensor, and is valuable to engineering application (Tai et al., 2004). The electronic nose (sensor responses analyzed by a neural network) achieved success similar to that obtained using the more usual fatty acid analysis by gas chromatography. Similar application in fatty acid analysis of soyabean oil is reported by Kovalenko et al (2006). An artificial neural network model is presented for the prediction of thermal conductivity of food as a function of moisture content, temperature and apparent porosity. (Sablani and Rahman, 2003).The food products considered were apple, pear, corn starch, raisin, potato, ovalbumin, sucrose, starch, carrot and rice. The thermal conductivity data of food products (0.012-2.350W/mK) were obtained from literature for the wide range of moisture content (0.04-0.98 on wet basis fraction), temperature (-42-130oC)and apparent porosity(0.0-0.7). Several configurations were evaluated while developing the optimal ANN model. The optimal model ANN consisted two hidden layers with four neurons in each layer. This model was able to predict thermal conductivity with a mean relative error of 12.6%,a mean absolute error of 0.081 W/mK. The model can be incorporated in heat transfer calculations during food processing. Rahman's model (at 0oC) and a simple multiple regression model predict thermal conductivity with mean relative error of 24.3%. An interesting application of

ANN for identification of organically farmed atlantic salmon from wild salmon is by analysis of stable isotopes and fatty acids is discussed by Molkentin et al (2007). Using isotope ratio mass spectrometry (IRMS), the ratios of carbon (δ 13C) and nitrogen (δ 15N) stable isotopes were investigated in raw fillets of differently grown Atlantic salmon (Salmo salar) in order to develop a method for the identification of organically farmed salmon. IRMS allowed to distinguish organically farmed salmon (OS) from wild salmon (WS), with δ 15N-values being higher in OS, but not from conventionally farmed salmon (CS). The gas chromatographic analysis of fatty acids differentiated WS from CS by stearic acid as well as WS from CS and OS by either linoleic acid or α-linolenic acid, but not OS from CS. The combined data were subjected to analysis using an artificial neural network (ANN). The ANN yielded several combinations of input data that allowed to assign all 100 samples from Ireland and Norway correctly to the three different classes. Although the complete assignment could already be achieved using fatty acid data only, it appeared to be more robust with a combination of fatty acid and IRMS data, i.e. with two independent analytical methods. This was also favorable with respect to a possible manipulation using suitable feed components. A good differentiation was established even without an ANN by the δ 15N-value and the content of linoleic acid. The general applicability in the context of consumer protection is recommended be checked with further samples, particularly regarding the variability of feed composition and possible changes in smoked salmon. Experimental measurements of the variation in the solid fraction during crystallization of lipid mixtures are often correlated in terms of the so-called Avrami model. Jose et al (2007) employed above model to describe measurements taken during the crystallization of blends of tripalmitin in olive oil at high concentrations. Although the blends appeared to behave ideally, the Avrami model failed to describe the experimental results over the entire range of tripalmitin concentration investigated. As an alternative to the description of lipid crystallization experiments, the use of continuous-time artificial neural network (ANN) approximators is proposed. ANN successfully reproduced the experimentally observed behavior for all temperatures and tripalmitin concentrations used. ANN based automatic grading and sorting systems for fruits and vegetables have been developed by various investigators. Saito et al (2003) have developed eggplant grading system using image processing and artificial neural network. The lighting conditions are discussed for taking color components of the eggplant image effectively. The shape parameters such as length, girth, etc. are measured using image processing. On the other hand, bruises of the eggplants are detected and classified based on the color information by using artificial neural network. Development of electronc nose for determination of fruit ripeness has been reported by Salim et al. (2005). A combination of machine vision and artificial neural network model for guava sorting which classify from size, weight and defect of guava has been described by Chokananporn and Tansakul (2008) and the system was evaluated by comparing with human sorting. Furthermore, the surface area of guava could be estimated from the artificial neural network model. The major diameter, intermediate diameter, minor diameter, and sphericity were used to classify the shape and used as the input parameters of the network. The sorting process was controlled by computer software which was well designed and created on visual basic 6.0. The experiments were carried out with fresh guava. The results from machine vision system were compared with those from human classifying capability. One hundred percent coincidence for the extra size and 73.3 percent coincidence for the class I and II size were obtained. For surface area estimation, the predicted surface area was foundto be nearly the same as that from the standard method. The lowest mean relative error (MRE) and mean absolute error (MAE) values were 0.15% and 0.39 cm2, respectively. Similar combination system for classification of beans is reported by Kilik et al (2007). Prediction of Milk shelf – life based on Artificial Predicting Neural networks and head space gas chromatographic data has been reported by Vellejo-Cordoba et al. (1995). Pasteurized milk was sampled during refrigerated storage at 4oC until termination of shelf life, as determined by sensory evaluation, sub samples were incubated at 24 +1oC for 18 hours prior to detection of volatiles by dynamic head space gas chromatograph (Cordoba & Nakai, 1994)). Several volatiles consisting mainly of aldehydes, ketones & alcohols were identified in milk. Not only increased peak areas of the compounds already present appeared in poorquality milk, new volatiles were also detected, including esters. Cross validation was used with 113 training sets, and 21 test sets. In PCR, the independent variables were the first 30 principal components and the dependent variable was flavor – based shelf life in days. The shelf life predictability of ANN was superior to PCR as indicated by carrying out regression analysis for experimental vs predicted shelf life and the squared correlation (r2) and the standard error of the estimate (SEE). The power of computational neural networks (CNN) for growth prediction of three strains of Salmonella as affected by pH level, sodium chloride concentration and storage temperature was evaluated by Herv's et al (2001). The architecture of CNN was designed to contain above three input parameters and growth as output parameter. The standard error of prediction (%SEP) obtained was under 5% and was significantly less than the one obtained using regression equations. Similar study by Zurera-Cosano et al (2005) reported an Artificial Neural Network-based predictive model (ANN) for Leuconostoc mesenteroides growth in response to temperature, pH, sodium chloride and sodium nitrite, was validated on vacuum packed, sliced, cooked meat products and applied to shelf-life determination. Lag-time (Lag), growth rate (Gr), and maximum population density (yEnd) of L. mesenteroides, estimated by the ANN model, were compared to those observed in vacuumpacked cooked ham, turkey breast meat, and chicken breast meat stored at 10.5°C, 13.5°C and 17.7°C. From the three kinetic parameters obtained by the ANN model, commercial shelf-life were estimated for each temperature and compared with the tasting panel evaluation. The commercial shelf life estimated microbiologically, i.e. times to reach 106.5 cfu/g, was shorter than the period estimated using sensory methods. Application of ANN for prediction of shelf life of green chilli powder (GCP) is reported by Meshram (2008).Green Chilli Powder (GCP) prepared by dehydration of Jwala variety of chilli in air–Radio Frequency (RF) combo dryer had 1.13% moisture content with 19% ERH. Danger and critical points were identified at 60.5 % and 63% ERH corresponding to 7.12% and 8.0% moisture content respectively. Storage study was carried out under ambient (25oC, 65% RH) and accelerated (38oC, 90% RH) conditions for GCP packed in Laminated aluminium foil (LAM) and Polypropylene (PP). Half Value Period (HVP) and shelf life at different combinations of temperature (T) and relative Humidity (RH%) for 100 g GCP pack was calculated based on WVTR (LAM =2.35, PP =4.16 units at 38oC,90% RH) and packaging constant. (Ranganna). Application of Artificial Neural Network (elite-ANN ©) for prediction of shelf life as function of T and RH% gave R^2 value >0.99 for both packings.

ANN in food processing

Various processing parameters are required to be monitored and controlled simultaneously, and it is quite difficult to derive classical structured models, on account of practical problems in conducting required number of experiments and lack of sufficient data. Possibility for application of ANN for optimizing the process parameters is an interesting area, with many potential applications. The effect of agglomerate size and water activity on attrition kinetics of some selected agglomerated food powders was evaluated by Hong Yan and Barbosa-Canovas (2001) by application of ANN. Investigation of the attrition of agglomerates is very important for assessing the agglomerate strength, compaction characteristics, and quality control. A oneterm exponential attrition index model and the Hausner ratio were used to study the effects of agglomerate size and water activity on the attrition kinetics of some selected agglomerated food powders. It was found that the agglomerate size and water activity played significant roles in affecting the attrition: the larger the agglomerate size and higher the water activity, higher was the attrition index under the same tap number. The Hausner ratio was well correlated with the attrition index at high tap numbers and might be used as a simple index to evaluate attrition severity for agglomerates. Knowing the effects of agglomerate size and water activity is very useful to minimize the attrition phenomenon during the handling and processing of agglomerated powders. Modeling and control of a food extrusion process using artificial neural network and an expert system is discussed by Popescue et al. (2001). A neural network model is proposed and its parameters are determined. Simulation results with real data are also presented. The inputs and outputs of the model are among those used by the human operator during the start-up process for control. An intelligent controller structure that uses an expert system and "delta-variations" to modify inputs is also proposed. A hypothesis on coating of food is put forward by Bhattacharya et al (2008), who have also discussed development of a system analytical model based on simulation studies and artificial neural network The process of coating of foods is a complex process due to the presence of a large number of variables, and unknown relationship between the coating variables and coating characteristics. Needs exists to develop a model that can relate the important variables and coating parameters that would be helpful in developing coated products. A system analytical model for coating of foods has been hypothesized. The model relates influencing variables to derived parameters that in turn relates the target coating parameters. The concentration of solids and temperature of coating dispersions are the examples of the influencing variables, whereas rheological parameters (apparent viscosity, yield stress, flow and consistency indices) are the derived parameters that finally decide the coating parameters such as total uptake, solid uptake and dimensionless uptake according to the hypothesized relations $y = f(x)$ and $z = g(y)$. The proposed hypothesis was initially examined by performing simulation studies conducted on steel balls (small and big) using sucrose solution and malt – maltodextrin dispersions at different concentrations (20-60%) and temperatures (5-80°C), and applying the theory of artificial neural network (ANN) for prediction of target parameters. The hypothesis was tested in actual system using corn balls and sucrose solution. The proposed analytical model has been employed to develop sweetened breakfast cereals and snacks. Application of ANN in baking has been studied out by few investigators. The bake level of biscuits is of significant value to biscuit manufacturers as it determines the taste, texture and appearance of the products. Previous research explored and revealed the feasibility of biscuit bake inspection using feed forward neural networks (FFNN) with a back propagation learning algorithm and monochrome images (Yeh et al 1995).

A second study revealed the existence of a curve in colour space, called a baking curve, along which the bake colour changes during the baking process. Combining these results, an automated bake inspection system with artificial neural networks that utilises colour instead of monochrome images is evaluated against trained human inspectors. Comparison of Neural Networks Vs Principal component regression for prediction of wheat flour loaf volume in baking tests has been reported by Harimoto et al. (1995). The objective here was to determine values of four parameters which minimize the standard error of estimate (SEE) between predictions of NN & actual, measured remix loaf volumes of the flour. Two hundred patterns (i.e. quality test results of 200 flours) were used for training the NN. The training tolerance specifies how close each output (remix loaf volume) of the network must be to the empirical response to be considered "correct" during training. The training tolerance is a percentage of the range of the output neuron. Networks with smaller tolerances require longer time to train. If a network is slow in learning, it is sometimes helpful to begin with a wide tolerance and then narrow tolerance. A back-propagation neural network has been developed by Ruan et al (1995) to accurately predict the farinograph peak, extensibility, and maximum resistance of dough using the mixer torque curve. This development has significant potential to improve product quality by minimizing process variability. The ability to measure the rheology of every batch of dough will enable online process control through modifying process conditions. Razmi Rad et al (2007) have shown the ability of artificial neural network (ANN) technology for predicting the correlation between farinographic properties of wheat flour dough and its chemical composition. With protein content, wet gluten, sedimentation value and falling number as input parameters six farinographic properties including water absorption, dough development time, dough stability time, degree of dough softening after 10 and 20 min and valorimeteric value as output parameters. The ANN model predicted the farinographic properties of wheat flour dough with average RMS 10.794. Indicating that the ANN can potentially be used to estimate farinographic parameters of dough from chemical composition. A neural network based model was developed for the prediction of sedimentation value of wheat flour as a function of protein content, wet gluten and hardness index (Razmi et al 2008). The optimal model, which consisted of one hidden layer with nine neurons, was able to predict the sedimentation value with acceptable error. Thus, ANN can potentially be used to estimate other chemical and physical properties of wheat flour. Ismail et al (2008) have compared chemometric methods including classical least square (CLS), principle component regression (PCR), partial least square (PLS),and artificial neural networks (ANN) for estimation of dielectric constants (DC) dielectric loss factor (DLF) values of cakes by using porosity, moisture content and main formulation components, fat content, emulsifier type (Purawave™, Lecigran™), and fat replacer type (maltodextrin, Simplesse). Chemometric methods were calibrated firstly using training data set, and then they were tested using test data set to determine estimation capability of the method. Although statistical methods (CLS,PCR and PLS) were not successful for estimation of DC and DLF values, ANN estimated the dielectric properties accurately (R2, 0.940 for DC and 0.953 for DLF). The variation of DC and DLF of the cakes when the porosity value, moisture content, and formulation components were changed were also visualized using the data predicted by trained network ANN is applied for prediction of temperature and moisture content of frankfurters during thermal processing (Mittal and Zhang, 2000). Lou, and Nakai (2001). Have discussed application of artificial neural networks for predicting the thermal inactivation of bacteria as a combined effect of temperature, pH and water activity. Linear Regression, NN & Induction Analysis to determine harvesting & processing effects on surimi quality is reported by (Peters et al 1996). Surimi production is highly technical process requiring considerable skill. Harvesting & Processing input combinations and product quality attributes for the pacific writing surimi industrial were collected and analyzed. Multiple linear regression (MLR), NN, & MS – Induction were used to determine significant variables in the industry. MLR incorporated time, temperature and date of harvest as the variables, whereas ANN could incorporate other significant variable factors intrinsic to the fish (moisture content, salinity, pH , length, weight) and processing variables (processing time, storage temp, harvest date, wash time, wash ratios) in addition to the above three variables. Most variables were highly interactive and non linear. The back propagation NN algorithm was used to relate the influences of the variables (inputs) and their effects on quality (output) as defined by gel strength the NN model was trained so that the model predication was = 10% of the actual value for all data points. Comparison of three analytical systems, MLR, NN, & MS –I showed that time from capture to final production, temp of storage and date of harvest were indicated to be critical to get desired gel strength by all systems. ANN & MS-I also identified fish weight and length, salinity & moisture of flesh as important processing parameters. In addition, NN analysis indicated flesh pH, wash ratios and geographic location were important factors that affect quality. NN and MS-I were effective computer based methods for analyzing large data sets of complex biological system. They were especially useful for determining factors that affect final product quality in a multi-process operation. A three-layer feed forward neural network was successfully applied by Paquet et al (2000) to model and predict the pH of cheese curd at various stages during the cheese-making process. An extended database, containing more than 1800 vats over 3 yr of production of Cheddar cheese with eight different starters, from a large cheese plant was used for model development and parameter estimation. Very high correlation coefficients, ranging from 0.853 to 0.926, were obtained with the validation data. A sensitivity analysis of neural network models allowed the relative importance of each input process variable to be identified. The sensitivity analysis in conjunction with a prior knowledge permitted a significant reduction in the size of the model input vector. A neural network model using only nine input process variables was able to predict the final pH of cheese with the same accuracy as for the complete model with 33 original input variables. This significant decrease in the size of neural networks is important for applications of process control in cheese manufacturing. Optimization of the process of extraction of soy-fiber from defatted soy-flour is reported by Gupta and Shastri (2005). Defatted soya flour (DSF) is a good source of proteins, which are extracted in alkaline medium. The concept of integrated processing of DSF involves simultaneous recovery of soya proteins and fiber, which find use in dietetic foods. Process needs to be optimized to solubiise maximum protein, which is recovered afterwards as Soya Protein Isolate (SPI), with minimum fiber disintegration, and maximum recovery. DSF (obtained from Rasoya Ltd. Nagpur) contained 40.3% protein and 25% fiber. Extraction of soy-fiber was carried out by alkaline extraction at 11 different concentration-time combinations with alkali concentration (range 0.1-0.5N) as variable I, and extraction time (range 0.5-1.5 hours) as variable II. Maximum recovery of the fiber after protein solubilization was the required output. ANN elite; software (Pandharipande &. Badhe,2003) was applied by selecting three hidden layers with 5 neurons, 0.9 learning rate and 0.001 back propagation error. Learning of the network was carried out using 9 data points from the experimental data, whereas remaining two data points were used for assessment of the learning status of the network. The comparison between the experimental and predicted results is given in Fig. (3)

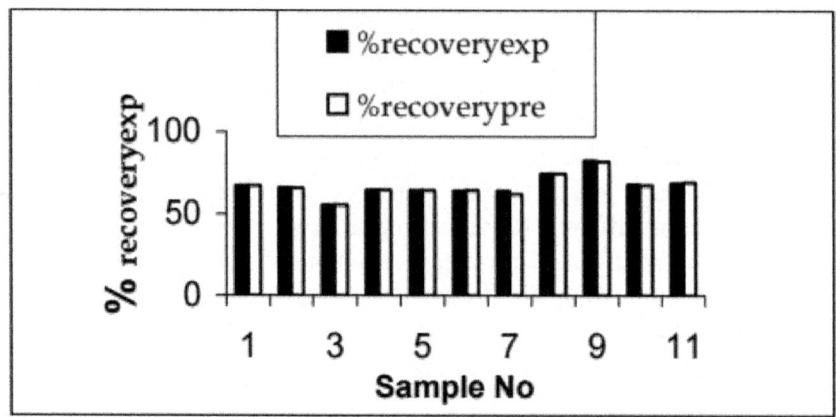

Figure 3: Experimental and predicted values for recovery of fiber from DSF

The optimum conditions predicting maximum percentage recovery under the above consideration were found to be 0.5 hrs extraction time with 0.5 N alkali(condition I) and 0.35N alkali for 60 minutes (condition II). Validity of the model was established by confirming the recovery under the selected combinations of alkali concentration and time which showed excellent correlation (R2=0.998) with the predicted values, Thus, it can be concluded that the developed Artificial Neural Network model has been used effectively as a tool in optimizing the process parameter for removal of fiber from DSF.

ANN IN THE FIELD OF BIOTECHNOLOGY

ANN can be a boon in the field of biotechnology in view of the complex nature of biocatalysts and microorganisms and their interactions with the environment. Prediction of models is usually very difficult on account of the lack of information about the physiological and biochemical constraints of biocatalysts, and their effect on physical phenomena like solubility of nutrients, oxygen transfer, and availability of water. ANN has the advantage that it can make accurate forecast even when the process behavior is non linear and data is unstructured. Since network training is fast, the method is suitable for on-line forecasting. Characteristic of the beer production process is the uncertainty caused by the complex biological raw materials and the yeast, a living organism. Thus, predicting the speed of the beer fermentation process is a non-trivial task. Data sets from laboratory-scale experiments as well as industrial scale brewing process were used to develop the neural network and descision tree. Simple decision trees were able to predict the classes with 95%–98% accuracy. Utility of these methods was checked in a real brewery environment. The neural network could, on average, predict the duration of the fermentation process within a day of the true value; an accuracy that is sufficient for today's brewery logistics. The accuracy of the decision tree in detecting slow fermentation was around 70%, which is also a useful result. (Rousu et al 1999). Beluhan and Beluhan (2000) describe estimation of yeast biomass concentration in industrial fed-batch yeast cultivation process with separate arificial neural networks combined with balance equations. Static networks with local recurrent memory structures were used for on line estimation of yeast biomass concentration in industrial bioreactor , and the inputs were standard cultivation state variables: respiratory quotient, molasses feed rate, ethanol concentration, etc. This hybrid approach is generally applicable to state estimation or prediction when different sources of process information and knowledge have to be integrated. Multivariate statistical methods namely, principal component analysis (PCA) and partial least squares (PLS), which perform dimensionality reduction and regression, respectively, are commonly used in batch process modeling and monitoring. A significant drawback of the PLS is that it is a linear regression formalism and thus makes poor predictions when relationships between process inputs and outputs are nonlinear. For overcoming this drawback of PCA, an integrated generalized regression neural networks (GRNNs) is introduced for conducting batch process modeling and monitoring. The effectiveness of the proposed modeling and monitoring formalism has been successfully demonstrated by conducting two case studies involving penicillin production and protein synthesis.(Kulkarni et al 2004). Application of neural network (ANN) for the prediction of fermentation variables in batch fermenter for the production of ethanol from grape waste using Saccharomyces cerevisiae yeast has been discussed by Pramanik (2004). ANN model, based on feed forward architecture and back propagation as training algorithm, is applied in this study. The Levenberg- Marquardt optimization technique has been used to upgrade the network by minimizing the sum square error (SSE). The performance of the network for predicting cell mass and ethanol concentration is found to be very effective. The best prediction is obtained using a neural network with two hidden layers consisting of 15 and 16 neurons, respectively. Online biomass estimation for bioprocess supervision and control purposes is addressed by Jenzsch et al (2006), for the concrete case of recombinant protein production with genetically modified Escherichia coli bacteria and perform a ranking. As the biomass concentration cannot be measured online during the production to sufficient accuracy, indirect measurement techniques are required. At normal process operation, the best estimates can be obtained with artificial neural networks (ANNs). Simple model-based statistical correlation techniques such as multivariate regression and principle component techniques analysis can be used as alternative. Estimates based on the Luedeking/Pirettype are not as accurate as the ANN approach; however, they are very robust. Techniques based on principal component analysis can be used to recognize abnormal cultivation behavior. All techniques examined are in line with the recommendations expressed in the process analytical technology (PAT)-initiative of the FDA. Badhe et al (2002) extended application of ANN to study hydrolysis of castor oil by Pancreatic lipase (Biocon India Ltd.) at 35 oC at pH 7.5 in immobilized membrane bio reactor to investigate the application of free and immobilized lipase for oil hydrolysis. Effect of three variables, e.g. enzyme concentration (range 0.1-0.5 ml), substrate concentration (Range 0.25 to 2.0g) and reaction time (range 2 –8 hours) on percent hydrolysis was investigated. Total 30 data points in the above mentioned range were subjected to training and validation using eliteANN software (Pandharipande & Badhe, 2003) with feed forward, sigmoidal activation function & delta learning rule. The topology of the system is described as in Table 1. ANN predictions were accurate (R^2 =0.998) for predicting the percentage hydrolysis of castor oil by lipase enzyme as a function of enzyme concentration, ratio of substrate to buffer concentration and reaction time.

Table 1: Topology of the ANN network applied for prediction for hydrolysis of castor oil using pancreatic lipase.

Number of neurons	input	3
	Output	1
First hidden layer		15
Second hidden layer		07
Training data points		26
Test		04
Learning rate		0.8

Cheese whey proteolysis, carried out by immobilized enzymes, can either change or evidence functional properties of the produced peptides, increasing the potential applications of this byproduct of the dairy industry. However, no information about the distribution of peptides' molecular sizes is supplied by the mass balance equations and Michelis Menten like kinetics. Sousa et al (2003) present a hybrid model of a batch enzymatic reactor, consisting of differential mass balances coupled to a "neural-kinetic model," which provides the molecular weight distributions of the resulting peptides.

ANN FOR PREDICTION OF ENZYME PRODUCTION

Mazutti et al (2009) have studies production of inulinase employing agroindustrial residues as the substrate to reduce production costs and to minimize the environmental impact of disposing these residues in the environment. This study focused on the use of a phenomenological model and an artificial neural network (ANN) to simulate the inulinase production during the batch cultivation of the yeast Kluyveromyces marxianus NRRL Y-7571, employing a medium containing agroindustrial residues such as molasses, corn steep liquor and yeast extract. It was concluded that due to the complexity of the medium composition it was rather difficult to use a phenomenological model with sufficient accuracy. For this reason, an alternative and more cost-effective methodology based on ANN was adopted. The predictive capacity of the ANN was superior to that of the phenomenological model, indicating that the neural network approach could be used as an alternative in the predictive modeling of complex batch cultivations. SSF is defined as cultivation of microorganisms on a moist insoluble substrate, which binds sufficient water to solubilize the nutrients. The desirable aw is 0.88 to 0.85 and the amount of water to be added is determined by the water binding capacity of the solid substrate. Although wheat bran is widely recommended ingredient in SSF, several other lignocellulosic agrowastes may be incorporated as inducers for specific products. (Deshpande et al 2008). On account of difference in water binding capacity of such varied substrates, it becomes necessary to optimize the amount of water for achieving maximum productivity. The system can be described as a unstructured model, on account of several undefined parameters and interactions. Possibility of application of ANN for prediction of extracellular enzyme production under SSF conditions was examined for several systems, specially to define optimum level of water in combination of solid substrate containing components with different water binding properties. Production of Pectin Trans Eliminase (PTE) by Penicillium oxalicum was carried out on wheat bran medium by incorporation of de-oiled orange peel (DOP), which was incorporated at different levels (range 25 – 75%) as first input parameter and levels of substrate: moisture ratio (range 2-3) as second input variable. Enzyme activity units /ml of crude enzyme extract (CEE) was first output parameters and specific activity (enzyme activity units/mg proteins) was second output parameter. ANN topology employed for the study had three hidden layers, each with 10 neurons, learning rate 0.9 and back propogation error =0.0014. The model was used for prediction of experimental conditions within the system framework for optimum enzyme production and the output predicted by ANN showed excellent concurrence with experimental results (Fig.4). Results clearly indicate that DOP is a good inducer, because increase in orange peel % increases the enzyme activity but to enhance the activity, it is necessary to increase moisture content simultaneously since orange peel has more moisture binding capacity. Optimum combination for high productivity as per the ANN analysis was found to be 60-65% DOP, with 90% moisture. (Yadav et al. 2003).

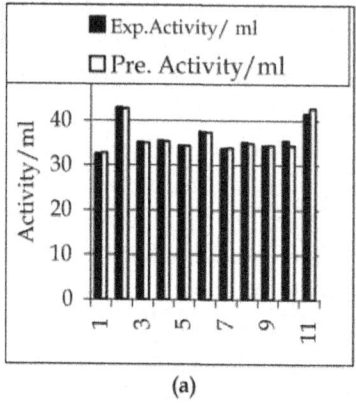 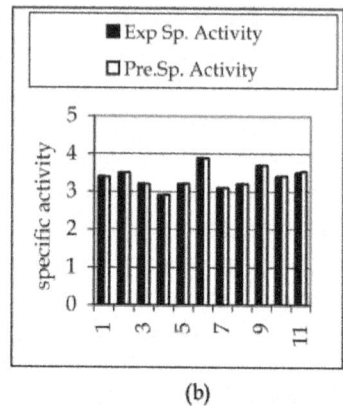

(a)   (b)

Figure 4: Experimental and predicted values for production of Pectin Trans Eliminase by Penicillium oxalicum on Wheat bran : Deoiled Orange Peel medium under SSF conditions (a) Enzyme units /ml of Crude enzyme extract (b) Specific activity

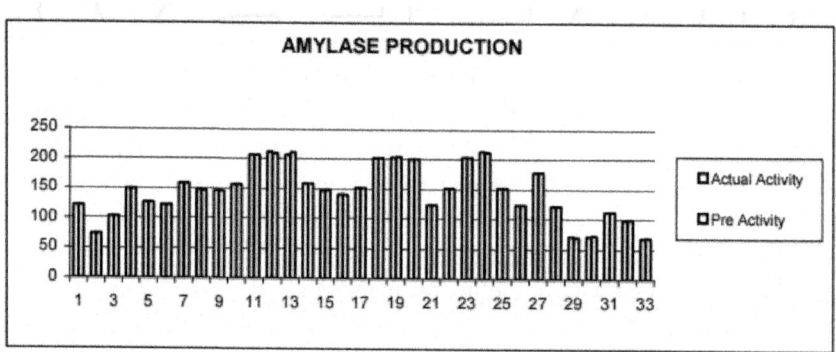

Figure 5: Experimental and predicted values for specific activity of Amylase produced by Aspergillus oryzae on sorghum grit and sorghum stalk medium under SSF conditions

Similar Study was carried out for production of amylase by Aspergillus oryzae by using combination of sorghum stalk and sorghum grits as substrate (Pandharipande et al. 2003). Sorghum stalk content varied between 0-100%, and the level of moisture varied between (30-70%) with total 33 data sets. Amylase activity units/ml in CEE, as well as the specific activity were the output parameters. The data was processed by eliteANN software (Pandharipande and Badhe, 2003), with three hidden layers of 20 neurons each, learning rate of 0.9, and back propogation error 0.0001. The experimental results are shown in Fig. 5. It was observed that the amount of inducer influenced the amount of water to be added. The optimum specific activity was obtained at inducer level 70% and moisture level 65% experimentally as against the predicted values of 85% inducer and 60% moisture. Application of ANN for prediction of cellulase and xylanase production by Solid State Fermentation (SSF) was studied using microorganisms Trichoderma reesei and Aspergillus niger (Singh et al. 2008, 2009). Experiments were performed with three variables on the production of xylanase and cellulase enzyme by T.reesei and A.niger by SSF. Total 60 different combination of wheat bran-sugarcane bagasse composition, water: substrate ratio and incubation time were selected as shown in Table 2.

Table 2: Range of variables selected for study of cellulase & xylanase production

Variable	Range	
	Low	High
1. Bran%	0	100
2.Water Substrate Ratio(v/w) W:S	1.875	3.125
3.Time of Incubation (hrs)	24	168

Experimental data was divided into two data series. First set, consisting of about 75-80% of the data points, named as 'Training Set'. It was used for training of ANN to develop independent models for xylanase and cellulase production, each containing three inputs (%wheat bran, W:S Ratio, and Hours of incubation) and one output (IU/ml), three hidden layers (10 nodes each) , learning rate 0.6 and final error 0.002. Adequacy and predictability of the model was tested by giving input parameters for the second data set named as 'Test set' and comparing the predicted and experimental values for T. reese (Fig. 6a & 6b,) and A.niger (Fig. 7a & 7b) respectively by using elite-ANN© software.

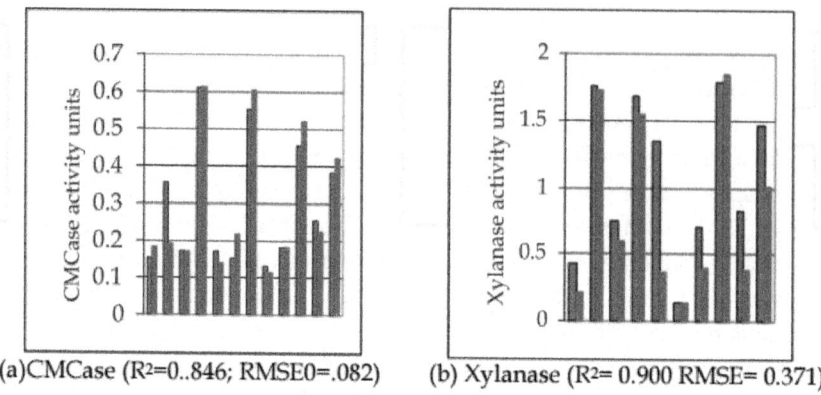

(a)CMCase (R²=0..846; RMSE0=.082)   (b) Xylanase (R²= 0.900 RMSE= 0.371)

Figure 6: Comparison between actual and predicted values of enzyme production for test data set of T.reesei (a) CMCase (b) Xylanase

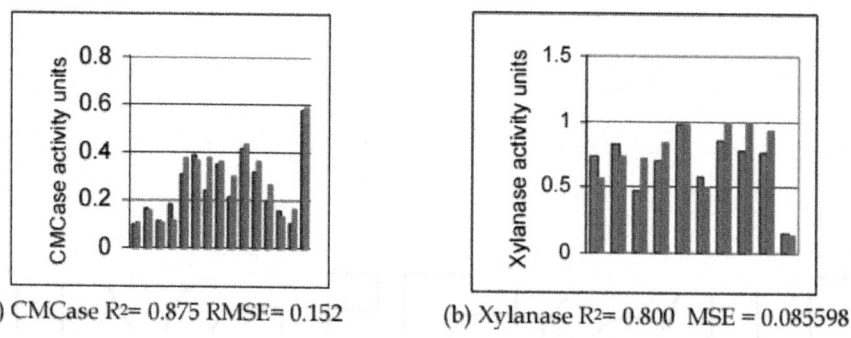

(a) CMCase R²= 0.875 RMSE= 0.152   (b) Xylanase R²= 0.800 MSE = 0.085598

Figure 7: Comparison between actual and predicted values of enzyme production for test data set of A niger (a) CMCase (b) Xylanase

Adequacy and predictability of the developed ANN mode 1 is judged by the comparison of the actual and the predicted values (Fig.6 &7), which show a satisfactory match as indicated by the correlation coefficients (0.0.90 & 0.81 for xylanase and 0.85 & 0.87 for cellulase) and root mean square error (0.35 & 086 for xylanase and 0.082 & 0.15 for cellulase) for T.reesei and A. niger respectively. Minor variations in the prediction may be due to complexity and inherent variability of biological system. (Pandharipande et al. 2007) Production of CMCase and Xylanase by A.niger and T.reesei under SSF condition is a function of %baggasse (which acts as an inducer) and aw (which supports the growth). Since wheat bran and baggasse differ in their water absorption capacity (WAC), proportion of water required to achieve desirable aw in combined substrate needs to be predicted. The observations indicate that CMCase and Xylanase production is optimum (>0.6 units) with greater than 50% baggasse, and ratio of water to substrate being 2.0. Thus it can be concluded that the model developed is validated for the given set & range of process conditions and can be used for the prediction of the enzyme activity at different combinations of parameters and selection of most appropriate fermentation conditions.

Table 3: ANN based predicted combinations for optimized production of enzymes by T. reesei

Xylanase				Cellulase			
Wheat Bran %	W : S Ratio	Hours	Predicted Activity IU/ml-	Wheat Bran %	W : S Ratio	Hours	Predicted Activity IU/ml-
90	2.75	168	1.769	55	2.5	120	0.527
65	2.5	156	1.491	50	2.25	144	0.497
60	2.5	168	1.387	45	2.25	120	0.443
55	2.25	120	1.524	40	2.75	120	0.426

Table 4: ANN based predicted combinations for optimized roduction of enzymes by A.niger,

Xylanase				Cellulase			
Wheat Bran %	W : S Ratio	Hours	Predicted Activity IU/ml	Wheat Bran %	W : S Ratio	Hours	Predicted Activity IU/ml
90	1.875	108	0.9503	80	1.750	108	0.5489
80	1.750	144	0.9741	75	1.875	120	0.5483
75	2.259	120	0.9027	70	2.000	136	0.4315
60	2.000	136	0.8836	65	2.250	144	0.4832

Above data was subjected to Response Surface Methodology, for Box Behnken model using second order regression equation obtained for the model expressed as follows:

$$y = \beta_0 + \beta_1 x_1 + \beta_2 x_2 + \beta_3 x_3 + \beta_{11} x_1^2 + \beta_{22} x_2^2 + \beta_{33} x_3^2 + \beta_{12} x_1 x_2 + \beta_{23} x_2 x_3 + \beta_{31} x_3 x_1$$

where x_1, x_2, and x_3 are inputs, y is the output, The Statistical analysis was done using Minitab1511. The correlation coefficient and MSE obtained by these two models is compared in Table 3 indicating suitability of both models.

Table 5: Comparison of ANN and RSM for prediction of cellulase and xylanase production by Solid State Fermentation (SSF)

Parameter	T.reesei		A,niger	
	Xylanase	CMCase	Xylanase	CMCase
$R^2$by ANN	0.9	O.846	0.8	0.875
R^2 by RSM	0.987	0.79	0.99	0.87
MSE ANN	0.371	0.082	0.856	0.152
MSE RMS	0.034	0.028	0.06	0.076

FUTURE PROSPECTS

Modern systems with diverse application areas demand expert & accurate calculations within a nick of time. For Such diverse and cutting-edge technology conventional systems have proved expendable and arduous. It is when the Artificial Neural Networks and Fuzzy Systems have proved their speed competitive potentials and expandability. In the last years several propositions for hybrid models, and especially serial approaches, were published and discussed, in order to combine analytical prior knowledge with the learning capabilities of Artificial Neural Networks (ANN). The intelligent modeling approach of models employing Artificial Neural Network in combination with other data analysis systems is able to solve a very important problem - processing of scarce, uncertainty and incomplete numerical and linguistic information about multivariate non-linear and non-stationary systems as well as biotechnological processes (Vassileva et al ,2000, Beluhan and Beluhan, 2000).

ACKNOWLEDGEMENT

Authors are thankful to Mr. S. L. Pandharipande for making available the ANN software and support for analysis of the experimental data. Authors express their gratitude towards Director, Laxminarayan Institute of Technology, Rashtrasant Tukadoji Maharaj Nagpur University, Nagpur for the encouragement and facilities provided at the institute.

REFERENCES

1.   Badhe YP, Joshi SW, Bhotmange MG, & Pandharipande SL (2002). Modelling of hydrolysis of castor oil by pancreatic lipase using artificial neural network Proceedings of National Conference on Instrumentation and Controls for Chemical Industries, ICCI 2002, Paper 1.2, 8-9th August 2002 Nagpur , organized by Laxminarayan Institute of Technology, Nagpur, India

2. Beluhan Damir, & Beluhan Sunica (2000). Hybrid modeling approach to on-line estimation of yeast biomass concentration in industrial bioreactor, Biotechnology Letters 22(8), pp. 631-635

3. Bhattacharya Suvendu, Patel Bhavesh K, & Agarwal Kalpesh (2003). Enrobing of foods: Simulation study and application of artificial neural network for development of products. Proceedings of International Food Convention "Innovative Food Technologies and Quality Systems Strategies for Global Competitiveness" IFCON 2003, Poster no TC- 32, pp 172, Mysore, December 2003.(AFSTI), Mysore India

4. Carapiso Ana I, Ventanas Jesus, Jurado Angela, & Garcia Carmen (2001). An Electronic Nose to Classify Isberian Pig Fats with different Fatty Acid Composition, Journal of the American Oil Chemists' Society, 78(4), pp. 415-418

5. Deshpande SK, Bhotmange MG, Chakrabarti T, & Shastri PN (2008). Production of cellulose and xylanase by T. reesei (QM9414 mutant), A. niger and mixed culture by Solid State Fermentation (SSF) of Water Hyacinth (Eicchornia crassipes), Indian Journal of Chemical Technology, 15(5), pp. 449-456

6. Eberhart, R. C., & Dobbins, R.W. (1990). Network analysis. In Neural Network PC Tools. A. Practical Guide, R.C. Eberhart and R.W. Dobbins (Ed.), Academic Press, San Diego, CA.

7. Gardner, JW, Hines, EL, & Wilkinson M (1990). Application of artificial neural networks to an electronic olfactory system, Journal Measurement Science and Technology, 1(5), pp. 446.

8. Gupta R, Pandharipande SL, & Shastri PN (2005). Optimization of the process of extraction of fiber from defatted soyflour using ANN, National Seminar on Global perspectives for India Food Industry by 2020- Food Vision 2020, organized by Laxminarayan Institute of Technology, Nagpur University, Nagpur.

9. Harimoto Y, Durance T, Nakai S, & Lukow O.M. (1995). Neural Networks Vs Principal Component Regression for Prediction of Wheat Flour Loaf Volume in Baking Tests, Journal of Food Science, 60(3), pp. 429–433

10. Herv's, C. G.,Zurera, Garcfa, R M., & Martinez J. A. (2001). Optimization of Computational Neural Network for Its Application in the Prediction of Microbial Growth in Foods Food Science and Technology International, 7: 159-163

11. Hong Yan, G.V., & Barbosa-Canovas (2001). Attrition Evaluation for selected Agglomerated Food Powders: The effect of agglomerate size and water activity, Journal of Food Process Engineering, 24(1), pp. 37-49

12. Hornik, K, Stichcombe, M, & White , H (1989). Multilayer Feed forward Neural Network are universal Approximate . Neural Network. 2, pp. 359-366 Huang Y, Kangas LJ, & Rasco BA. (2007). Applications of artificial neural networks (ANNs) in food science. Crit. Rev Food Sci. Nutr. 47(2), pp. 113-26

13. Huiling Tai, Guangzhong Xie, & Yadong Jiang (2004). An Artificial Olfactory system based on Gas Sensor Array and Back-Propogation Neural Network, Lecture Notes in Computer Science, Advances in Neural Networks, Vol. 3174, pp. 323-339

14. Igor V. Kovalenko, Glen R. Rippke, & Charles R. Hurburgh (2006). Measurement of soybean fatty acids by near-infrared spectroscopy: Linear and nonlinear calibration methods Journal of the American Oil Chemists' Society, 83(5)

15. İsmail Hakkı Boyacı, Gulum Sumnu, & Ozge Sakiyan (2008). Estimation of Dielectric Properties of Cakes Based on Porosity, Moisture Content, and Formulations Using Statistical Methods and Artificial Neural Networks, Food Bioprocess Technol 2(4), pp. 353-360

16. Jenzsch, Marco, Simutis, Rimvydas, Eisbrenner, Günter, Stückrath, Ingolf, & Lübbert, Andreas (2006). Estimation of biomass concentrations in fermentation processes for recombinant protein production, Bioprocess and Biosystems Engineering, 29(1), pp. 19-27

17. Jose Alberto Gallegos-Enfante, Nuria E.Roha Guzman, Ruben F.Gonzalez-Laredo, & Ramiro Rico- Martinez (2007). The kinetics of crystallization of tripalmitin in olive oil: an artificial neural network approach Journal of Food Lipids, 9(1), pp. 73–86

18. Kılıç, K., Boyacı, İ.-H., Köksel, H., & Küsmenoğlu, İ. (2007). A classification system for beans using computer vision system and artificial neural networks. Journal of Food Engineering, 78, pp. 897–904.

19. References and further reading may be available for this article. To view references and further reading you must this article.

20. Kulkarni Savita G., Chaudhary Amit Kumar Nandi , Somnath Tambe, & Kulkarni Bhaskar D. (2004). Modeling and monitoring of batch processes using principal component analysis (PCA) assisted generalized regression neural networks (GRNN), Biochemical Engineering Journal, 18(3), pp. 193-210

21. Lenz, J, Hofer, M, Krasenbrink, J, B, & Holker U (2004). A Survey of Computational and physical methods applied to solid state fermentation, Applied Microbiology and biotechnology, 65(1), pp. 9-17

22. Lopes, M.F.S., Pereira C. I., Rodrigues F.M.S., Martins M. P., Mimoso

M.C., Barros T. C., Figueiredo Marques J. J., Tenreiro R. P., Almeida J. S., & Barreto Crespo M. T. (1999). Registered designation of origin areas of fermented food products defined by microbial phenotypes and artificial neural networks. Appl. Environ. Microbiol., 65, pp. 4484–4489

23. Lou, W., & Nakai, S. (2001). Application of artificial neural networks for predicting the thermal inactivation of bacteria: A combined effect of temperature, pH and water activity. Food Research International, 34, pp. 573–579

24. Mazzuti Marcio M, Corrazza Marcos L, Filho Francisco Maugeri, Rodrigues Marai Isabel, Corraza Fernanda C, & Triechel Helen (2009). Inulinase production in a batch bioreactor using agroindustrial residues as the substrate: experimental data and modeling, Bioprocess and Biosystems engineering, 32(1), pp. 85-95

25. Meshram C.N. (2008). Studies on dehydration of agro based products using radio frequency dryer, M.Tech (ChemTech.) Thesis sumitted to Nagpur University

26. Mittal G.S., & Zhang , J. (2000). Prediction of temperature and moisture content of frankfurters during thermal processing using neural network, Meat Science, 55(1), pp. 13-24

27. Molkentin Joachim, Meisel Hans, Lehmann Ines, & Rehbein Hartmut (2007). Identification of Organically Farmed Atlantic Salmon by Analysis of Stable Isotopes and Fatty acids, European food Research and Technology, 224 (5) pp. 535-543

28. Pandharipande, M.S., Pandharipande, S.L., Bhotmange, M.G., & Shastri P.N. (2003). Application of ANN for prediction of Amylase production by Aspergillus oryzae under SSF conditions; Proceedings of International Food Convention"Innovative Food Technologies and Quality Systems Strategies for Global Competitiveness "IFCON 2003", Poster no PD 37, pp. 259, Mysore, December 2003.AFST(I), Mysore, India

29. Pandharipande S. L., & Badhe Y.P.(2003). Software copyright for 'elit-ANN' No. 103/03/CoSw dated 20/3/03

30. Pandharipande S.L. (2004). Artificial Neural Networks, Central Techno Publications, Nagpur.

31. Paquet, J., Lacroix, C., & J. Thibault J, 2000. Modeling of pH and Acidity for Industrial Cheese Production, J Dairy Sci., 83, pp. 2393–2409

32. Peters, G., Morrissey, M., Sylvia, G., & Bolte, J. (1996). Linear Regression, Neural Network and Induction Analysis to Determine Harvesting and Processing Effects on Surimi Quality, Journal of Food Science, 61, pp. 876–880

33. Pramanik K., (2004). Use of Artificial Neural Networks for Prediction of Cell Mass and Ethanol Concentration in Batch Fermentation using Saccharomyces cerevisiae Yeast, IE (I) Journal.CH, 85, pp. 31-35

34. Popescu Otelia, popescu Dimitrie, wilder Joseph, & Karwe Mukund (2001). A New Approach To Modelling and Control of a Food Extrusion Process Using Artificial Neural Network and an Expert System, Journal of Food Process Engineering, 24(1), pp. 17-36

35. References and further reading may be available for this article. To view references and further reading you must this article.

36. Razmi-Rad E., Ghanbarzadeh B., Mousavi S.M., Emam-Djomeh Z., & Khazaei J. (2007). Prediction of rheological properties of Iranian bread dough from chemical composition of wheat flour by using artificial neural networks, Journal of Food Engineering, 81(4), pp. 728-734

37. Razmi-Rad, E., Ghanbarzadeh B., & Rashmekarim, J. (2008). An artificial neural network for prediction of zeleny sedimentation volume of wheat flour. Int. J. Agri. Biol., 10, pp. 422–426

38. Rousu, J., Elomaa, T., & Aarts, R.J.(1999). Predicting the Speed of Beer Fermentation in Laboratory and Industrial Scale. Engineering Applications of Bio-Inspired Artificial Neural Networks, Lecture Notes in Computer Science, 1607, pp. 893–901

39. Ruan R, Almaer S, & Zhang J (1995). Prediction of Dough Rheological Properties Using Neural Networks, Cereal Chem., 72(3), pp. 308-311

40. Rumelhart, D.E., Hinton, G.E., & Williams, R.J.(1986). Learning internal representations by error propagation. Parallel distributed Processing: Explorations in the Microstructure of Cognition, 1, MIT Press, pp. 318–362

41. Sablani Shyam S., & M. Shafiur Rahman (2003). Using neural networks to predict thermal conductivity of food as a function of moisture content, temperature and apparent porosity, Food Research International, 36,pp. 617–623

42. Salim Siti Nordiyana Md, Shakaffl Ali Yeon Md, Ahmad Mohd Noor, Adom Abdul Hamid, & Husin Zulkifl (2005). Development of electronic nose for fruits ripeness determination, 1st International Conference on Sensing Technology, November 21- 23, 2005 Palmerston North, New Zealand pp. 515-518

43. Singh Aruna, Tatewar Divya, Shastri P.N., & Pandharipande S.L. (2008). Application of ANN for prediction of cellulase and xylanase production by Trichoderma reesei under SSF conditions, Indian Journal of Chemical Technology, 15(1), pp. 53-58.

44. Singh Aruna, Tatewar Divya, Shastri P.N., & Pandharipande S.L. (2009). Validity of artificial neural network for predicting effect of media components on enzyme production by A. niger in solid state fermentation. Asian Journal of Microbiology Biotechnology Environmental Science, 11(4), pp. 777-782

45. Sousa Ruy, Resende Mariaam M, Giordano Raquel L.C., & Giordano Roberto C, (2003). Hydrolysis of cheese whey proteins by alcalase immobilized in agarose gel particles, Applied Biochemistry and Biotechnology, 106(1-3)

46. Vallejo-Cordoba B., Arteaga G.E., & Nakai S. (1995). Predicting Milk Shelf-life Based on Artificial Neural Networks and Headspace Gas Chromatographic Data, Journal of Food Science, 60(5), pp. 885–888

47. Vassileva S., B. Tzvetkova B., Katranoushkova C. and Losseva L.(2000) Neuro-fuzzy prediction of uricase production Bioprocess and Biosystems Engineering 22,(4), pp 363-367

48. Wongsapat Chokananporn, & Ampawan Tansakul (2008). Artificial Neural Network Model for Estimating the Surface Area of Fresh Guava, Asian Journal of Food and AgroIndustry, 1(3), pp. 129 – 136

49. Yadav Sangeeta, Shastri N.V., Pandharipande S.L., & Shastri P. N. (2003). Optimization of water and DOP level for production of pectin trans eliminase by Penicillium oxalatum under SSF condition by Artificial Neural Network, Proceedings of International Food Convention"Innovative Food Technologies and Quality Systems Strategies for Global Competitiveness" IFCON 2003, Poster no FB 28, pp. 48, Mysore , December 2003.(AFSTI) Mysore, India

50. Yasuo Saito, Toshiharu Hatanaka, Katsuji Uosaki, & Hidekazu Shigeto (2003). Neural Network application to Eggplant Classification, Lecture Notes in Computer Science, Vol. 2774, pp. 933-940

51. Yeh Jeffrey C.H., Hamey Leonard G. C., Westcott Tas, & Sung Samuel K.Y. (2005). In Proceedings of the IEEE International Conference on Neural Networks 2005, pp. 37--42, {IEEE}

52. Zhang, Jun, & Chen, Yixin (1997). Food sensory evaluation employing artificial neural networks Sensor Review, 17(2), pp. 150-158(9)

CITATION

CHAPTER 1

Rashid Abafita Abawari, Microbiology of Keribo Fermentation: An Ethiopian Traditional Fermented Beverage, DOI: 10.3923/pjbs.2013.1113.1121

CHAPTER 2

Bokulich NA, Bamforth CW, Mills DA (2012) Brewhouse-Resident Microbiota Are Responsible for Multi-Stage Fermentation of American Coolship Ale. PLoS ONE 7(4): e35507. doi:10.1371/journal.pone.0035507

CHAPTER 3

Kim Y, Yoon S, Lee SB, Han HW, Oh H, Lee WJ, et al. (2014) Fermentation of Soy Milk via Lactobacillus plantarum Improves Dysregulated Lipid Metabolism in Rats on a High Cholesterol Diet. PLoS ONE 9(2): e88231. doi:10.1371/journal.pone.0088231

CHAPTER 4

Camarasa C, Sanchez I, Brial P, Bigey F, Dequin S (2011) Phenotypic Landscape of Saccharomyces cerevisiae during Wine Fermentation: Evidence for Origin-Dependent Metabolic Traits. PLoS ONE 6(9): e25147. doi:10.1371/journal.pone.0025147

CHAPTER 5

Favier M, Bilhère E, Lonvaud-Funel A, Moine V, Lucas PM (2012) Identification of pOENI-1 and Related Plasmids in Oenococcus oeni Strains Performing the Malolactic Fermentation in Wine. PLoS ONE 7(11): e49082. doi:10.1371/journal.pone.0049082

CHAPTER 6

Sun X-y, Zhao Y, Liu L-l, Jia B, Zhao F, Huang W-d, et al. (2015) Copper Tolerance and Biosorption of Saccharomyces cerevisiae during Alcoholic Fermentation. PLoS ONE 10(6): e0128611. doi:10.1371/journal.pone.0128611

CHAPTER 7

Y. Widyastuti, R. and A. Febrisiantosa, "The Role of Lactic Acid Bacteria in Milk Fermentation," Food and Nutrition Sciences, Vol. 5 No. 4, 2014, pp. 435-442. doi: 10.4236/fns.2014.54051.

CHAPTER 8

Scheinemann HA, Dittmar K, Stöckel FS, Müller H, Krüger ME (2015) Hygienisation and Nutrient Conservation of Sewage Sludge or Cattle Manure by Lactic Acid Fermentation. PLoS ONE 10(3): e0118230. doi:10.1371/journal.pone.0118230

CHAPTER 9

Oshoma CE, Greetham D, Louis EJ, Smart KA, Phister TG, Powell C, et al. (2015) Screening of Non- Saccharomyces cerevisiae Strains for Tolerance to Formic Acid in Bioethanol Fermentation. PLoS ONE 10(8): e0135626. doi:10.1371/journal.pone.0135626

CHAPTER 10

M. Assohoun, T. Djeni, M. Koussémon-Camara and K. Brou, "Effect of Fermentation Process on Nutritional Composition and Aflatoxins Concentration of Doklu, a Fermented Maize Based Food," Food and Nutrition Sciences, Vol. 4 No. 11, 2013, pp. 1120-1127. doi: 10.4236/fns.2013.411146.

CHAPTER 11

Csutorás, C. , Rácz, K. , Nagy, G. , Hudák, O. and Rácz, L. (2014) Large Scale Experiments on the Investigation of the Effect of High Concentrations of Aflatoxin B1 on the Fermentation of Different Wines. Journal of Agricultural Chemistry and Environment, 3, 41-47. doi: 10.4236/jacen.2014.32B007.

CHAPTER 12

M. Alberto, M. Perera and M. Arena, "Lactic Acid Fermentation of Peppers," Food and Nutrition Sciences, Vol. 4 No. 11A, 2013, pp. 47-55. doi: 10.4236/fns.2013.411A007.

CHAPTER 13

Madhukar Bhotmange and Pratima Shastri (2011). Application of Artificial Neural Networks to Food and Fermentation Technology, Artificial Neural Networks - Industrial and Control Engineering Applications, Prof. Kenji Suzuki (Ed.), ISBN: 978-953-307-220-3, InTech,

INDEX